Wilderness Preservation and the Sagebrush Rebellions

WILDERNESS PRESERVATION AND THE SAGEBRUSH REBELLIONS

by

William L. Graf

Rowman & Littlefield Publishers, Inc.

ROWMAN & LITTLEFIELD PUBLISHERS, INC.

Published in the United States of America
by Rowman & Littlefield Publishers, Inc.
8705 Bollman Place, Savage, Maryland 20763

British Cataloging in Publication Information Available

Library of Congress Cataloging-in-Publication Data

Graf, William L., 1947–
Wilderness preservation and the sagebrush rebellions / by William
L. Graf.
p. cm.
Includes bibliographical references.
1. West (U.S.)—Public lands. 2. Wilderness areas—Government
policy—United States. I. Title.
HD243.A17G72 1990
333.78'16'0973—dc20 89–70230 CIP

ISBN 0–8476–7420–7 (alk. paper)

5 4 3 2 1

Printed in the United States of America

∞™ The paper used in this publication meets the minimum requirements of American National Standard for Information Sciences—Permanence of Paper for Printed Library Materials, ANSI Z39.48–1984.

For Kenny,
who taught the value of the individual

Contents

List of Figures

This book outlines a particular view of the western public lands, the view of a geographer who is a geomorphologist, an earth scientist specializing in river mechanics. Like its subject, the book is not strictly geography or history, science or public policy, politics or biography. Perhaps the beauty of this story is that it occupies the border lands of all these areas, drawing on each intellectual source region for part of its interest, and becoming an important story only when all the areas are connected.

In the following pages I have focused on wilderness areas for four reasons. First, these remnants of relatively undisturbed public lands are continually at the center of debates about public land policy in the late twentieth century. Second, the evolution of American land policy over the past century, as it affected wilderness areas, provides a window for viewing the forces that shaped more general public land policies. Third, an understanding of the natural processes affecting the western United States is incomplete without accounting for the influences of public lands generally, and of wilderness areas in particular. Finally, wilderness areas are components of the American map representing all that is best about a nation that is so incredibly wealthy that it can afford to set them aside; a nation so wise that it can make a divisive but collective decision in favor of preservation.

This book is a story told in four episodes. Each episode includes federal initiatives regarding the public lands and a western rebellion against the initiatives. This story has heroes and villains, though the exact identity of characters depends on the reader's point of view. This story includes idealism, deception, humor, victories, defeats, sweeping landscapes, and boardroom and congressional intrigue. This story includes people and places that are famous, and some that are obscure but important. In the end, this story is one way to explain why there are almost 90 million acres of American wilderness.

My understanding of this story has grown slowly under the influence of many people. In addition to the many students who have offered constructive comments and useful information during my courses on federal public lands at the University of Iowa and Arizona State University, I owe a debt of special gratitude to my wife, partner, and colleague, Professor Patricia Gober. Her contributions and patience in seemingly endless discussions about people, places, and policy made an important difference in the final product.

The land where I write this Preface is administered by the Bureau of Land Management, an agency that has decided not to recommend that the area be included in the National Wilderness System. Porphyry copper in the shelving rocks on the opposite canyon wall have an irresistible lure for decision makers whose deliberations have brought out the conflicts between development and preservation. But from my vantage point, I can see in the blue, hazy distance the Mazatzal, Four Peaks, and Superstition Wilderness Areas, places where values other than copper have prevailed. Be it copper, timber, grass, or solitude, each area of the public lands offers something of value. There is no such thing as an empty place.

White Canyon Wilderness Study Area,
Arizona

Introduction

In the final years of the twentieth century, the conflict between wilderness preservation and commodity users is the most powerful and pervasive force in shaping public policy for America's federal lands. One-third of the nation's surface is administered by the federal government and owned collectively by the people, but the geographic distribution of that land is concentrated in the West (Figure 1). Virtually all of the remaining American wilderness is administered by the federal government, but its preservation and management has been the source of divisive conflict among the nation's citizens and their representatives. This book is the story of those conflicts and of how the wilderness areas were reserved.

The term "sagebrush rebellion" is used in the following pages to represent organized resistance in the West to federal public land policies. The term was coined in the 1970s, but it is equally applicable to events that transpired almost a century before. The sagebrush rebellion of the 1970s and 1980s was not a new phenomenon, but rather a new episode in a lengthy history of such actions. The preservation of natural areas is a relatively new concept in law that is the product of an equally lengthy history of philosophical evolution. Wilderness preservation represents, along with free public schools and national parks, a unique contribution of American culture.[1] To understand the recent conflicts between wilderness preservation and the sagebrush rebellion, it is necessary to explore similar previous events. In their sum, the rebellions and the efforts at preservation that stimulated them have shaped the geography of the nation's public lands.

Because most of the public lands are in the western part of the nation, and because most of the resistance to wilderness preservation has come from the same area, the West (exclusive of Alaska, which is a special case) is of special importance to the story of wilderness preservation and the sagebrush rebellions. The region might be called

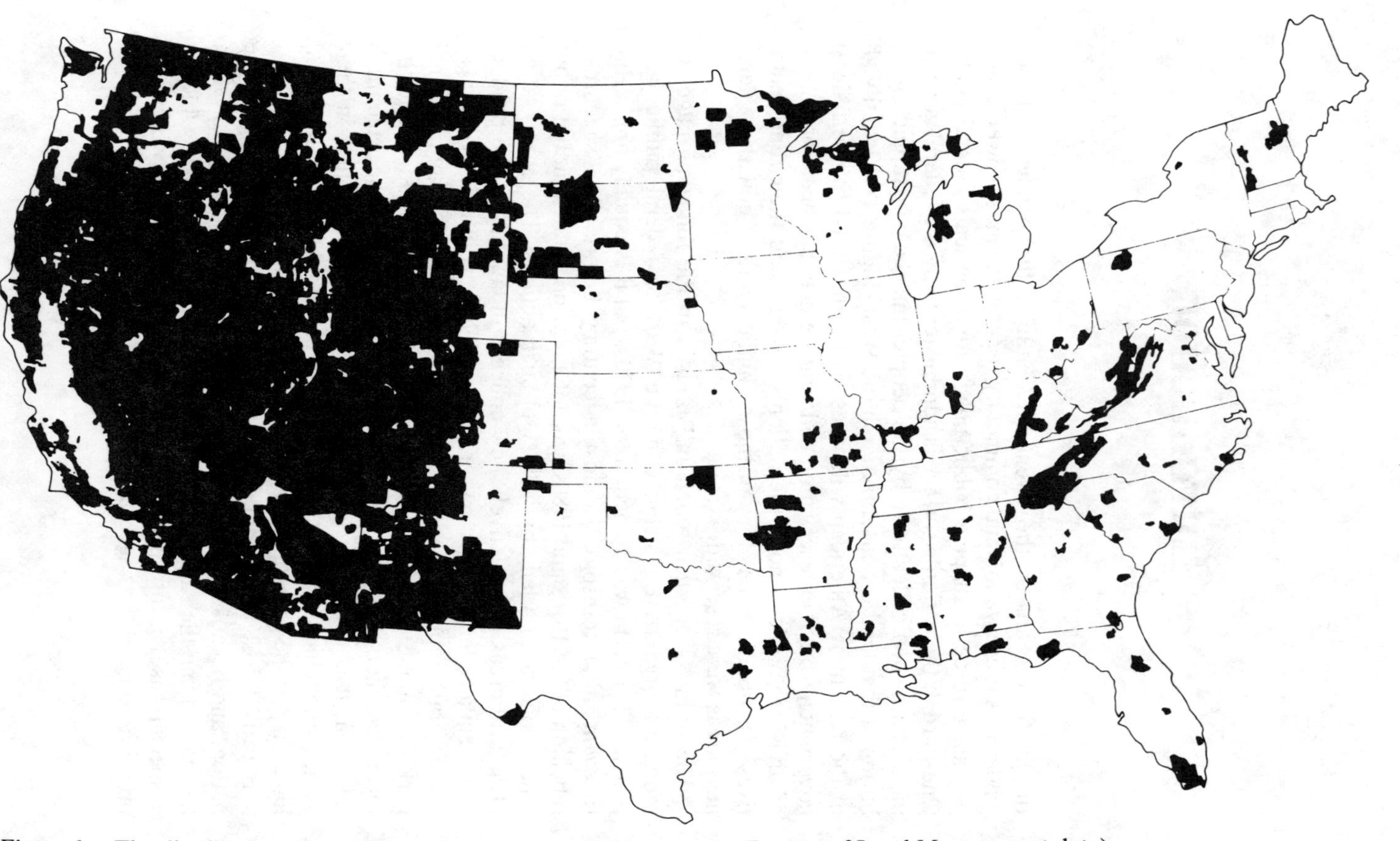

Figure 1. The distribution of federal public lands in the lower 48 states (Bureau of Land Management data).

''The'' West, because it is most similar to the image that most modern Americans have of the ''old West'' where the frontier process ended (Figure 2).[2] Nevada, Utah, and Arizona form the heartland of this ''sagebrush empire,'' which also includes western Colorado and Wyoming, and parts of New Mexico, Montana, and Idaho.

The people of this region and their culture are distinctive and separate from surrounding regions in spite of the rails, interstate highways, and jet routes that crisscross the nation. The population of these states is less dense than anywhere else in the country except Alaska. In 1980, for example, Nevada had only 8 persons per square mile, while Utah had 17, and Arizona 24.[3] By comparison, states east of the Rocky Mountains that are primarily rural, and that by European standards are sparsely populated, seem to the Westerner to be bursting at the seams with people. In 1980, forest-covered Maine had 34 persons per square mile, agricultural Iowa had 54, and Tennessee in the rugged Appalachians had 109.

The story of wilderness preservation and the sagebrush rebellions is therefore a story about the West, and about how the West has been perceived by the rest of the nation. The story begins shortly after the Civil War with the adventures of a geologist and geographer, John Wesley Powell, who sought to fill one of the last remaining blank spaces on the American map. Ironically, the end of the story is the preservation of part of that blank space as wilderness.

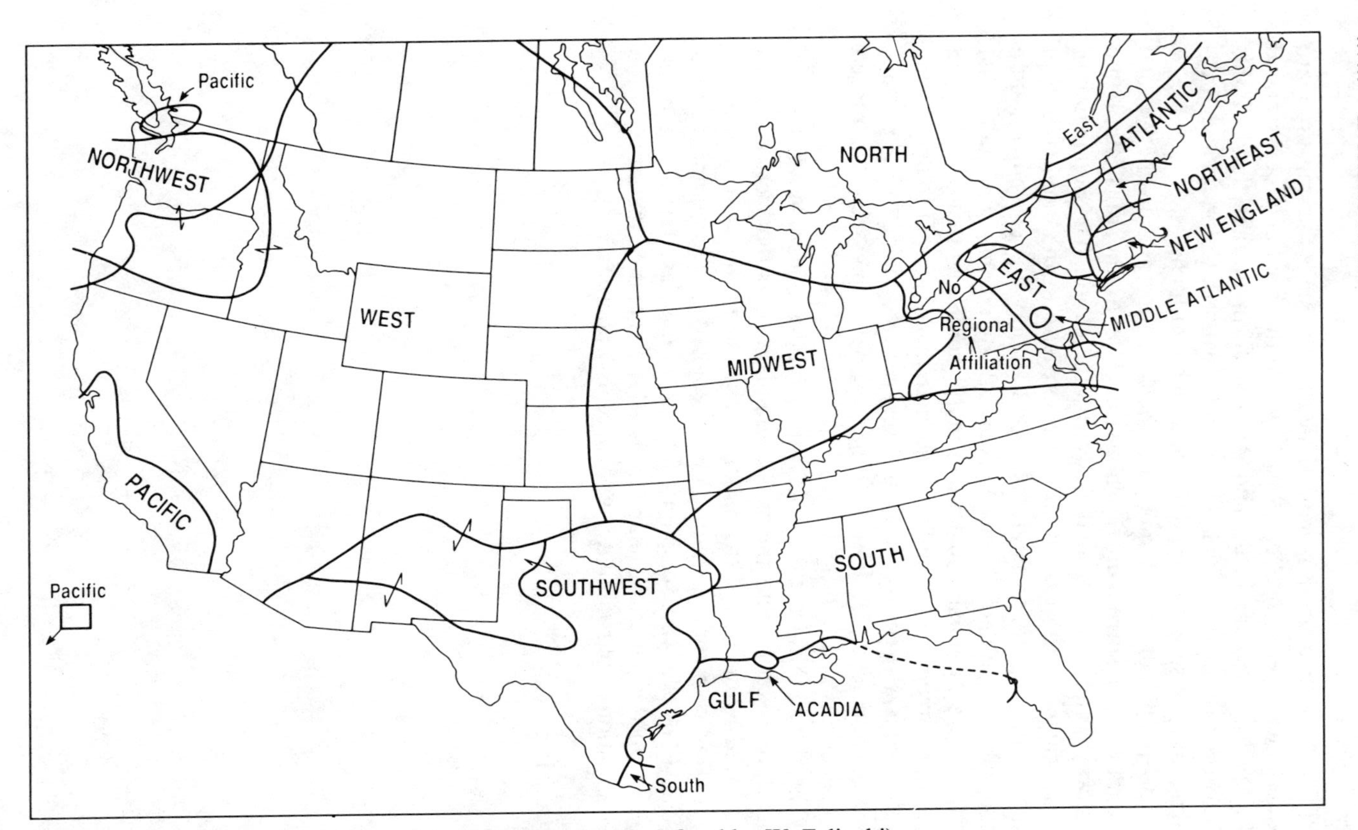

Figure 2. The vernacular regions of the United States (as defined by W. Zelinski).

Part I

The First Rebellion: Irrigation Lands

Chapter 1

Surveying the West

After the Civil War, western public lands were one of America's most vital resources, yet little was known of them, and they appeared on maps as vast blank spaces. John Wesley Powell, an Illinois college professor, set out from the town of Green River, Wyoming, to explore the interior West along the Green and Colorado Rivers in May, 1869.[1] Using boats brought from Chicago on the just-completed transcontinental railroad, Powell and nine associates (mostly trappers and frontier bums) entered a wilderness from which they were not expected to return.[2]

Powell was a thoughtful man who rose from a boyhood on a Wisconsin farm in the 1840s to become a leading scientist by the end of the century.[3] He had been a school teacher, an artillarist (losing his right arm after a wound at Shiloh), college professor, explorer, geographer, geologist, ethnologist, and bureaucrat. His supreme self-confidence displayed itself in his exploration of wilderness rivers to the bureaucratic maneuvering that led to the first sagebrush rebellion. His death was reported to the outside world during his 1869 expedition, but after ninety-nine days in the canyons he triumphantly completed the first traverse of the Green and Colorado Rivers, including the Grand Canyon. Powell became a celebrity during a period of political scandals and economic reversals when the American public desperately needed a hero.

Powell was also a consummate image manager, and he parlayed his exploits into congressional funding for a second, more scientific expedition down the rivers in 1871. He formalized his work into the United States Geographical and Geological Survey of the Rocky Mountain Region, which Congress viewed as a means to better understand the public lands and waters of the Colorado River.[4] Under Powell's direction, the Survey mapped the arid lands of Utah and northern Arizona and began scientific studies that remain landmark beginnings in geology, geomorphology, and ethnology. Powell's survey also successfully competed for funds against rival surveys.[5] In 1879, all the surveys were combined to form the U.S. Geological Survey under the uncertain leadership of geologist Clarence King.[6] Within a year, Powell replaced King and guided the fledgling agency through its formative years and intense controversies.

Shortly before he took over the combined surveys, Powell outlined the conclusions of almost a decade of field observation in the public domain of the West in a report with the unassuming title *Report on the Lands of the Arid Region of the United States, With a More Detailed Account of Lands of Utah.*[7] His report marshalled the scientific evidence of his surveys to prove that water would limit the use of public lands in the intermountain area.[8] The initial limited edition was rushed through printing to demonstrate the productivity of his survey and to serve as ammunition in the efforts to reform land laws and corrupt practices in the management of federal lands. Recognizing the significance of the report, Congress authorized a second edition in 1879 of 5,000 copies.[9] After more than a century, the arid lands report was still being reprinted as the most insightful examination of the land and water problems of the West.

Powell's report was a summary, with extensive examples, of his statements on public land management made to congressional committees, philosophical societies, the first Public Lands Commission, the National Academy of Sciences, and later in the popular press.[10] He defined the arid region as the area west of the one-hundredth meridian (midway across the Great Plains) where mean annual rainfall was less than 20 inches. Of this region, Powell contended that only a small amount was irrigable, a statement hotly disputed by expansionist congressional representatives from the states and territories involved. Powell also called for survey and classifications of the lands, recognizing portions of the public domain that would be used for mineral extraction, logging, irrigation, or grazing.[11] A century later, similar classification proposals for wilderness stimulated a fourth sagebrush rebellion.

Most of the arid lands report was devoted to the analysis of irrigation and pasture lands. Powell explained that most small streams in the West had been diverted for irrigation of high-altitude croplands. The more productive lowlands, with their rich alluvial soils, required projects for irrigation canals and dams that were so large that they were impossibly expensive for individuals or groups of farmers. Powell feared that if private companies developed the large lowland rivers, monopolies would follow, and that a few monied individuals would gain control of the productive areas. He proposed instead that the federal government undertake such projects, or that cooperative irrigation districts formed by groups of individual farmers should control construction of large irrigation works and distribute the benefits.

Powell also made radical proposals regarding survey methods and grazing lands. He proposed that the familiar rectangular survey be replaced by methods ensuring that each parcel of property fronted on a stream. In recognition that the western arid environment could not support dense herds of livestock, he proposed that the land laws be changed so that an individual could claim at least 2,560 acres for a grazing homestead rather than the then-accepted 160 acres.

With an eye toward his congressional lobbying efforts, Powell included in his report two sample bills that would implement his proposals. Neither bill survived burial under an avalanche of western criticism. Nonetheless, his document was one of the most visionary reports ever published by the federal government, and within a few decades almost all his suggestions were put into practice.[12] The period between publication and implementation was so rife with controversy that it qualifies as the first sagebrush rebellion.

Powell's proposals and his efforts to bring them into practice drew the lines of the first rebellion because they triggered a massive western response to the suggestion that the federal government should control irrigation development. Powell's emphasis on cooperative efforts was antithetical to the emphasis on the individual entrepreneur of the times, and to advocate massive federal assistance of irrigation projects was political heresy in the 1870s. By advocating allotments of the public domain in blocks of thousands of acres for grazing, Powell turned against the prevailing legislative wisdom, which was couched in more limited terms. Finally, by suggesting small blocks of irrigation lands to stimulate private development, he aligned himself against the congressional political leaders who were supported by land irrigation speculators dealing in large tracts. At the time of its publication, his report gave a clear view of reality and a look to the future.[13]

When Powell's report appeared in 1878, management of the American land resource was out of control despite numerous legislative attempts to bring sanity to the use of the nation's major source of capital.[14] The first sagebrush rebellion took place against a chaotic backdrop. From the birth of the nation until the 1860s, acquisition of additional federal lands was a major component of federal land policy (Figure 3). In the intermountain West, land that was the focus of all the sagebrush rebellions came into federal possession by three routes. The core of the sagebrush country was obtained by cession from Mexico at the conclusion of the Mexican-American War in 1848.[15] The northern perimeter of the region was obtained from Great Britain in the Oregon Compromise of 1846.[16] Finally, after realizing that the primary rail route between the South and southern California remained in Mexican control, abashed American negotiators arranged to buy what is now southern Arizona from Mexico in the Gadsden Purchase of 1853.[17] With these additions, the western American empire was complete, and the "Manifest Destiny" of the United States to occupy the continent from one ocean to the other was satisfied. Because the federal government was left in possession of much land but little capital, a policy of land disposal accompanied the acquisitions. A series of land laws resulted that were intended by Congress to promote settlement, but the result was waste, corruption, and greed on a scale seldom witnessed in American history.

The original public lands (or public domain) were created when the Continental Congress pursuaded several of the original colonies to cede their western land claims to the government of the newly formed United States.[18] Begun in 1780, the process involved lands east of the Mississippi River and was complete by 1802. When the states west of the Mississippi River were admitted to the union, they were carved from lands administered by the federal government. Each state agreed to relinquish all claims to nonappropriated (that is, unsettled) lands within its borders, an arrangement destined to be an issue in the fourth sagebrush rebellion of the 1970s.

The public lands were surveyed by the federal government by a method ill-suited to the western environment.[19] In 1785 the Land Ordinance established a system of rectangular surveys that was effective in the relatively well-watered, flat terrain of the midcontinent. When extended to the rugged arid West, the system was the source of problems for settlers, because it divided the land surface into rectangles of standard shape and size. In the early 1800s, a series of laws were enacted by Congress relating the cash sale of public land parcels and making the standard parcel size 80 acres. A family might subsist on an 80-acre farm in Ohio, but later in Iowa the size was problematic, and in arid southern Idaho an 80-acre farm could not sustain a family. In 1812 the General Land Office was established with responsibility for

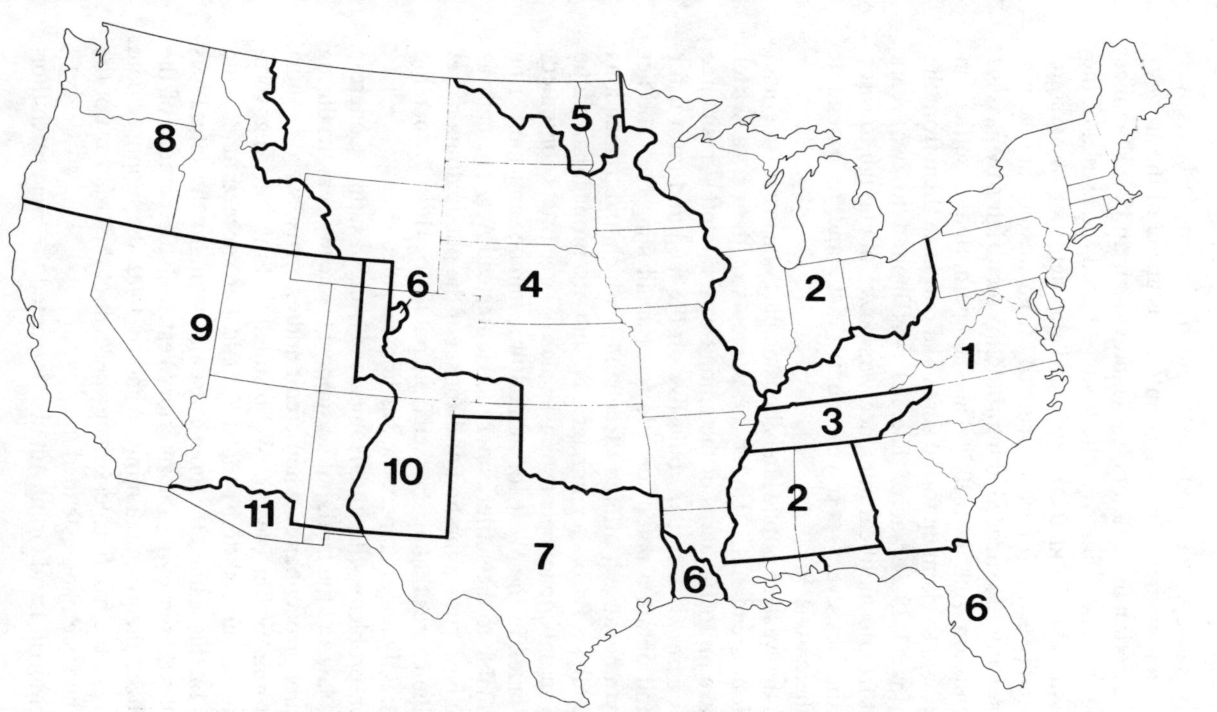

Figure 3. Acquisition of land by the United States for the lower 48 states. 1) The original thirteen states. 2) State cessions to the United States. 3) North Carolina cession to the United States (1790) and United States to Tennessee (1806 and 1846). 4) Louisiana Purchase (1803). 5) Red River Basin of the North. 6) Treaty with Spain (1819). 7) State of Texas. 8) Oregon Compromise with Great Britain (1846). 9) Cession from Mexico (1848). 10) United States purchase from Texas (1850). 11) Gadsden Purchase from Mexico (1853) (Bureau of Land Management data).

surveys, sales, and maintenance of records.[20] Throughout the period of settlement in the intermountain West, the General Land Office and its local district offices administered the transfer of land from public to private ownership.

The orderly survey, sale, and settlement of the public lands was frustrated by frontiersmen who ranged into areas ahead of the surveyors. Otherwise known as squatters, these individuals and their families frequently cleared and planted public lands, awaited the arrival of the Land Office surveyor, and then demanded the rights of ownership. Their representatives in Congress, primarily from western and southern states or territories, succeeded in protecting their rights by passage of the Pre-emption Act in 1830.[21] This act permitted a settler on unsurveyed tracts to claim up to 160 acres and purchase them from the government for $1.25 per acre. The act was limited to two years duration, but frontier political influence was powerful enough to force renewal of the act every two years until 1840, and to extend it indefinitely thereafter.[22]

After 1862 the western public lands became part of a great American myth: the Homestead Act.[23] Under the precepts of the act, a settler could purchase up to 160 acres of the land after six months' residence for $1.25 per acre. Or a settler could live on the land and farm it for five years, and obtain a patent or title for a small filing fee. The act became a myth because it offered "free land," which had tremendous appeal to landless European immigrants, but its propaganda value exceeded its actual impact because cash sales of large tracts continued. The richest lands in the West for agriculture, mining, and lumbering were not available to the settler under the Homestead Act. The act was probably the best-known method of disposal of the public domain, yet of the 1.1 billion acres that had passed from public to private ownership by the 1970s, only about 25 percent was homesteaded.[24]

The major problem with the Homestead Act was that Congress provided the law and the land, but no support for needy persons during the early years of occupancy when credit and expertise were in short supply. Historian Walter Prescott Webb stated that the federal government bet the settler 160 acres that he could not succeed at dryland agriculture. In this game, the house usually won. In the late 1800s when Powell was surveying the interior West, only 35 percent of the settlers finalized their original filings,[25] and by 1890, the supposed date of the closing of the frontier, less than 4 percent of the area west of the Mississippi had been homesteaded.[26]

The 1862 Morrill Land Grant Act also impacted the West by authorizing the federal government to grant lands to the individual states. Proceeds from these lands were to be used to support state education systems, so that individual parcels became known as school sections.[27] In the sagebrush rebellions, a common goal of western representatives was to have additional federal lands ceded to the states, but the management record of the states for lands they received previously was inconsistent. Nevada sold all of its lands into private ownership, while most other states managed them for maximum financial return to support state education. All western states were not created equal.[28] Utah received 14 percent of its total area as state lands, while Colorado received only 7 percent. By the 1980s many states had not yet completed the selection of lands to be transferred. In Arizona 190,000 acres were "owed" by the federal government to the state.[29]

As the frontier swept westward into ever drier portions of the continent, concepts about land management developed in the East faltered. A tract of 160 acres was insufficient to support a family in much of the western plains and beyond.[30] Congress created a series of laws in the 1870s to make adjustments for the arid western conditions by increasing the acreage available to the homesteader providing proof of newly planted trees (Timber Culture Act, 1873) or developed irrigation (Desert Land Act, 1877). The Timber and Stone Act (1878) provided for the transfer of nonagricultural lands for less than the price of a log per acre. The problems plaguing the Homestead Act continued, and the failure rates for settlers were disheartening. Eventually, the Timber Culture, Desert Land, and Timber and Stone Acts accounted for only about 3 percent of all disposed federal land.[31]

Failures of legitimate settlers were compounded by the illegal activities of land speculators.[32] In order to obtain title to tracts larger than those specified by the law (but required for survival), filings were frequently made on contiguous land parcels in the names of all members of the family, sometimes including tracts deeded to the family dog or a favorite horse. Ranchers required hired hands to file for lands and then provided for immediate transfer of title from the employee to the ranch owner. In early years these services were accomplished for a fee received by the individual filing the claim, but later the practice was so common that the price for a tract of 640 acres was a glass of beer.

The specific provisions of the land laws were circumvented. Trees required by the Timber Culture Act were planted but not cared for. The Desert Land Act required claimants to bring water to their lands, intending to stimulate irrigation development. Unscrupulous speculators, however, filed for nonirrigable lands and brought water in kegs, poured it on the ground, and then collected signatures of witnesses who duly testified that water had been brought to the land.[33] In other cases, when dwellings were required on claimed lands, tiny houses mounted on wheels were moved9 from tract to tract for certification.

In the intermountain region, control of water courses was critical, so that ranchers, their family members, and their employees filed claims for the areas around the few water sources.[34] They then effectively controlled surrounding unclaimed lands. Acquisition of lands in tremendous quantities for sale in subdivisions or for cattle ranches was so common that the "robber baron" was a familiar character in the 1870s and 1880s. Lewis Mumford referred to the raid on western lands where millions of acres of public land passed into the control of a relatively few individuals as a party of gluttons, the "Great Barbeque."[35]

Widespread success of illegal methods could not succeed without the active participation of public officials in the General Land Office. Local officials in district offices, swamped by mountains of paperwork, either could not police the process or were willing participants in fraud. In the central office in Washington, money changed hands frequently in return for swift approval in questionable cases. Some lawyers made careers of manipulating federal land officers and processes, and they openly advertised their expertise in newspapers.

A century later, critics would charge that Defense Department officials frequently retired from public duties to employment by weapons contractors for whom they had rendered favors while in office.[36] In the 1880s, the roles were similar but the players were land officers who retired to positions in railroad companies and land speculation firms.[37] In 1885 the commissioner of the General Land Office reported that the public interest had been subverted in "every branch of the service."[38]

Mismanagement and fraud in the administration of the public lands reached such shameful proportions that in 1879 President Hayes appointed the first Public Lands Commission to investigate the problems and suggest solutions.[39] James A. Williamson, commissioner of the General Land Office, was the commission chair. Other members were Clarence King (director of the newly formed Geological Survey), Powell (because of his widely read report on arid lands), Alexander T. Britton (a land lawyer), and Thomas Donaldson (a prominent Philadelphian). Clarence Dutton, a geologist with wide experience in the intermountain West, served as secretary. After a year of documentary research and travel, the commission found that the general principles of the government land policy were sound, but they recommended wholesale changes in the General Land Office. They recommended that the sale of lands in the arid West be in much larger parcels, and proposed new land laws to prevent the abuses of special provisions such as in the Timber Culture and Desert Land Acts. The commission also recommended that the public lands be classified into use categories to better manage and market them. A century later, a similar classification scheme for wilderness lands would trigger another rebellion. The report of the commission appeared in 1880, but it languished for over a decade without meaningful response.

Although disposal of the public domain was the watchword of America's land policy in the 1870s and 1880s, the period also saw the development of the new dimension of preservation, a concept destined to evolve into a national wilderness system. When the advance parties of the Hayden Survey, one of Powell's competitors, returned from the Yellowstone Plateau of northwest Wyoming in 1871, they brought stories that were hard to believe.[40] It was possible to catch fish from a lake and drop them directly into a nearby spring for cooking. Geysers roared with amazing regularity. The landscape was a fairyland of colored formations with weird shapes.[41] But even as the surveyors arrived, entrepreneurs were operating their businesses in the Yellowstone River Valley. The surveyors, geologists, and geographers concluded that the spectacular landscape should not pass into the hands of private exploiters, but that it should remain in public ownership. Aided by sterling photographs of the little-known area by William Henry Jackson and spectacular oil paintings by Thomas Moran,[42] Hayden convinced Congress to set aside the region as the world's first national park in 1872. There were no appropriations for management of the new park, and preservation of scenic parts of the public domain continued to be a minor part of public policy, but from these beginnings would one day grow the world's first wilderness system.

The state of California had established a policy of landscape preservation by setting aside Yosemite State Park in 1852.[43] The preserve was small and encompassed only the floor of a limited glacial valley in the High Sierra, but it represented an important beginning: the land was ceded by the federal government to the state for preservation purposes.[44] When lax administration of concessionaires and touring companies permitted the degradation of the area, public outcries were orchestrated by an increasingly well-known literary figure, John Muir. Ultimately, the area returned to federal jurisdiction. In 1890 the park was augmented by forest lands surrounding the original park zone in the valley, and Yosemite joined Yellowstone as one of the crown jewels of the evolving park system. Also in 1890, Sequoia and General Grant (later renamed King's Canyon) National Parks were carved from the public domain, and a preservation thread was permanently established in American land policy.

The preservation thread was strengthened by more general reform movements in the nation. The abuses of the management system for public lands became a rallying point for reformers in the 1880s who were concerned about the drift of the American economy into the control of industrial monopolies. The actions of these reformers were a trigger mechanism that initiated a reaction in the form of a western sagebrush rebellion. The reformers wanted to ensure that the entire federal government, not just the General Land Office, served the individual and protected the opportunity for individual enterprise. Part of the reform movement stemmed from the corruption of the federal government that extended throughout the Congress and into the White House, especially during the Grant and Hayes administrations. The railroad monopolies were especially powerful, and they directly affected the land disposal process because Congress routinely granted public lands to them for sale.[45] The proceeds from the sales were to finance railroad construction. In the eyes of farmers and shippers, the railroad owners were the villains of the century, an opinion shared by social reformers of eastern urban centers.[46]

The efforts of reform-minded politicians to democratize and popularize the land disposal process paralleled efforts to preserve some parts of the public domain from any disposal. The opening of the West for settlement led to the development of a popular image of vast open spaces, spectacular scenery, abundant wildlife, and noble savages. The literary works of Francis Parkman and the paintings of George Catlin led to a romantic conception of the continent west of the Mississippi that became common by the 1840s.[47] Thomas Cole added his voice to those calling attention to the loss of wilderness in Europe, and to its survival in the United States.[48] Even Horace Greeley of "Go West, young man" fame wrote in 1851 that in contrast to Europeans, Americans should "spare, preserve, and cherish some portion of our primitive forests."[49] Eastern editorialists began to make the case for preservation in the 1850s and 1860s.

At the end of the nineteenth century, the most ardent spokesman for the developing preservation thread was John Muir, a true zealot for the preservation perspective.[50] Though he detested writing, the general public eagerly devoured every line from Muir's reluctant pen, and he single-handedly sparked a preservationist attitude among people in influential positions. In the Geological Survey, many young researchers and surveyors embraced Muir's doctrine, and although they were not generally in the forefront of the preservation controversies (except for Powell), they were in important supporting roles.[51] They were additionally effective because of their wide field experience and romantic attachment to the spectacular western landscapes that were the subjects of Muir's writings.

Preservation often encountered western opposition because of water development interests. To control the large streams would require massive capital investments that were impossible for individuals.[52] Large corporations bought many of the best dam sites on medium-sized streams and began delivering water over long distances.[53] In many parts of the arid West, the monopoly of water threatened to eliminate the individual farmer, and Powell's report, written by a farmer's son, provided a blueprint for control of water monopolies. Carl Schurz, the secretary of interior who received Powell's report, was strongly supportive of the reform suggestions.[54]

Powell's report was forwarded to the National Academy of Sciences for comment in April 1878.[55] The president of the academy was Joseph Henry, an early supporter of Powell's exploration of the Colorado Plateau. Powell was so appreciative of the support that he named a mountain range after his benefactor. Henry and his successor, O. C. Marsh, supported Powell's proposed reforms. In December 1878, the academy reported favorably on the concepts of the arid lands report, but the House Committe on Public Lands, considering reform legislation, was divided concerning the best course of action.[56] Representatives including P. S. Wigginton (California) and Abraham Hewitt (Colorado) backed the reforms, but dissension on the committee prevented legislative action.

Early in 1879, reform congressmen avoided the controversy in the public lands committees in the House and Senate by attaching a land provision to an appropriations bill. Most western representatives opposed the reforms as expected, and the rhetoric set a pattern for subsequent sagebrush rebellions. One westerner targeted Powell as the major foe of development interests, and darkly charged him as a "charlatan in science and intermeddler in affairs of which he has no proper conception."[57] Undeterred, Powell continued his testimony before public groups and congressional committees, frequently illustrating his points by using huge multicolored maps or by drawing diagrams on a chalkboard.[58] He obviously enjoyed himself. Almost out of sheer exhaustion, Congress finally approved a weakened version of the reforms. The new law directed that the western surveys should be consolidated into the new U.S. Geological Survey and that the office of Land Commissioner be created, but there were to be no adjustments in the basic land laws. Powell's crusade had been temporarily delayed.

The importance of the potential irrigation lands increased in the early 1880s because they were the focal points of planned development. Western politicians and developers contended that only simple irrigation projects were needed to supplement water supplies in marginal agricultural environments. They argued that the term "desert"[59] was inappropriate for the West, and they organized numerous train tours of the region for reporters and public officials to prove the productivity of the allegedly arid zone. The tours invariably were timed for May or early June when moisture was most abundant.

This new attention to western development also focused attention on the issue of land fraud. Fraud in land dealings continued to be so common that in the West it was unremarkable. By 1883, settlers rarely could secure title to lands without first paying illegal claimants or corrupt land officials.[60] William A. J. Sparks, land commissioner in 1885, reported to Congress that fraud was rampant in all western states and territories as well as throughout the central General Land Office. The *New York Times* averaged one editorial on public land questions per week for three years beginning in 1885, reflecting a rising symphony of demands for reform.

In frustration, Commissioner Sparks finally suspended all cash land sales, and on December 23, 1885, he established a board of review to investigate land office records and claims by settlers. Meetings were held throughout the western states and territories to protest what was viewed as arbitrary and capricious dictates from Washington. The first sagebrush rebellion was heating up. Some meetings were organized by farmers, but most apparently were the work of land lawyers and speculators who were suddenly out of business. A *New York Times* editorial dryly noted that the sentiment of many westerners might be gauged by the lack of similar protests regarding fraud.[61] The rebels charged that fraud did not exist, and that it was the product of overly active imaginations of special agents of the land office who collected their data from the windows of moving trains. Finally, Sparks's superior, Secretary of Interior Lucius Lamar, ordered reopening of cash sales, but the board of review investigated almost 25,000 land filings and canceled 75 percent of them as fraudulent.[62]

The land management decisions about western areas were made

more difficult by unpredictable environmental conditions. Conditions in the period from 1886 to 1888 were especially severe. Whereas the early 1880s had been mild and relatively well-watered in the West, the late 1880s were a pronounced dry period,[63] and ranges that had supported thousands of cattle literally dried up and blew away. Cattle managers who ran huge herds on the public domain and harvested public grass at no cost began to cut their herds, but not quickly enough. To complicate the arid conditions, winters were unusually cold and windy, and entire herds were lost in the cruel winter of 1886–1887. In the high plains, spring melt streams carried so many cattle carcasses that the rivers appeared to be filled with floating cordwood.[64] It had been assumed that cattle could forage during the entire year, but after the late 1880s, supplemental feeding was seen as a prerequisite for successful operations. The collapse of the range economy exerted additional pressures in the first sagebrush rebellion since consolidation of bankrupt operations demanded an ever freer reign by entrepreneurs on public lands.

As the changing environment forced operational changes, the case for land reform was embraced by an increasing number of public opinion leaders. Speculators on the western frontier scrambled to obtain as much land as possible before the laws were changed. In 1886, preemption entries leaped 500 percent over the previous year.[65] Land Commissioner Sparks stirred the hornet's nest of western congressional representatives to action by his investigations, and he was attacked as having a vendetta against trans-Mississippi River interests. Led by Congressman James Laird of Nebraska, the westerners finally succeeded in forcing Sparks's resignation, and the reformist secretary of interior, Lucius Lamar, soon followed. President Grover Cleveland appointed a lackluster Henry Villas of Wisconsin as interior secretary, and he appointed S. M. Stockslager of Illinois to the hot seat of land commissioner. Stockslager continued the efforts of Sparks in exposing corruption, and the friction between eastern reformers and western developers continued.

The Senate gained an ardent spokesman for western development when Nevada returned William Stewart to Washington on a platform of free coinage of silver and improved irrigation programs. "Big Bill" (as he was known by constituents and colleagues) wasted little time in setting the wheels of Congress in motion: on February 13, 1888, he initiated a Senate resolution asking the U.S. Geological Survey to report on the possibility of creating an Irrigation Survey. As the puppet of the developers, such an organization in the bureaucracy could represent a force for western expansion and could provide support for the expansion of irrigation by making available the necessary technical expertise. A wary Stockslager wrote that he was "unable to see any urgent necessity for it,"[66] but Powell as director of the Geological Survey saw the suggested Irrigation Survey as a way to implement his cherished dream of scientifically rational development. The two vitriolic personalities of Powell and Stewart combined forces, and on March 20, 1888, the Senate approved a resolution calling for the creation of an Irrigation Survey.

Through a summer of bitter debate, the authorization of the Irrigation Survey occupied the attention of Congress. Senator Stewart and other western representatives tried to claim credit for the concept, but some members of Congress saw the new survey as an additional step in the march of John Wesley Powell toward all-encompassing power in the administration of the nation's natural resources. Senator Preston Plumb (of Nebraska) rightly pointed out that the concept of an irrigation survey had been proposed by Powell a decade earlier in his arid lands report. Plumb added his personal assessment of Powell: "He expects someday to be incorporated into the Constitution of the United States as an amendment."[67]

Some reformers were concerned that if the survey were established, speculators would buy potential dam sites at low prices from the government and then sell to dam builders at high prices. Recognizing this possibility, Representative George Symes (Colorado) inserted into the authorizing bill the proviso that lands potentially susceptible to irrigation development were to be reserved from public sale until the most likely sites for construction were identified. Western interests, fearful that such a sweeping reservation of public lands would result in a lockup that would slow development, tried to remove the provision. They were successful only in adding a statement that the president could open irrigation lands to settlement at his discretion, and the entire bill was enacted on October 2, 1888.

The legality of the legislation was soon tested. On August 5, 1889, the acting land commissioner stopped all cash sales in the West and invalidated all patents (deeds) issued after approval of the irrigation survey. From the standpoint of the land law reformers, it all appeared quite sensible. Stop all land sales, scientifically survey and classify the resources, and then dispose of them in an orderly, democratic fashion. The procedure was amazingly similar to that proposed for potential wilderness areas almost a century later, when the Federal Land Policy and Management Act of 1976 directed the Bureau of Land Management to halt all development on potential wilderness lands until they could be analyzed for wilderness and competing values. The argument against the reservation plans was the same in the 1880s as it was in the 1970s: the West was being economically hog-tied by a restrictive federal land policy.

Powell's ally, Big Bill Stewart, belatedly realized that the proposed survey would require years of expensive research, while economic growth in the western states would be suspended. He had originally envisioned a brief analysis of potential dam sites and quick release of lands for development. Gadflies like Congressman Hilary Herbert (Alabama) continually insisted that the irrigation survey was the scheme of a maniacal director (Powell), that the survey was not feasible, and that no one was going to develop the irrigation sites anyway because of the tremendous expense. The government was taking the first step in a potential billion-dollar investment, Herbert claimed, since only the federal government could amass the needed capital. Herbert, Stewart, and other antifederalists mobilized their political forces with clearly defined objectives: eliminate Powell, dismantle his survey, and release the federal lands for unrestricted development. After ten years of germination, the first sagebrush rebellion was in full flower.

Chapter 2

The Big Thinkers

An understanding of the first sagebrush rebellion depends on more than an appreciation of its history, and the environmental processes of the arid western landscape. The clash between federal controls and western efforts at unrestricted development were, at root, the clash of personalities and philosophies. An exploration of the backgrounds of the opinion leaders in the Irrigation Survey problem illustrates how the big thinkers on both sides of the question tried to implement contrasting grand designs for the public lands. Definable groups include the leading sagebrush rebels (mostly senators), reformers (mostly executives and congressmen), the "experts" of the Public Lands Commission and Geological Survey, and the preservationists.

The biggest thinker among the senators, William N. Stewart, was also an astute politician who represented well the sentiments of the residents of the intermountain region. Big Bill Stewart was the prototype sagebrush rebel.[1] Elected senator from Nevada in 1864–1875 and 1887–1904, he was the most powerful political force for the economic development of his state, and his activities showed a willingness to pursue federal assistance to that end. His flamboyant oratory and aggressive personality endeared him to his constituents and made him effective in the Senate, where he usually got what he wanted. His ability to establish the Irrigation Survey by means of attaching a rider to a Sundry Appropriations bill to avoid the Public Lands Committees was typical of his success. In his legal practice, he represented mining companies, and during his tenure in Washington he was a friend of the mineral industries, ensuring the passage of favorable mining laws and encouraging the development of federal expertise in economic geology. His integrity was unquestionable.

Stewart's interest in irrigation was as logical as his interest in mining. Nevada's economy was irrigation-dependent, and during the 1880s it lagged behind other sagebrush states in irrigation development. In the Irrigation Survey, Stewart saw a quick means of identifying developable reservoir sites, and by having the federal government do the work, he could avoid chronic funding shortages at the state level. He saw Powell as a useful tool, but at first was apparently not familiar with Powell's democratic opinions or his bureaucratic expertise. The two men seemed to strike a respectful friendship during the months leading up to the formation of the Irrigation Survey, and they enjoyed traveling in the West together during the summer of 1888 to collect information about water conditions.

The friendship was doomed. Powell was a theorist, an activist who was concerned with implementing abstract ideas. Stewart was the classic western politician, interested in action, development, growth, and tangible economic progress. While Powell was most comfortable with an evening of billiards and scientific discussion among pioneering geologists and anthropologists, Stewart was more likely to enjoy an evening of hard political bargaining over whiskey and cigars. When it became obvious that Powell intended to scientifically survey the reservoir sites with complete topographic mapping, Stewart's impatience emerged. His impatience was further taxed with the Land Office ruling that suspended all entries until the surveys were complete. With characteristic vigor, Stewart attacked his one-time friend personally, hoping to destroy the Irrigation Survey he had created, thus freeing lands for immediate development.

Stewart's attacks took the same approach that would be used by all subsequent sagebrush rebels in attacking their federal opponents. Realizing that Powell's scientific position was secure, he went to work on Powell's bureaucratic efforts. "There is no man in this country who is his superior as a lobbyist or who can better organize and control Congress," Stewart told his colleagues.[2] "He uses the vast appropriations . . . where they will do the most good, in procuring more appropriations."[3] Long after the controversy was ended, Stewart retained his bitter feelings toward Powell and thought of him as a "scientific wiseacre" who masqueraded as an expert, trying to tell the true experts, western developers, how to do their own business.

Stewart proved to be an unbeatable enemy for Powell because the senator was not a simple, crass, money-hungry developer. He had a heroic vision of the developing West, and whether, as a rough-neck lawyer in the mining camps or as a widely respected Senate leader, he was a lifelong activist for western economic progress. His craggy face and flowing beard were familiar in the nation's capitol wherever major decisions were made, and his interactions with presidents spanned the period from Lincoln to Theodore Roosevelt. He had a history of advocating decentralized authority, and when it appeared to him that Powell was amassing undue amounts of power in the directorship of the Geological Survey, he relentlessly pursued his one-time friend. Powell had an easy task when fending off critics who had only financial interests in the irrigation question, but with Stewart he was set against an idealist like himself.

Stewart's most strident ally in the Senate was Henry M. Teller from Colorado. Teller began his law practice in Illinois, but he settled in Central City, Colorado, a mining area known as the richest square mile on earth. Before statehood, Central City was the economic capital of the territory, and Teller's expertise in mining and irrigation law made him a familiar figure in the area.[4] When Colorado became a state in 1876, he was elected the state's first U.S. senator. His wide, square face, friendly smile, and bushy eyebrows were a fixture in the Senate into the twentieth century.

At first, Teller's political career appears inconsistent, but it can be best summarized as reflecting the actions of a political liberal except in those questions relating to economic development interests. When Colorado needed a fine legal and business mind to organize a railroad, Teller proved to be a shrewd promoter and president of the Colorado Central Railroad.[5] Simultaneously, he provided legal counsel for the eastern railroad monopoly of Jay Gould. As interior secretary in President Arthur's cabinet, he was concerned about the status of Indians, and he directed the Geological Survey to inventory reservation lands for marketable coal in order to provide economic security for them. During his same term as interior secretary, however, he arranged the lease of reservation grasslands to cattlemen at scandalously low prices of two cents per acre per year.[6] And although he frequently attempted land law reform legislation in the Senate, as interior secretary he also supported such spurious projects as federal land give-aways to railroads by declaring a 50-mile "right-of-way" on either side of the tracks, and the granting of suspect patents or deeds through the Land Office.

In terms of general political philosophy, Teller was more similar to Powell than to Stewart. Like Powell, he was the son of a farmer, and he taught school. Both were classic Jeffersonian democrats who viewed cities as dens of political inequity, saw great danger in industrial monopolies, and considered the individual yeoman farmer the country's salvation. But when Powell's Irrigation Survey and the General Land Office suspended western economic expansion with a freeze on the disposal of federal lands, Teller's votes were predictably aligned with the sagebrush rebellion. He was a powerful political ally of Stewart and others who sought destruction of federal restrictions.

Stewart and Teller enjoyed the enthusiastic support of many of their fellow senators in their rebellion, but the most significant were Gideon S. Moody (South Dakota), Preston B. Plumb (Nebraska), and James Laird (Nebraska). As representatives of the Great Plains, their interests were understandably piqued when Powell included their states in the "arid zone," an action they feared would discourage capital investment and that was contrary to their own experiences. While Powell was claiming that agriculture *as practiced in the eastern states* was not possible in the arid zone, South Dakota and Nebraska were producing prodigious quantities of cereal grains in the late 1870s and 1880s without the aid of irrigation. A long-term record shows that both Powell and the senators were correct, because drought in the early 1890s and periodically thereafter showed that for a few years there was sufficient moisture, but interruptions in the supply could be severe and lengthy. Nonetheless, in an impressive display of political control of the perceived natural environment, Powell produced an updated version of his arid lands map, and excluded most of Nebraska and South Dakota from his arid zone.[7]

Gideon Moody was a lawyer who came to prominence in the Dakota Territory and later in South Dakota because of his position as general counsel for the Homestake Mining Company, the leading corporate entity in the gold fields of the Black Hills. His service in the Senate in the late 1880s and early 1890s was focused on irrigation and mining matters, and he was one of Powell's most vocal critics. Moody was distrustful of the federal "experts," and he questioned the need for topographic surveys by scientists he felt were merely pompous professors. "You cannot satisfy an ordinary man by any theoretical scheme or by science," he observed with characteristic sarcasm.[8] In reference to the reservoir mapping efforts of the Irrigation Survey, Moody observed "our people in the West are practical people, and we cannot wait until this geological picture and topographical picture is perfected."[9] Moody and his constituents were interested in immediate action, not rational, well-planned (and thus inevitably delayed) action.

Senator Preston B. Plumb was also a key figure in the first sagebrush rebellion, but his input was more from a personal perspective than from a philosophical one.[10] He simply disliked Powell and did not approve of his survey activitites. When western senators claimed credit for originating the concept of an irrigation survey, it was Plumb, sensitive to Powell's empire-building activities, who pointed out that Powell was the true instigator.[11] In 1890, Plumb initiated what was to become a standard sagebrush rebellion effort: he proposed the transfer of all unappropriated lakes and rivers to state control. It did not succeed.

Congressman James Laird (Nebraska) was a vocal opponent of reform of land laws.[12] While Stewart, Teller, Plumb, and Moody concentrated their efforts on Powell and his Irrigation Survey, Laird took on the reform-minded land commissioners. When W. A. J. Sparks began publishing evidence of fraud in the annual Land Office reports of the mid-1880s, Laird mounted a counteroffensive that also set a pattern for subsequent sagebrush rebellions. He ensured a reduced budget for the General Land Office that removed most of the special agents who investigated land fraud. A special agent who persisted in prosecution of survey frauds was summarily dismissed as a result of congressional pressure, and funds for resurvey of lands in disputed areas were eliminated. Although not strictly representing the intermountain region, Laird's rhetoric was well received there when he charged that Commissioner Sparks was interested only in a reformer label for political purposes. He magnanimously charged that Sparks had a "vendetta against the best interests of all the territory beyond the Mississippi River."

Laird's sanctimonious bubble was later burst, however, when it was publicly revealed that his own record was tainted by fraud.[13] A company formed by Laird's brother had improperly obtained control of more than twelve miles of land bordering a Kansas stream. A profitable stock operation developed from these illegal beginnings and much of Laird's antireform activity appeared to be self-serving.

The primary target of the sagebrush rebels was John Wesley Powell. By the time of the Irrigation Survey controversy, Powell was widely known. After his adventures in the Colorado Plateau country, he wrote a best-selling book about his adventures, uncharacteristically sacrificing accuracy for drama.[14] He toured the country to give public lectures on his exploits and the wonders of the Colorado River system, and he marketed a lengthy series of stereoscopic photographs (actually taken by J. K. Hillers and E. D. Beaman) of the then little-known region. His popular success was matched with success in scientific circles. His concepts of landscape change and river processes influenced physical geography and physical geology for over half a century, and his pioneering work among primitive Indian tribes are cornerstones of American ethnology.

His definitions of water resource problems and solutions in the arid lands were not well received at the time of the initial publication, but by the early twentieth century, almost all his suggestions had been adopted, and he was revered as the father of modern reclamation. The reservoir behind Glen Canyon Dam, completed in 1962, was named in his honor, and although some misinformed later environmentalists claimed that Powell would have decried the reservoir that destroyed a vast region of scenic beauty, it is more likely that he would have been gratified with developments. He was the country's leading thinker on reclamation, and though he had an appreciative eye for scenic beauty, his writings reveal a strongly developed utilitarian view of nature. He advocated the construction of dams in Boulder Canyon, the eventual site of Hoover Dam.[15]

Much of Powell's political effort exercised from his position as director of the Geological and Irrigation surveys was motivated by an antimonopoly perspective. Like Senator Stewart, he saw the nation developing its greatest expression of democracy in the success of the individual farmer, and he saw his arid lands proposals as the only means of ensuring the survival of the independent farmer. His statement to the House Committee on Irrigation was typical of his views and pinpoints his political motivation for land reform: "I suppose that it is well known to the committee that the desert land privilege and the timber culture privilege are the great agencies by which the public domain passes from the possession of the Government into the hands of capitalists and corporations."[16]

Powell preferred to implement his reform ideas in a scientific fashion, and he felt that a topographic map showing an accurate portrait of the landscape was a prerequisite to successful public land management. When the Irrigation Survey was authorized, most of the money was spent on topographic mapping, a decision that seemed scientifically logical to Powell. Stewart and his associates saw the topographic mapping as a waste of time and money, a diversion from the important business of reservoir site identification, and further evidence of Powell's megalomania. Significantly, a century after Powell's arid lands report, many land management problems could still be traced to the lack of adequate topographic maps.

At the time of the completion of his arid lands report, Powell received valuable support from Secretary of Interior Carl Schurz.

While Powell came to his reformist views on land management by way of scientific analysis, Schurz came to similar views by way of politics. Schurz fled his native Germany because of his involvement in the 1848 revolution.[17] After living in exile in Switzerland, he migrated to the United States and helped organize the Republican party. In 1869 he became the first German-American to be elected to the Senate, based on his support from the German-speaking population of St. Louis. He was a liberal, outspoken critic of corruption, even within his own party, and had a special personal goal of implementing a merit-based civil service system. In public land affairs he was especially concerned with timber, and he consolidated what meager federal control was available when he was appointed secretary of interior by President Hayes in 1877.

Schurz's political opinions on land reform favored the independent farmer or rancher over the corporations and suspected monopolies. He was the man to implement the political philosophy of Henry George, a well-known social commentator of the 1870s.[18] Both George and Schurz viewed the federal government as a protector of the private citizen, and were alarmed at the corporate raids on the public domain. George advocated an almost socialistic approach to land legislation, and Schurz frequently proposed radical corrective measures during his tenure as secretary of interior. George summed up the feeling of both men: "A generation hence our children will look with astonishment at the recklessness with which the public domain has been squandered. It will seem to them that we must have been mad."[19]

After Schurz's stint as secretary of interior, he continued to be active as a liberal reformer in the highest levels of American politics, and he continued his interest in public lands and timber management. He was a focal point of debate and attracted much criticism because of his activism. He was witty but fought with documented facts, irony, and sarcasm. He rarely turned down an excuse to debate, either face-to-face or through the newspapers. His German background was frequently used against him, but he rarely had difficulty in deflating his critics, usually because of their obvious financial interests in the issues.[20] For example, Schurz easily undercut criticism of his timber policies by Senators James G. Blaine (New York) and Timothy O. Howe (Wisconsin), both of whom were financially and politically dependent on lumber companies.[21]

William Andrew Jackson Sparks was even more aggressive than Schurz in the exposure of corruption in the General Land Office.[22] Appointed as Land Office commissioner in 1885 by Secretary of

Interior Lucius Q. C. Lamar, Sparks generated a series of annual reports that documented the fraud in land dealings. His reports were effective because the criticism appeared in the publications of the very agency he was attacking. Sparks instituted operating procedures in the General Land Office that closely adhered to the legal responsibilities of the office, but these procedures resulted in a drastic slowdown in the transfer of federal lands into private control. Economic and political pressures to speed up the process were soon brought to bear on Sparks and his superior, Secretary Lamar. When William F. Villas replaced Lamar in 1887, the pressure increased and Villas dismissed the crusading land commissioner the following year. Although Lamar and Villas expressed the intention of instituting more rapid processing of land claims by replacing Sparks, both secretaries had commercial interests. Lamar had connections with land-hungry railroads and the Villas family was extensively involved in Wisconsin lumbering and land speculation.

Sparks had political views that were similar to Powell's and Schurz's. He was known as a Jeffersonian democrat and served four terms in the House of Representatives, frequently campaigning on a platform of land reform, railroad regulation, and improvement in the civil service. When he was appointed land commissioner, he had returned to private life in Illinois from government service, so he had few political connections to impair his purge of the land office. His independence in large part explains his success in exposing and documenting fraud, despite a hostile Congress and press. Western editorialists were especially vehement in their condemnation of his efforts, which they perceived as unnecessarily restrictive and biased against the cattle industry. *The Wyoming Daily Sun's* comment was typical in spirit if not in form: "Thou shalt have no other gods than William Andrew Jackson Sparks, and none shalt thou worship. Thou shalt not raise cattle upon the land neither sheep nor any living thing but only corn the same as in the State of Illinois."[23]

Many of the important reforms in the federal approach to the public lands took place during the first administration of President Grover Cleveland, who was elected on a reform platform in 1884.[24] His record as a consistent reformer as mayor of Buffalo and then governor of New York provided his base of popular support. Sparks had Cleveland's direct support in the purge of the land office as well as in the recommendation for eliminating the land laws for disposal. Cleveland indicated that if Congress was unable to come to grips with the land problems, then control of the public domain should go to the state.[25]

This suggestion was not pursued during the first sagebrush rebellion, but in the twentieth century it would become a central theme. It appears that Cleveland suggested the transfer more as a political ploy to stimulate Congress to enact reform legislation rather than as a bona fide proposal.

Reformers in the executive branch were generally faced with a hostile Congress, where the entrenched interests of resource developers and their monopolies prevailed. In the Senate there was almost no support for altering land laws that enriched the timber, mining, ranching, and water companies, but there were a few individuals in the House of Representatives who occasionally challenged the power of Senators Stewart and Teller and their clique. Foremost among the reformers were Abraham Hewitt of Colorado and John H. Reagan of Texas.

When Powell's arid lands report first appeared in 1878, Hewitt's support was critical because he was a member of the House Public Lands Committee, and he could blunt the opposition. Antireform efforts were led by another Coloradan, Thomas M. Patterson, who was a Denver lawyer who made a fortune in real estate investments.[26] Hewitt was successful in continuing congressional investigations of land laws, and throughout the first sagebrush rebellion he resisted the efforts of western resource developers to derail reforms.

John H. Reagan was an able ally of Hewitt in the House of Representatives whenever land reforms became an issue. As with many who sought reforms in the first sagebrush rebellion, Reagan was from a family background centered on small farms. Later, as a senator, Reagan was an ardent supporter of the Irrigation Survey, and he was instrumental in securing passage of the October 2, 1888 bill, which brought the rebellion to a head. Whenever Stewart launched one of his attacks on Powell, Reagan spoke for the defense, but Stewart was clearly the more powerful of the two. One of Reagan's strategic weaknesses in the debate was that as a representative of Texas, he was from a state with no federal lands.

An analysis of the personalities influencing the course of the first sagebrush rebellion would be incomplete without recognition of the role of Hilary A. Herbert, congressman from Alabama. Herbert was a strident advocate of economy in government and somewhat of a gadfly for the sagebrush rebels as well as for the reformers. His drawn face and piercing eyes belied a subtle sense of humor. After lengthy testimony before a House Committee on Irrigation where Powell had inundated the committee with maps, figures, and detailed analyses,

Herbert wearily moved "that the meeting adjourn until Saturday, and it be understood that Maj. Powell commences talking at half past 10 o'clock, whether there is anybody to talk to . . . or not."[27] Herbert constantly corrected the testimony of witnesses as well as his congressional colleagues, and he had the aggravating capability of being consistently correct on matters of fact. He brought a sense of realistic perspective to the first sagebrush rebellion that was sorely needed.

The first Public Lands Commission, appointed by President Hayes in 1879, could only be successful if all sides of the public lands dispute were adequately represented. The commission's failure to stimulate resolution of the controversy was probably the product of the bias toward reform in its membership. Powell was a member and successfuly secured the commission's support for his arid lands proposals. Clarence King, as director of the newly formed Geological Survey, was a logical appointment to the commission, but his reform opinions were well known. James A. Williamson, as commissioner of the General Land Office, was also a logical appointment to the commission, but he had been installed in the Land Office by then Secretary of Interior Schurz with the obvious intent of reforming the office. In addition, Williamson's record was somewhat checkered.[28] Despite his reformist exterior, he was involved in two questionable cases. He had staked a settlement claim in Utah seven years before his appointment as land commissioner, but was refused a patent (deed). When he became commissioner, the decision was reversed. Also, after leaving the Land Office he directly entered the service of a corporation he supposedly regulated, the Atlantic and Pacific Railroad Company. Another commission member, Thomas Donaldson, appeared to be a political neutral whose primary function was a gatherer of statistics.

Of the five commission members, the antireformists had only one representative, Alexander Britton. Independently wealthy, Britton was a land lawyer who assisted in codifying public land laws. His firm, Britton and Grey, was especially effective in handling claims that encountered difficulties in obtaining approval from the General Land Office because some legal clerks in the office were Britton's relatives.[29] A claimant in need of "hurry-up" processing by the Land Office was usually wise to invest in the services of Britton and Grey to secure the desired federal approvals. Britton was outnumbered on the Public Lands Commission, and the commission's report was an advocacy document for reform rather than a politically useful tool for compromise. As a result, it failed to bring about an effective solution to the controversy.

One problem in the controversy was a lack of accurate knowledge about the western environment and about the physical-nature processes that acted in that environment. The U.S. Geological Survey occupied a crucial position, therefore, as the only organization qualified to answer questions about the development of western resources.[30] Although Powell clearly assumed an advocacy position in the rebellion, the agency he headed contained most of the available impartial experts.

By the time the first sagebrush rebellion had developed into the impass over the irrigation lands, the survey was almost a decade old. Under Powell's direction it had grown into a research organization modeled after the German science institutes of the era.[31] Despite the opposition of Representative Herbert, it became the focal point of basic research in the earth sciences and produced most of the advances in U.S. geology and geography in the late 1800s. When questions arose as to the stability of climate in the arid West, and it was uncertain whether or not the advancing agricultural frontier stimulated increases in rainfall, the survey was the agency that was called upon to resolve the issue. Powell called on his close friend G. K. Gilbert to investigate the Great Salt Lake as an indicator of climatic change. The result was a classic work of environmental research that did not totally resolve the issue, but which is still used as a major scientific reference.[32]

Gilbert's role in the first sagebrush rebellion was important but indirect. He was so competent as an administrator that Powell left the survey virtually in his control and was thus free to fully participate in the ongoing political processes.[33] Powell spent prodigious amounts of time lobbying congressmen, speaking at public gatherings, and testifying before committees, all while controlling the major government research agency through Gilbert as his surrogate. Gilbert was purely a researcher (and reluctant administrator) who had absolutely no interest in politics. His entire career as a pioneering geologist and physical geographer was highly productive, and his work added luster to the young Geological Survey, further strengthening Powell's base of bureaucratic support.

Powell and Gilbert were close personal friends and, along with another geologist, Clarence Dutton, they dominated early research in the intermountain area. Dutton was similar to Gilbert in that he was apolitical, especially so because he was a captain in the Army on leave to civilian agencies.[34] Dutton served as secretary of the first Public Lands Commission in his closest brush with politics. Dutton, like Gilbert, was a perceptive scientific researcher who made it much easier for Powell to allow the survey to operate on its own while he was off to the political wars.

Although most earth scientists viewed the survey to be productive of useful research, approval was far from universal among members of Congress. Senators Stewart and Teller saw the survey (and the associate Irrigation Survey) as an empire constructed by a power-mad director. Senator N. A. Gorman expressed the feelings of many congressmen when he testified that the survey was "the asylum of all the scientists in the country who have nothing else to do," and "the final step of the young men who are graduating at the colleges of the country.[35] Reservations such as these were central to the controversy of the first sagebrush rebellion because, after October 2, 1888, lands were not available for disposal until this organization completed its analysis, and many political critics had no faith that the surveys of irrigation lands were being expeditiously completed.

By the late 1880s and early 1890s, the philosophy of land preservation was building a group of followers toward a political critical mass. The movement was largely embodied in one man, John Muir.[36] Throughout most of the 1880s, he produced no significant additions to his published literature, but spent most of his time and effort managing his wife's ranch in Martinez, California. His wilderness writings of earlier decades were well known, especially in the urban East, and it appeared to Robert Underwood Johnson, editor of the *Century* magazine, that the literary market was overdue for another Muir infusion.[37] It took more than a year, but Muir was finally pursuaded to take up the crusade for Yosemite, and his articles, supplemented by Johnson's editorials, mobilized Eastern public opinion in favor of federal preservation of an enlarged area around Yosemite Valley. Both authors railed against the commercial interests who were prominent in the main Yosemite Valley, and the issue fit neatly into the magazine's anti-monopoly format.

The Yosemite controversy was a reflection of the irrigation and public lands issues of the first sagebrush rebellion because the opposition to change was manifest in commercial entities with financial interests in the outcome. Muir had a well-developed dislike for monopolies, and at first he felt that the Southern Pacific Railroad, owner of a spur line that served the park area, led the opposition to federal preservation efforts. Leland Stanford, head of the railroad, was non-committal, but when he was replaced by Collis Huntington, the issue moved to resolution, in large part because of the behind-the-scenes support of the Southern Pacific political machine.[38] The railroad's support could not be too obvious because suspicious observers expected only self-serving actions from the railroad lobby. In the end, even Muir expressed gratitude to the railroad.

Except for this single instance, the preservationists had little impact on the first sagebrush rebellion, but the groundwork was laid for a phenomenal growth in the movement. Muir was an effective writer, and his almost religious attitudes toward environmental preservation were slowly attracting interest among opinion leaders and decision makers. After 1890 he appeared more frequently in public, and was an even better speaker than a writer. His approach to nature was radically different from the scientists of the Geological Survey, but they respected his naturalist experience and frequently read his work. On a personal level, however, scientist and evangelist did not mix well. Gilbert, who helped host Muir during a Washington visit wrote "Muir has gone and I think everyone is relieved . . . his talk, which never ceased, became very tedious. I don't believe I have any use for him."[39]

A preservationist more appealing to the scientists was George Perkins Marsh.[40] Marsh was the first modern American physical geographer, and in his book *Man and Nature* he outlined in scientific and historical terms the intricate dependencies of human society on the natural world.[41] During diplomatic assignments to Turkey and Italy, he toured many Mediterranean countries and concluded that the downfall of many ancient civilizations was rooted in inadequate land management. Accelerated erosion and soil loss resulting from overgrazing and reckless lumbering had destroyed the productive landscapes upon which civilizations were ultimately based, and Marsh warned his fellow Americans that they were on a similar course to disaster. His arguments were well reasoned and fortified with unassailable facts gathered from his wide travels. His book, first published in 1864 and revised in 1874, was widely read. Muir read and was influenced by the work,[42] and the Survey scientists apparently accepted many of his conclusions.

In addition to wise management and conservation of marketable resources, Marsh urged wilderness preservation as a part of public land policy. It would seem a century later that political activists borrowed directly from Marsh when they explained the reasons for wilderness preservation. He wrote that it was desirable to preserve part of the American landscape in its primitive condition as a "museum for instruction . . . a garden for recreation, and an asylum where indigenous trees, . . . plants, . . . and beasts may dwell and perpetuate their kind."[43] More than any other single person, Marsh can be identified as the originator of American conservation,[44] and his influence pervades the preservation arguments in all the sagebrush rebellions. When his sense of science and history were combined with the sense of the wilderness religion of John Muir, the preservation movement was about to bloom. The bloom was too late to have an impact on the first sagebrush rebellion, so that in the late 1800s the commercialization of the American public lands policy was almost total.

Chapter 3

The Mythical Land of Gilpin

As the grand strategists planned their intricate maneuvers in the first sagebrush rebellion, the object of the controversy, the environmental resources of the sagebrush region, remained little known or understood. Accurate maps of the region were unavailable until the geographical and geological surveys began to produce results in the 1870s, but after their efforts many significant blanks remained in the regional picture. Rivers flowed across the maps in wiggling lines that might or might not approximate reality, and entire mountain ranges did not appear on many supposedly accurate renditions. Western promoters were undeterred, however, and where known geography was lacking, some happily supplied imaginary terrain. William Gilpin, an early promoter of Colorado, advised eastern investors of the productive plains and meadowlands of the arid region, and fabricated an entirely imaginary kingdom in Colorado, complete with fictitious landscape.[1] This mythical land of Gilpin was a simple extension of more limited flights of fancy by other developers infected with the disease of fabrication referred to by one historian as "Gilpinism."[2]

Potential townsites were especially attractive as foci for investment, and when none were handy, they were invented. Stimulated by reports of Powell's pending journey through the Green and Colorado River systems, Samual Adams set out on a similar mission from the mining camp of Breckenridge in the Colorado Rockies.[3] In four small rowboats with few provisions, Adams and ten associates managed to struggle down the Blue River and some unknown distance down the Colorado River (then known as the Grand River). After carrying the boats more often than riding in them, they walked back to civilization, apparently without even passing the modern site of Grand Junction. Adams knew from reports and common sense that the Colorado and Green Rivers eventually joined at a confluence, so he began promotion of a town at the meeting of the two rivers even though he had never seen it. Grandiose plans were laid for Junction City, and accommodation was made for a busy commercial center located at the obvious crossroads of transportation in the region. Adams could envision the well laid out streets and stately homes, and could almost hear the whistles of profitable steamboats. Gilpinism, even in its most highly developed form, was no competition for reality though, and plans for Junction City evaporated like dew on a hot desert morning. The confluence of the Green and Colorado Rivers lies in narrow, steep-walled canyons over 1,000 feet deep. The only flat land is on a few temporary sandbars and beaches, all inundated by spring floods. In 1964 it became a part of the wilderness section of Canyonlands National Park.[4]

In the 1870s and 1880s, almost all of the land in the sagebrush region was in the public domain, so that ignorance about its characteristics was doubly hazardous. Private and corporate investors from the eastern United States and Europe had no reliable source of information about the areas of their investments, and depended solely on their trust of the promoters. The second hazard was to the nation's public policymakers who were charged with making laws and executive decisions with little or no information. Public documents concerning the region were sketchy accounts of explorers such as John C. Fremont (somewhat of a Gilpinite himself) or the more accurate but incomplete accounts of the early surveys of Powell, Wheeler, Hayden, and King. The published works of these reconnaissance surveys are filled with geologic, anthropologic, and biologic observations that reflect the interests of the survey leaders and were included in spite of the wishes of Congress, which paid the bills to obtain extensive and accurate evaluations of the public lands. Not only was basic geography of the sagebrush region a mystery, but its major resources of climate, water, land, and minerals were known only at a speculative level.

One of the major problems in management of the western environment was the mystery of its climate. In 1810, when Zebulon Pike reported on his brief reconnaissances in 1805 and 1807 of what is now central Colorado, he labeled the area as the Great American Desert, and the name stuck.[5] His impressions were reinforced by similar conclusions reached by Major Stephen H. Long in 1820. For decades the land beyond the prairies was assumed to be agriculturally unproductive and of little value. This impression was slowly changed, however, as farmers pushed the frontier of agriculture westward after the Civil War. As they moved into new areas, they found that crops could be produced in this supposed desert, and by their own observations they concluded that "rain follows the plow."[6] This axiom held that because civilization brought turnover of the soil, deep-seated moisture was released to the air, and that it returned to earth as precipitation useful for crops. By the 1870s and 1880s enthusiasm for agriculture in the area once thought of as desert was so great that farms were being extended into marginal areas. Farms were abandoned and reoccupied regularly when available moisture fluctuated. When Powell commissioned G. K. Gilbert's study of the Great Salt Lake in an attempt to determine whether climate was indeed changing, the only answer he received was that on a short-term basis more precipitation was occurring in the sagebrush region than previously, but that such a change might be part of an unknown cycle, and might only be temporary.[7]

The concept that settlement stimulated rainfall was probably a reflection of the general optimism of the country in the 1870s.[8] The idea received credence through frequent repetition by eastern journalists, especially Bayard Taylor of the *New York Tribune* and Samuel Bowles of the *Springfield Republican*. Railroad and real estate promoters inflated precipitation records to attract settlers. Ferdinard Hayden, whose survey was eventually combined with others to form the U.S. Geological Survey, supported the idea of human-induced rainfall. The most ardent prophet of climatological bliss was probably Charles Dana Wilber, a Nebraska land speculator. Along with his close friend Samuel G. Aughey, professor of Natural Sciences at the fledgling University of Nebraska, Wilber concocted a complete treatise concerning the beneficial influences of civilization on rainfall, which he published in 1881.[9]

In his book, Wilber railed against eastern conspirators who he thought were trying to thwart the westward progress of the frontier. Weather observers and scientists were dismissed as wiseacres, kid-gloved experts, closet philosophers, charlatans, and quacks. He contended that the eastern scientists seemed to feed on gloomy prognostications of the western climate "for a reason similar to that which calls for stimulants." Data could also be easily dismissed: "notwithstanding the fact that the meteorological observations taken at U.S. forts show but a slight increase between the time prior to and subsequent to settlement, I still firmly believe that the increase of rainfall in actual inches is considerable." Railroad promoters obviously wanted potential settlers to believe so too, and filled the final pages of Wilber's book with advertisements.

One of the prevaricators of facts who was the subject of Wilber's scorn was William B. Hazen, a military officer with extensive experience in western weather observations. Hazen, a close friend of Presidents Garfield and Hayes, was appointed Chief Signal Officer in the War Department in 1880, and managed the newly formed Weather Bureau. In his popular publications he ruthlessly exposed railroad and land promoters who exaggerated true conditions. He supported his contentions of climatic variability and cyclic change with measurements from military forts, and in an era when dendrochronology (the study of tree rings as evidence of climatic change) had not even formally begun, he noted that "the annual layers of timber show this change of seasons [drought cycle] to be the regular order."[10]

By the 1980s, the development of computerized techniques and extensive collections of samples permitted a reconstruction of regional drought conditions using tree-ring evidence.[11] The results indicated that although drought conditions were probably not widespread during Pike's reconnaissances of 1805 and 1807, Long's observations in the 1820s were probably made during an exceptionally dry period.[12] The tree-ring evidence suggests that during his expedition the western interior was experiencing a drought that was the most severe in over 200 years and was worse than the one in the 1930s. Recovery of the region, especially in the 1870s, led observers to their optimistic conclusions and to an unwarranted distrust of Long's powers of observation.[13]

By 1980 almost a century of precipitation records could provide some answers to the questions posed by the earlier researchers. Rainfall in the sagebrush region has been remarkable as much for its variability as for its scarcity (Figure 4). For example, at Salt Lake City in the heart of Powell's arid region, annual rainfall has varied from a maximum of 23.7 inches in 1875 to a minimum of 8.7 inches in 1979. Radical fluctuations have occurred from one year to the next, and in one case, a well-watered year with 18.4 inches was followed by a drought year with only 9.0 inches. Classifying Salt Lake City as part of Powell's arid region, where annual rainfall is less than 20 inches, has been quite sensible over the period of record from 1875 to 1980. However, in one three-year period, the annual rainfall was less than 10 inches and in another nine-year period, the annual rainfall exceeded 20 inches.

Salt Lake City's precipitation record shows evidence of long-term cycles as suspected by Hazen and Gilbert.[14] Generally, the odd-numbered decades (1870s, 1890s, and so forth) were characterized by declining annual precipitation, while the even-numbered decades

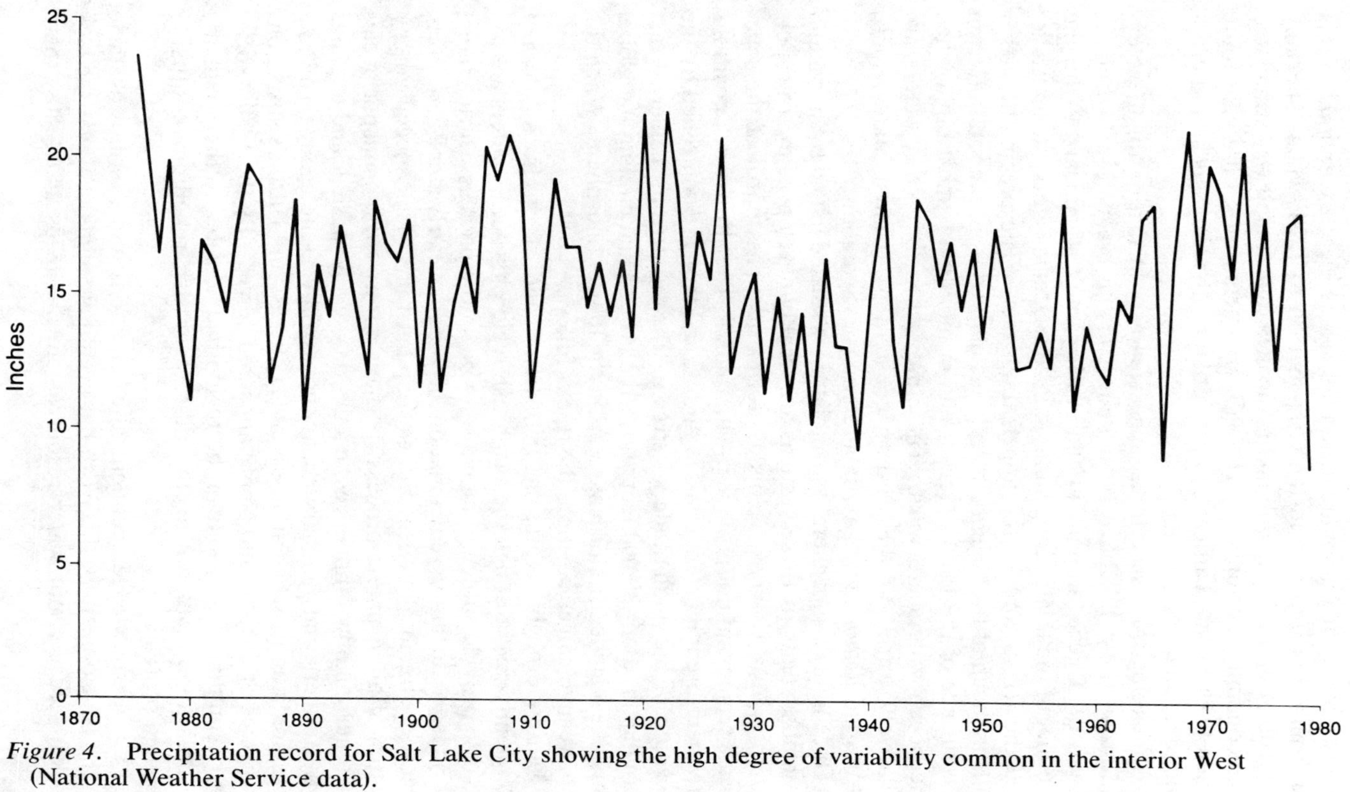

Figure 4. Precipitation record for Salt Lake City showing the high degree of variability common in the interior West (National Weather Service data).

showed increasing amounts of rainfall. There were exceptions to this generalization, most importantly from 1886 to 1890, when especially dry conditions aided in debunking the idea that rainfall was stimulated by economic development. Hazen's arguments took on added force when farmers and cattlemen went broke by the thousands, and crop- and rangelands desiccated.

Climatological research has suggested an explanation for these cycles that could not have been known in the era before high-flying aircraft.[15] Cyclic drought in the American West appears to be the product of the relationships among Pacific Ocean waters, storm tracts, and the jet streams of the upper atmosphere. The precipitation that falls in much of the sagebrush region is derived from evaporation from the surface of the Pacific Ocean, and evaporation is most rapid when the ocean waters are warm. Therefore, when the ocean waters are relatively warm off the western coast of the United States, more atmospheric moisture is available through evaporation.

Once in the atmosphere, this moisture is carried inland by storm and frontal systems. If these systems are weakly developed, the transportation is also weakly developed, storms may pass north of the sagebrush region and not provide it with rainfall. The storm track positions are therefore important considerations, and they are controlled by the position of the midlatitude jet stream. The jet stream, a river of fast moving air about 30,000 feet above the ground and circling the globe, acts like a guide for storm systems in the lower atmosphere. When the jet stream changes position, the tracks follow.[16]

During drought years, the jet stream and storm tracks do not frequently pass through the sagebrush region, or ocean waters are too cold to provide sufficient moisture to begin the process. During moist periods, a fortuitous combination of warm ocean waters and storm track/jet-stream alignment brings frequent rains. This special combination of atmospheric process, perhaps controlled by sunspot cycles or other unknown influences, occurs on a cyclic basis,[17] and explains why farmers and ranchers could successfully argue against Powell's dire predictions, while at other times Gilbert's and Hazen's suspicions of cycles of disaster and plenty appeared correct. By the 1980s, social and economic institutions in the region had adjusted to these climatic fluctuations,[18] but during the first sagebrush rebellion, confusion reigned supreme.

Lack of knowledge about climatic conditions was a serious obstacle in the development of the West, but even more significant was the lack of knowledge about water resources. In the 1870s and 1880s, there were no measurements of river discharges and their variation, even though such data were essential to efficient planning for irrigation agriculture. With the establishment and funding of the Irrigation Survey, Powell began the training of the hydrologists so desperately needed in the West. His friend Clarence E. Dutton and F. H. Newell made rapid progress with their students, but the environment was a formidable foe. When Dutton reported on his efforts to measure discharges on the Salt and Gila Rivers in central Arizona, he wrote, "It is not easy to secure men who are willing to banish themselves for months at a time from all human intercourse and remain alone in one of the cruelest deserts of the world merely to watch the rise and fall of a river."[19]

Even without precise data, much irrigation development took place in the sagebrush region between the end of the Civil War and 1890. The development was stimulated by the voracious appetites of agriculture and mining for water. At first, agricultural production in the region was primarily for local consumption by army troops assigned to quell restive Indians or for support of mining communities.[20] Gold and silver, followed by copper, stimulated great demand for agricultural products that could be produced only by irrigation. In addition, ore processing required large amounts of water to separate waste materials from valuable metal, so that mining ventures stimulated the construction of dams and waterways.[21]

The importance of irrigation to the economic development of the interior West led to a popular movement in support of irrigation that took on almost evangelical fervor.[22] The unquestioned leader of this movement was William E. Smyth, a Massachusetts-born journalist.[23] During the first sagebrush rebellion, and from his position as editor of a Nebraska newspaper, he was an ardent speaker for the unconfined development of public lands. Later, he founded a long-running Irrigation Congress and a major irrigation journal. At one of the meetings of the Irrigation Congress, Smyth proposed a unique combination of private and government capital to irrigate a million 40-acre farms on the Western public domain.[24] Powell pricked Smyth's irrigation balloon by pointing out the lack of suitable land or water for such a fantastic scheme, and a furious debate ensued. A Mexican delegate declared it was the only bullfight he had ever seen in the United States.[25]

At the root of the issue was lack of knowledge about the available land and water resources. Smyth and the irrigators relied on optimism and faith in engineering technology, but Powell based his estimates on field experience and knowledge of the limitations of the physical environment and its processes. He was not a pessimist, but simply a realist, and history has proven him right. The irrigators scoffed at Powell's extention of his estimate of 2.8 percent of Utah as irrigable to the entire region including Utah, Wyoming, Colorado, New Mexico, Idaho, and Montana. By 1974, with almost all economically irrigable lands receiving water from small private dams to huge federal structures, the total was 2.7 percent.[26]

Symptomatic of the political and economic pressures superimposed on the lack of knowledge of the entire region were those brought to bear in the case of the El Paso Dam. Immediately upstream from El Paso, Texas, the Rio Grande passes through a narrow gorge of resistant rock with two ideal dam sites. Major Anson Mills, an army officer in the area, first promoted the idea of an El Paso dam in the narrows, but he later became involved as an investigator in an area near Fort Seldon.[27] There, W. H. Llewellyn and his associates proposed to build a diversion work that would take all the water from users downstream, including Mexican farmers who had used the water for decades. In Washington, Mills testified against the scheme, and Llewellyn swore revenge by blocking the construction of Mills's pet project, the El Paso Dam.

The El Paso structure had many attractive features. Powell reported it as a showcase for the Irrigation Survey, there was an agreement with Mexico to divide the benefits, and Congress had appropriated the needed money.[28] But working against it was a layer of quicksand 30 to 50 feet deep at the dam site, and siltation rates so great that the river would fill any reservoir at the narrows within 150 years of the dam's closure. Llewellyn and his friends obtained a charter from the territory of New Mexico to build a dam at Elephant Butte, a site upstream from El Paso near Las Cruces, New Mexico. Their dam would take all the water from the Rio Grande and leave the El Paso site dry. After Supreme Court trials and several decades, a dam was finally built at Elephant Butte, but the effort was a public one. As for El Paso, what little water passes Elephant Butte Dam also passes El Paso because the gorge was never dammed. Just as Powell had warned, without defined irrigation districts and basinwide planning, the private and local interests would prevail over public and national interests.

Cattlemen have played important roles in all the sagebrush rebellions and have been primary forces in the groups urging the release of federal land restrictions. The western cattle industry passed through a series of definable stages in the period after the Civil War. First was the open range period, which extended from about 1865 to 1880. From the core areas of ranches in Texas, herds fanned north and west in search of free grass.[29] The cattle grazed on public land, and cattlemen paid no fees to the federal government, although local governments who could catch up with the itinerate herdsmen assessed some taxes.[30] Many small outfits and a few large ones raced with each other to take as much grass as quickly as possible. The open range system collapsed because overgrazing destroyed some ranges and allowed brush to replace nutritious grasses on others. Violence, stemming from conflicting claims to lands that in fact were publicly owned, led to the demise of the system.

After 1880, the industry became dominated by companies largely financed by eastern and European capitalists who viewed the industry as an opportunity for speculative adventures.[31] For a relatively small investment and with a dependable field manager, the capitalist, like his cattle, grew fat at public expense. The large companies operated on unfenced ranges until the hard winter of 1886–1887, and subsequent drought years forced managers to scale back their operations, to fence portions of the range, and to institute supplemental winter feeding.[32] Fences or not, however, the land being used was publicly owned and without substantive controls, so that the entire system represented a form of public subsidy.

Some farsighted ranchers saw the long-term implications of protective measures on public lands.[33] James E. Tuttle, a Wyoming operator and occasional visitor to the famous "Cheyenne Social Club," argued with his colleagues that the days of the big herds were limited, and that overgrazing was destroying the foundation resource for the cattlemen. Arnold A. Mowry, a well known and widely respected cattleman, gave a brief preview of subsequent sagebrush rebellions when he suggested that the public domain should be released from federal jurisdiction to individuals, who he predicted would treat the land and its grass with more respect if it were their own.

Overgrazing illustrated the pervasive ignorance about western resources. When cattlemen entered many areas in the sagebrush region, they found valleys with flat, grassy floors, but in the late 1800s and early 1900s almost all western valleys were eroded by the streams flowing through them. Deep trenches or arroyos were created in some valleys, and in other cases valleys were stripped of sediments, leaving only barren rocky surfaces.[34] Much of the destructive erosion took place after the introduction of cattle in large numbers. As the cattle grazed the native grasses and trampled new seedlings, the exposed surfaces permitted rainfall to run off quickly instead of slowly seeping into the turf. Once collected in channels, the large quantities of water eroded deep trenches, and once begun, the erosion undermined floodplains along the channels. Lush valley floors that were productively planted in crops were partially or totally eradicated, although climatic changes and engineering works may have contributed to the catastrophic erosion.[35] A century later, the erosion damage still scarred public as well as private land.

Miners played an important role in the development of the sagebrush region because they preceeded other Americans in settlement of Nevada, Arizona, Idaho, and Colorado. When federal surveyors climbed unnamed peaks to make their readings, they invariably found the broken whiskey bottles of unknown prospectors who already had been there.[36] Much of the early prospecting, because it occurred before scientific surveys, was hit-and-miss and depended on subtle indicators such as slight color changes, unusual breaks in slopes, and even vegetation changes. Many mining ventures were not made for obtaining metal, but rather for purposes of development and sale of stock and securities. The major federal law controlling prospecting on public lands, passed in 1872, was essentially designed to give prospectors free rein. The federal government was not so much concerned with obtaining funds from leases of mining rights as it was with stimulating a home-grown mining industry. The public therefore received few direct benefits from the minerals on public lands.

Federal surveyors were sensitive to coal deposits and recorded their observations whenever such fuels were found, but many surveyors were not geologists, and some were corrupt. Private companies hired their own specialists to identify valuable coal lands, and then sought to gain control of the lands by any means available, usually with an effort to avoid lease charges. For example, near Trinidad, Colorado, 61 preemption entries were made on public lands in 1873 for a total of 10,000 acres of "agricultural" lands.[37] All the entrymen and witnesses were later proved to be fakes. The lands were paid for in government script, a form of federal IOU, in the name of A. C. Hunt, a former governor of Colorado. Federal land agents discovered too late that the lands were more than simply agricultural: they contained large quantities of coal. Eventually, all the lands were controlled by the Colorado Fuel and Iron Company.

In the Trinidad case, the investors, even if they operated illegally, knew the nature of the resource. Others were not so lucky. In the late 1800s, the prevailing scientific opinion concerning copper deposits was that if an outcrop of copper-bearing rock appeared on the surface, the copper-rich zone widened as depth increased downward from the surface.[38] Thus, if a small outcrop was located, a flurry of financial activity was sure to ensue as investors sought to obtain a part of the anticipated profits. Unfortunately, in almost all copper deposits in the western United States, the ore body becomes more narrow with depth, so that small outcrops were not underlain by massive deposits. In many cases the paper profits generated by stock trading exceeded the profits from the sale of actual metal, and the public purse benefited from neither.

Thus it was that for many Americans the region between the Rockies and the Sierras was truly the land of Gilpin. To them it was a treasure chest owned by no one and waiting for anyone with the key of unbounded optimism. The climate was assumed to be changing to one increasingly favorable to agriculture; water resources were available to irrigate the lands where climate was uncooperative; cattlemen saw free resources for unlimited expansion; and miners saw a mother lode under every outcrop. Those few individuals who were aware of the true conditions either worked for private corporations or, if they were in the employ of public agencies, were muzzled by political opposition in and out of government.

A major aspect of the first sagebrush rebellion that prevented its speedy and efficient resolution was that the object of the controversy—the environment—was poorly known and poorly understood by rebels as well as the federal agencies. When Powell received his appropriations for the Irrigation Survey, he recognized the need for accurate knowledge and immediately initiated an effort to intensively survey potential reservoir sites. But while Powell's minions scoured the western mountains in search of likely sites, and while the General Land Office held up all land sales in anticipation of the results, the sagebrush rebels began to work on the other side of the continent in the nation's capital.

Chapter 4

Resolution and the Seeds of Discontent

In 1890, the activities of the first sagebrush rebellion took place mostly in congressional hearing rooms. William Stewart had mounted a broad attack on the Geological Survey and the Irrigation Survey.[1] During Senate hearings, he had tried to discredit the topographic mapping activities of the Irrigation Survey as unnecessary, unauthorized, wasteful, and a pet project of the director.[2] Later, he and most other western senators tried to legislatively undo the authorization of the Irrigation Survey, but its authorization, through sundry appropriation bills, made the survey hard to kill. It never had an official birth.

Stewart and his allies, especially Gideon Moody, attacked Powell personally, charging him with every misconduct possible, including misappropriation of funds and failure to follow the directives of Congress. Every scientific controversy Powell had been involved with was resurrected and explained as evidence of his lack of ability. Stewart argued that Powell was a megalomaniac, attempting to withdraw vast areas of the public domain without regard to the professional opinions of others.

In order to demonstrate to the rebels who remained at home that the fight was being carried forward, Stewart and other congressional representatives established an Artesian Survey in the Department of Agriculture. The new agency was to aid in the development of artesian water sources, and its creation was a deliberate evasion of the Geological Survey, which was clearly on record as believing that such wells were useful only in very limited areas. The Artesian Survey flourished briefly, but it was doomed to a short and unproductive life. The farm lands watered by groundwater would have to await the development of effective deep pumping systems, which were decades in the future.

Opponents of the Irrigation Survey attempted a direct repeal of the legislation that authorized withdrawal of lands in the arid region until survey of dam sites was finished, but they were unable to do so, partly because they did not have the support of President Harrison. Failing in this legislative approach, they resurrected the question of proper definition of the arid zone. The purpose of part of this effort was to undermine Powell's position, but it was also a well-orchestrated advertising campaign for western development. Gilpinism was in full force in the halls of Congress as testimonials frequently extolled the productivity of the lush western landscape. Powell, to no avail, countered with his now-familiar arguments on the lack of adequate moisture.

The apparent point of contention focused on the question of topographic surveys. Stewart pointed out that engineers had no trouble building dams in many areas without Powell's topographic surveys. "I wish to say there is not an experienced engineer in America I have met with who agrees with the Director," stormed Stewart.[3] There were only limited numbers of potential dam sites, Stewart contended, and in many cases the topographic surveys were useless because they resulted in the selection of dam sites that were structurally unsound. The question boiled down to which should come first, the site for the dam or the delineation of the drainage area that supplied the water. A minor but significant detail was that many congressmen could not understand how money they had appropriated for irrigation survey work was being spent on topography. They recognized that topographic information was useful, but could not accept that it was necessary.

Although the bitter disagreement over the utility of topographic surveys appeared to be the crux of the controversy, the true issue was a philosophical one surrounding the role of the federal government in the management of the public lands. Should the federal government divest itself of the public domain as quickly as possible to stimulate a rush to development, or should it manage the environment in support and subsidy of the individual farmer or irrigator? Stewart's position was consistently one of open and free development of the resources at all possible speed and without interference from Washington. Powell's position was consistently one of support for governmental assistance in a slower more orderly development dictated by scientific analysis.

On July 16, 1890, the vote was taken, and the sundry appropriations bill passed Congress with no funds for the continuation of the Irrigation

Survey. A specific repeal of the arrangements to withdraw lands for survey was included. All land claims made during the controversial period were declared valid. Powell's position was completely demolished, and the sagebrush rebels had won. It was the beginning of several years of decline for the survey, and western development rushed into the water business. Dam and canal companies sprang up like weeds throughout the arid region: most died after short lives.

The significance of the elimination of the Irrigation Survey was that water development in the West would proceed without overall direction or plan, and vast investments would be wasted on inefficient water delivery systems. The lack of basinwide planning left a legacy of confusion and legal conflict that would last a century. The failure to adequately classify lands brought about the irrigation of marginal areas to the detriment of other, potentially more productive areas. Finally, the ideas espoused by Powell were eventually enacted anyway at a later time when costs were greater, conflicts more difficult, and established wasteful practices were even more firmly entrenched.

A review of the events of the first sagebrush rebellion reveals clearly why the rebels won and Powell's approach to federal land management lost. First, Stewart was the leader of one of the most powerful group of men in the Senate. When he spoke, he spoke not only for himself, but he articulated the positions of influential men such as Teller and Moody. These were hardly newcomers to the art of congressional debate, and they brought to bear vast experience in unraveling legislative and bureaucratic entanglements. They fought well and were accustomed to winning.

Another strength of Stewart and his allies was that they accurately represented the general opinions of the citizens who elected them. The vast majority of westerners viewed Powell's schemes as foolish designs by a power-hungry, elitist scientist who by happenstance turned up in a powerful bureaucratic position. Western residents saw themselves as the last vestiges of the American frontier spirit, which was motivated by a "full speed ahead" attitude. They were not willing to wait to begin their life's work while a bearded, one-armed old man in Washington frittered away the taxpayers' money on seemingly useless surveys.

Finally, and perhaps most importantly, Stewart and the other sagebrush rebels may have appeared to be motivated at times by the lure of quick profits, but they had a vision of western development as the final expression of America's Manifest Destiny. To them, not only would the nation stretch across the continent, but it would develop into the greatest civilization of the modern world. They saw the West as a part of that great culture, and their consistently overoptimistic views of resources were simple extensions of the optimism of the country. And that country had been built, in their view, with little help from the federal government. Powell did not recognize that these views were honestly held, and he challenged the integrity of his major opponents.[4] He characterized the controversy as a "fight against the speculators pure and simple. I am doing what I can to prevent moneyed sharks from gobbling up the irrigable lands."[5]

From a perspective of several decades after the events, it is also obvious that Powell lost the irrigation controversy in part through problems with his own side of the issue. First, Powell's forces were simply outnumbered. Although Senators John H. Reagan (Texas), Arthur P. Gorman (Maryland), and James K. Jones (Arkansas) supported his position throughout the debate, the sagebrush rebels in Congress were more numerous and more powerful. The indifference to the controversy on the part of the majority of eastern and midwestern congressmen and senators ensured that the western representatives won the congressional voting contests.

Another problem was that the executive branch in general and the Geological Survey in particular lacked skilled debaters. Although Powell was an articulate spokesman for a rational scientific approach to the management of federal lands, he had no one else to call upon to support his testimony. G. K. Gilbert, the brilliant researcher, was reduced to a mumbling nonentity in the intense congressional hearings, and his invovled, long-winded explanations brought confusion to the listeners rather than insight. Powell finally stopped inviting Gilbert to accompany him to the hearings, so that one of the most intelligent individuals in federal science was relegated to administering the survey while Powell played politics. WJ McGee (who preferred the name "WJ" to his full name of William Jackson) was Powell's brightest protégé, but he was too inexperienced in the intricacies of congressional debate to be of much help. McGee, a dapper young man with a cookie-duster mustache, would make his contributions to federal conservation during the region's sagebrush rebellion.

In retrospect, Powell made a strategic error in placing so much emphasis on topographic surveys. Such surveys were at the root of his plans for orderly western development, but by allowing them to become the focal point of controversy, he allowed the debate to stray from the true issue of the proper role of the federal government in land management. The error was compounded by his lengthy bureaucratic history of supporting the topographic surveys at the expense of other programs, so that he appeared to be self-serving when he seized on the irrigation question as a vehicle for their continuation. Also, he simply overestimated the importance of the topographic surveys in dam site selection. Later experience showed that the suitable dam sites had to be identified first from an engineering perspective, and then topographic surveys would be required for only a limited number of basins. With Powell's approach, all basins had to be surveyed. Powell did not even have the support of his own employees on this issue. When his long-time friend and associate, Clarence Dutton, was called to testify, Dutton clearly indicated the inappropriate nature of extensive topographic surveys for dam site selection. Shortly thereafter, Dutton returned to service in the Army after almost twenty years of being "on loan" to Powell's survey. He and Powell never worked together again.

The basic nature of the problem at hand also worked against Powell. Stewart and the sagebrush rebels were espousing an extremely simple course of action that was easily understood by the public: no government interference with "business as usual." Powell's proposals, however, were a radical departure from established practices, and scientific opinions were complex and based on data not generally known or appreciated by the public. Realizing this, Powell tried to overcome the problem through a miniature public education program of his own, but the job was too much for one person. All his writings for popular publication and public-speaking engagements were still the work of only one man, and in the end he failed to reach the general public.

Finally, Powell's greatest sin was no sin at all: he was ahead of his time. He was a reformer just before reform was becoming acceptable; he was a conservationist before the term was coined; and he believed in arid-land reclamation before it became recognized as a national goal. Powell's intelligence and vitality were great enough to hasten reform, conservation, and reclamation, but his mistakes and his opponents were enough to prevent their realization until the twentieth century.

With the Irrigation Survey eliminated, western rebels played their trump card, a May 1890 bill to cede all unappropriated lakes and rivers to state or territorial control. Reformers defeated the bill, but its philosophy reappeared in every subsequent sagebrush rebellion. Recognizing public pressure to end land fraud and congressional pressure to simplify the chaotic land laws, several representatives mounted a campaign to revise the entire system of federal land disposal. Lewis E. Payson (Illinois) and William S. Holman (Indiana) assembled a divergent group of supporters for a General Revision Act designed to eliminate the most fraudulent aspects of land laws. Up to this time Republicans had generally been on the side of the rebels and Democrats had voted for reform, but bipartisan sponsorship of the bill with the backing of President Harrison ensured its ultimate success. The public, aroused by editorialists in their demand for change, could no longer be ignored. So it was that the Republican party, long-time supporter of land laws that were easily subverted, eventually shared the credit for their repeal.

The General Revision Act, finalized in 1891, repealed the Preemption Act, Timber Culture Act, and the auction sale of public lands.[6] It made major alterations to the remaining land laws, including the Homestead Act, to reduce corrupt practices. The concept of classification of public lands, a logical precursor to land use planning, was debated but did not survive to the final form of the act. Section 24 of the act attracted little attention outside a limited circle of congressmen and lobbyists, but later it assumed immense importance to federal land policy. The section authorized the president to set aside areas of the public domain as forest reserves in order to protect and manage the timber resource. This forest reserve section signaled a dramatic change for public land policy away from mere disposal of lands to active management, in part as a response to the problems associated with the irrigation lands.

The act contained the seeds of discontent, however. Unbeknownst to those who voted for the bill, this minor deflection of the course of federal policy from disposal to management would later be an indicator of a complete change in direction that would eventually eliminate the disposal function altogether, and resistance to that change would trigger another sagebrush rebellion. The General Revision Act signaled an end to the first rebellion over irrigation lands, but it also signaled the beginning of the second rebellion over forest lands.

Although the script would be similar for the second rebellion, the players were destined to be different. The land itself was different. Widespread erosion had forever changed the mountain sides and valley floors. The stirrup-high grass in western meadows was a story related by mountain men, trappers, and guides, to be contrasted with overgrazed wastelands in the 1890s. The original grasses were being replaced by juniper, Russian thistle, and cheat grass. Floods, once restrained by beaver dams and forest-covered slopes, roared unimpeded down the desolate canyons and valleys.

The cultural landscape of the 1890s was changing too. The U.S. Census Bureau, in a little noted comment, stated that the frontier was no longer evident in the distribution of the American population.[7] Rutted wagon roads and steel rails split the sagebrush region into ever-smaller pieces of wilderness, which were further diminished by mining camps that blossomed and died according to the whims of an emerging national economy and the vagaries of geology.

The combatants of the first rebellion went on to other, if not bigger, things. Big Bill Stewart and Hilary Herbert continued their relentless attacks on the U.S. Geological Survey, and extracted the revenge of victors as they reduced its budget by two-thirds. The Survey released most of its scientists, once labeled wiseacres, and shrank almost to oblivion before beginning a resurgence with the new century. In 1894, Powell resigned as director to the satisfaction of the leading rebels. Among those arguing for increased appropriations for the reemerging survey at the end of the century were Stewart and Herbert.

Stewart and Teller, the leading rebels, served in the Senate well into the new century, and sponsored a long list of legislation to improve life in the West, including enlightened treatment of Indians and more accurate surveys of mineral wealth. Herbert was also instrumental in the rise of the United States as a major naval power. Gideon Moody, vitriolic spokesman for the rebels, spent a successful decade as general counsel for some of the largest mining companies in the North and West, and died in the gentle climate of southern California in 1904. George Symes, bombastic proponent of the scheme to irrigate forty million acres of the West, retired to the California wine country to write of the glories, half realized, of a great civilization built on irrigation canals.

Most of those who opposed the rebels fared well too. Major Anson Mills, his cherished El Paso Dam a dream never implemented, amassed a fortune from his patents on a woven cartridge belt adopted by the Army. William Sparks and his boyhood friend, S. M. Stockslager, returned to successful midwestern law practices after their tumultuous years as commissioners of the General Land Office. John H. Reagan, champion of the Irrigation Survey in the Senate, returned to Texas and became the state's powerful railroad commissioner. William S. Holman, one of the originators of the forerunner of the General Revision Act, served in the House of Representatives until his death in 1897, and became the foremost Jeffersonian Democrat of his day. His colleague in proposing wholesale revision of the land laws, Lewis E. Payson, survives in spirit: Payson, Arizona, a small town nestled in the pine-covered mountains, was named after him.[8] G. K. Gilbert, author of the research on the climatic change evidence of the Great

Salt Lake and Powell's able assistant, was released from administrative duties and returned to his beloved research. He became the nation's most successful geologist and geographer.

John Wesley Powell did not fare as well as his fellow reformers. After his forced resignation from the Geological Survey, he continued his anthropological work for the Smithsonian Institution and laid the groundwork for modern anthropology in the United States. But failing health and a broken spirit led to a decline from which he never recovered. After the first sagebrush rebellion, he aged quickly and the fire seemed to go out of him. He socialized less frequently with Washington friends, and he seemed to be adrift, a captain without a ship. In 1902, Congress passed into law the Newlands Act, which began a massive federal reclamation effort in the West. The act embodied all of the major principles concerning arid-land water management that Powell had proposed over twenty years before, but Powell was too old and sick to celebrate.

He had not lost all sense of humor, however. During a period when medical authorities believed that brain size was directly related to intelligence, Powell made a friendly wager with his young aide, WJ McGee, that Powell's brain was the larger of the two.[9] Each left appropriate instructions, and after their deaths, Dr. Edward Spitzka announced the results of his comparative measurements: the director had won. The doctor concluded that Powell was endowed with a superior brain, and that "what is more, he used it well."

Part II

The Second Rebellion: Forest Lands

Chapter 5

Forests Without Trees

When Powell and other early explorers attempted to survey the inter-mountain region of the American West, their work was frequently halted by poor visibility resulting from an early form of air pollution. Smoke from forest fires obscured the terrain, often for days at a time. When Clarence King's group worked in the Rocky Mountains of Colorado, they found forest fires to be so extensive that they suspended survey work during an entire summer,[1] and when Powell mapped the forests of Utah in his analysis of arid lands, he found as much as a third of the once forested lands had burned over.[2] Reports of these extensive losses of valuable timber resources, combined with equally rampant fraud on timber lands of the public domain, led to the prospect of American forests without trees and the inclusion of a forestry section in the General Revision Act of 1891. While the general act signaled the end of the first sagebrush rebellion, it led eventually to a second rebellion as the implications of reserving forest lands from entry became apparent.

The forestry part (Section 24) of the General Revision Act was the product of the efforts of a few individuals and the maneuvers of opposing interest groups. The story of Section 24 and the second sagebrush rebellion began with a seemingly harmless lecture in 1873 by a New York physician and amateur statistician, Franklin B. Hough. He presented his lecture to the highly regarded American Association for the Advancement of Science: the title was "On the Duty of Governments in the Preservation of Forests."[3]

Hough made his historical plea for an active role of government in forest resource management based on his experiences while working on the 1870 U.S. Census. He noted that timber production was declining in the northeastern and upper midwestern states because of the exhaustion of the standing timber in those areas. His statement was spurred by the first stirrings of state activities in forestry as well, and ten states enacted legislation encouraging the planting of trees (Figures 5 and 6).[4] As a widely read individual, he was also influenced by Marsh's *Man and Nature*, just as the irrigation specialists had been influenced before the first sagebrush rebellion. Hough, like Marsh, commented with alarm on the obvious parallels between the loss of timber in Europe and in the United States.[5]

Hough was a persuasive speaker, and the American Association for the Advancement of Science responded to his lecture by appointing a committee to lobby the federal government in the matter of forest protection.[6] The committee, headed by Hough, journeyed to Washington for an intensive lobbying effort. They visited Joseph Henry (John Wesley Powell's benefactor at the Smithsonian Institution), Commissioner of Agriculture Frederick Watts, and President Grant. Their labors paid dividends in the form of a draft joint resolution prepared by General Land Office Commissioner Willis Drummond and Secretary of Interior Columbus Delano. Armed with a statement of support from Grant and the tactical aid of Congressman Mark A. Dunnell (Minnesota), Hough approached Congress with intentions of creating an Office of Forestry. Congressional efforts at economy doomed the attempt from the beginning.

Hough tried again, however, and he was successful in 1876 when his continued association with Congressman Dunnell produced a minor legislative coup. A bill to designate a federal forestry specialist expired in the House Public Lands Committee, but the ever diligent Dunnell attached a rider to an obscure bill as appropriate money for seeds to be distributed by the Department of Agriculture. In addition to the funds for the seeds, the final bill contained $2,000 for the appointment of a forester, the first association between American forestry and the Department of Agriculture.[7]

Hough wanted to ensure that the best man (whom he considered to be himself), would be appointed to the new position and Dunnell assured him that he would be made commissioner of forestry.[8] The appointment was made, and with support from Agriculture Secretary Watts, Hough's work produced the first of three major reports in 1878.[9] This monumental *Report Upon Forestry* identified the major problems facing the U.S. timber economy. Because of the strong role of private property rights, wasteful practices went unchecked; there were no provisions for forest lands in the public land laws; there were no

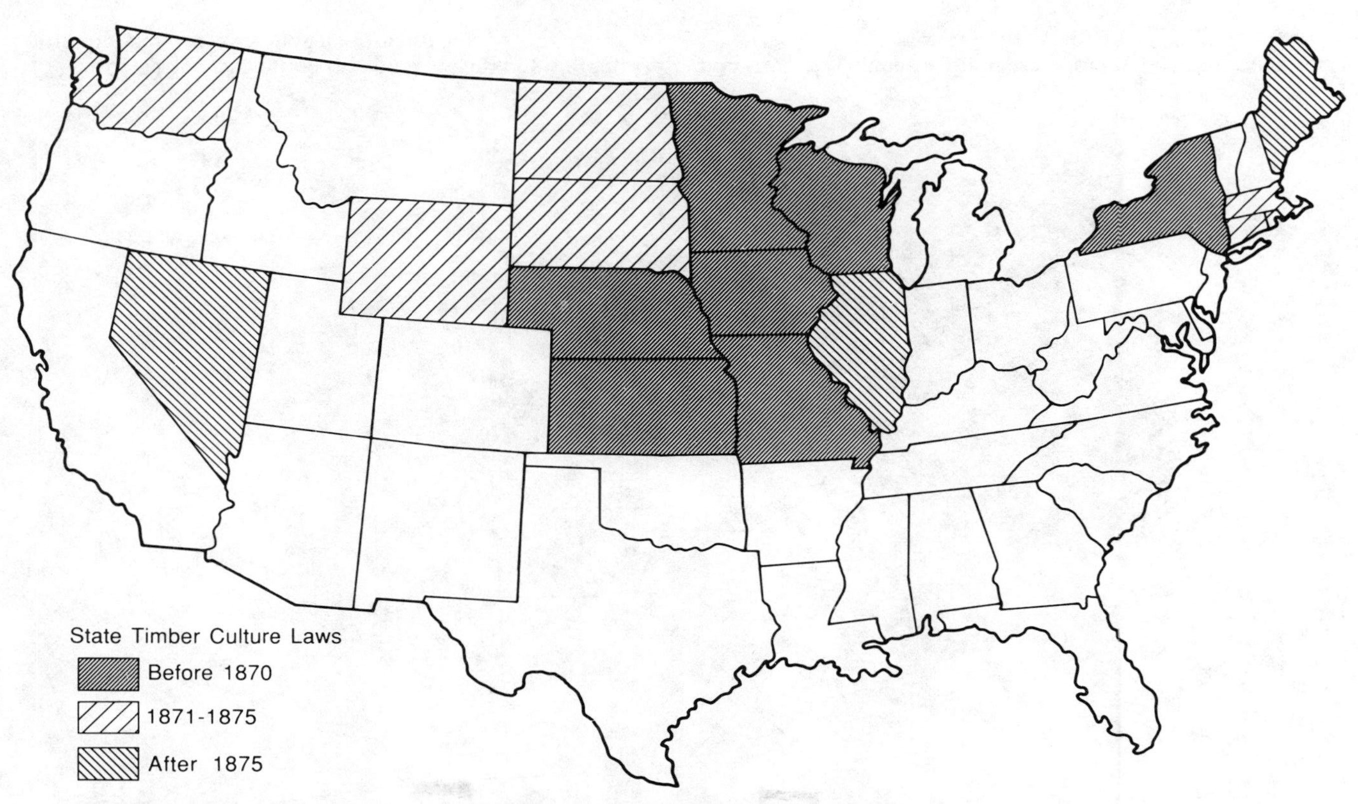

Figure 5. The distribution of states with timber culture laws (data from J. Ise).

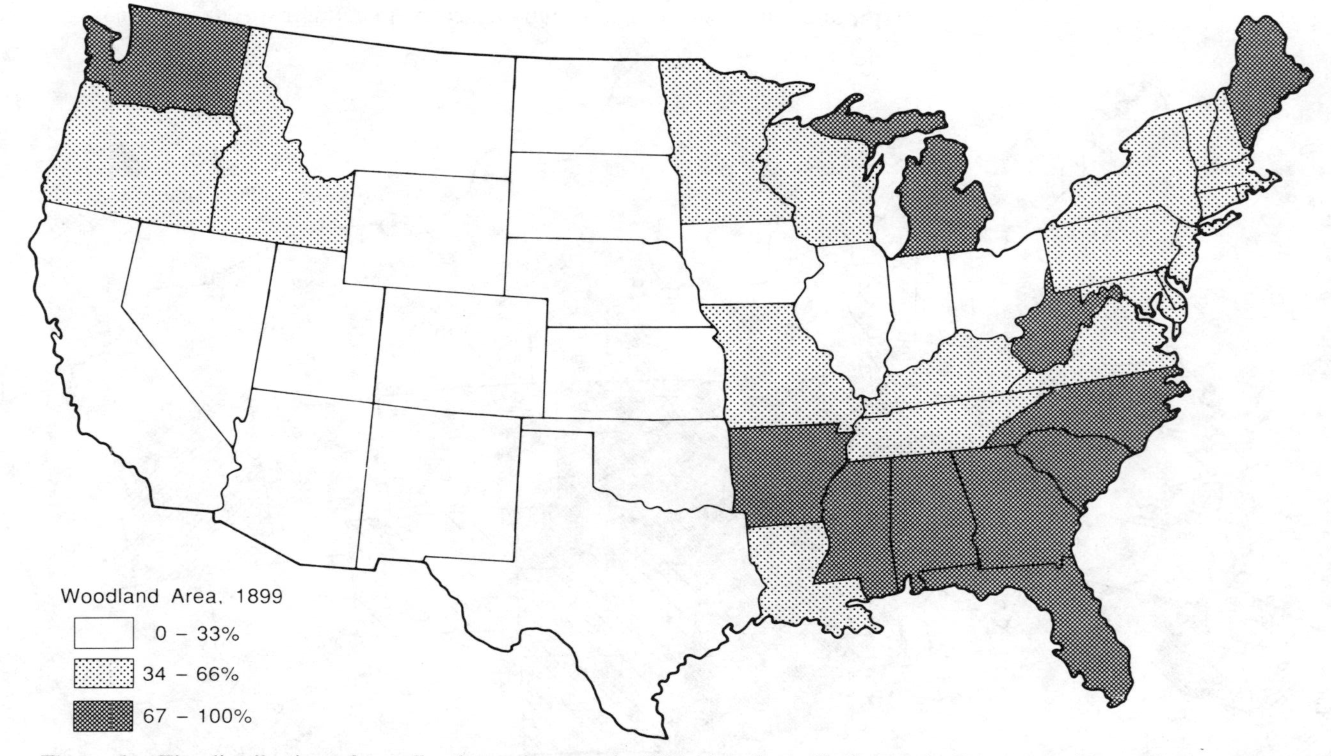

Figure 6. The distribution of woodland area by state in 1899 (data from H. Gannett). Compare with the distribution of state timber culture laws shown in Figure 5.

trained foresters or any attempts to halt theft of public timber; and timber markets were subject to radical price fluctuations. In a particularly insightful section, Hough wrote that America's "pioneer mentality of viewing natural resources as being infinite was largely at fault."[10]

Subsequent volumes of the *Report Upon Forestry* were effective in showing why and how the federal government should become involved in forest protection. Volume II, published in 1880, was a scientific examination of the prominent role of timber in America's export market, and it led to the establishment of an Office of Forestry (with Hough in charge, of course) in the Department of Agriculture.[11] Volume III, published in 1882, provided a review of corrective measures and served as a blueprint for federal activities that would come in the following decades.[12] Hough's forward-looking suggestions included the withdrawal of timberlands from the public domain, establishment of management regulations, and harvesting through leases to individuals and companies. He took a tack directly opposite that of R. W. Raymond, a General Land Office commissioner who noted the pervasive timber thefts and concluded that the "entire standing army of the United States could not enforce the regulations. The remedy is to sell the lands."[13] A similar suggestion surfaced again four years later.[14]

The scientific community took a major role in this early period of forestry development. Research into forestry problems rapidly intensified, and in the year when Volume III of the *Report Upon Forestry* appeared, the meeting of the American Association for the Advancement of Science contained 87 papers on the subject, compared to only one (Hough's) just nine years before.[15] In 1873, William Henry Brewer gave the first academic lecture on forestry in the United States at Yale University, and a year later he published the first comprehensive review of the resource.[16] In 1875, the American Forestry Association was formed in Chicago, and throughout the remainder of the century it was an influential scientific organization with direct and powerful ties to the highest levels of the federal government.[17]

After so much progressive change in a conservative political climate, Hough was flirting with bureaucratic disaster. In 1883 after numerous clashes with the new development-oriented secretary of agriculture, George B. Loring, he was demoted to the lowly status of "agent." Hough was particularly rankled by the secretary's appointment of special interest men as foresters without Hough's approval. One such appointee received Hough's harshest judgement: "a rebel and a Democrat."[18] In 1884, Hough died, but he left a rich legacy that would eventually grow into a professional forest service, a system of national forests, and the world's first wilderness system, all founded on an obscure rider to a little-noticed appropriation for seeds.

The leadership vacuum created by the demise of Hough was filled in 1886 when Bernhard E. Fernow was appointed chief of the Division of Forestry.[19] Fernow, a German immigrant and staunch Republican, was appointed by Democratic President Grover Cleveland primarily because of his European training as a forester and because of the aggressive support of the American Forestry Association. As Fernow was settling into his new office, the first sagebrush rebellion was brewing over irrigation lands, and he could easily see the volatile politics surrounding the general question of public resource management. He managed to make progress, though, and was able to establish congressional statutory recognition of the Division of Forestry. With the aid of Senator Eugene Hale (Maine), he proposed a strong antifraud bill for timber lands, but Senator Preston B. Plumb (Nebraska) took time out from his attacks on Powell's Irrigation Survey to ensure that the bill never progressed beyond the committee stage.

In 1889 it was clear that the plight of timber resources in the public domain was reaching crisis proportions.[20] The American Forestry Association's Law Committee requested an audience with President Harrison to address the issues. Committee members Fernow, Nathanial Eagleston (chief forester prior to Fernow), and Edward O. Bowers (of the General Land Office) made their case, but Harrison was unimpressed, and he took no action.

Recognizing the futility of influencing policy through the executive branch, the Forestry Association lobbied Congress in 1890 and 1891.[21] The efforts, combined with pressures related to the chaotic state of the federal land laws, resulted in the passage in both houses of Congress of a General Revision Act, but differences between the two bills had to be resolved in a conference committee.[22] The conference committee was dominated by congressmen interested in conservation: Senators Walthall (Mississippi) and Pettigrew (South Dakota); Congressmen Payson (Illinois), Pickler (South Dakota), and Holman (Indiana). Only the ubiquitous Senator Plumb (Nebraska) had a record of anticonservation votes. As the conference committee deliberated on the two bills to be resolved into a single one acceptable to both houses, forestry conservationists noticed that although there was no forestry provision in the legislative package, they had control of the conference committee.[23]

The American Forestry Association Law Committee influenced the process by returning to the executive branch and lobbying Secretary of Interior John W. Noble.[24] Powell also joined in the effort to convince Noble to lend support to an effort to add a forestry section to the bill,[25] although Fernow later recalled that at one point frustration on the conservation issue caused Powell to grump that "the best thing to do for the Rocky Mountain forests is to burn them down."[26] Unsuccessful in his bid to temporarily reserve reservoir lands, Powell saw in the forestry section a chance to reserve valuable watersheds to prevent excessive lumbering that contributed to catastrophic floods and erosion. Convinced that the forestry question was one more loose end that could be tied up by the general revision bill, Noble recommended to the conference committee the addition of Section 24.

The conference committee added the section that gave the president the authority to set aside forest reserves from the public domain, thus preventing the designated lands from passing into private ownership.[27] The primary rationale for the reserves would be to protect the timber resource and to preserve the quality of the watersheds involved. The supporters of the measure viewed the section as a means of preserving timber resources and of preventing their wasteful and rapid harvest. The conference committee approved a revised bill that included the new section, and both houses approved the measure. It was soon signed into law by President Harrison.

Although it was not noted at the time, the procedure by which Section 24 was added was illegal.[28] According to congressional rules, conference committees were not permitted to add new provisions to bills. Otherwise, small conference committees could undo the work of the full Congress. Somehow Section 24 was overlooked in the process, and it was signed into law. American forestry continued its erratic legislative history with its first forester paid by funds appropriated for a seed bill and its forests reserved by an illegal addition to a major land law. In later years, when the significance of Section 24 became apparent, several individuals claimed credit for its initiation, including Senator R. E. Pettigrew (South Dakota)[29] and Robert Underwood Johnson,[30] the crusading editor of the *North American Review*. At the time, however, only the American Forestry Association was willing to claim the then dubious honor.

In view of later controversies, the immense power conferred on the president, and the implications for closing significant portions of the public domain to use, it seems surprising that the General Revision Act of 1891 (including Section 24) received such uncontested approval in Congress. Several factors allowed the bill and its phantom forestry section to slip through the system.[31] The bill was a long, difficult document, and it is unlikely that any congressman was completely familiar with its provisions. Because of the bill's complexity, it was impossible to change one clause without setting off a chain reaction of related adjustments. The major points of the bill had been debated many times before, so the views of most congressmen were already on record. The final approval process occurred at the end of the session when many votes were hurriedly taken, and there was little time for scrutiny of individual bills. It appears that no one in Congress recognized the vast powers conferred on the president by Section 24. Most members, if they gave the section any thought at all, envisioned the president withdrawing from the public domain minor amounts of high-quality timber lands and a few select drainage basins important for water supply. None envisioned the eventual development of an extensive system of federally owned forests.[32]

Two major problems with Section 24 immediately appeared when President Harrison proclaimed the first forest reserve (Yellowstone) and then built the reserve system to 13 million acres by the end of 1892. First, there was a lockup of resources by withdrawing significant amounts of western land from potential development, the same problem that stimulated the first sagebrush rebellion. Second, there was no provision for management of the new reserves. Unable to secure a complete legislative package that included reservation of the lands and an apparatus for management, the conservation interests had obtained only half their goal. With the passage of the 1891 law, forests were only preserved, set aside without prospect of productive use. Western opposition to the forest reserve process was swift, and by 1893, the second sagebrush rebellion was a feature of the American political landscape.

Early opposition to the policy of reserving the public domain for forest and water protection came from clearly defined groups in the West, such as lumber companies who were bitterly opposed to forest preservation because they saw the reserves as a deliberate attack on their industry.[33] Many of the largest companies depended on timber from the public domain as a stable supply for their saw mills. Land was purchased legally from the public domain, but much was also obtained fraudulently by manipulating land laws.[34] Because most of the land laws pertained to agriculture lands, representatives of the timber companies first had to attest to the falsified claim that the land was obtained under one of the many land laws designed to stimulate settlement. The land was then turned over to the company for cutting, and, when cut, was usually abandoned and, eventually, taken over by local governments in lieu of back taxes. Without agricultural potential and without timber, however, the lands had no market value.

During the 1880s and 1890s, railroad companies usually remained on the fringes of the forest controversies because they owned large tracts of forest.[35] When Congress granted the railroads public lands as a stimulus for rail expansion, the companies came into possession of prime forest lands in the mountains of all the western states. The Northern Pacific Railroad in particular became involved in the timber business because of its extensive holdings in Montana, Idaho, and Washington. Smaller companies subverted the federal process by obtaining title to public lands by promises to haul freight and passengers. In the case of the Union River Logging Railroad, the only passengers carried were timber cruisers and lumberjacks, and the only freight carried was lumber stolen from lands that were abandoned along with the rails after the timber was cut.[36]

Mining companies generally opposed the policy of forest reservation on two counts.[37] First, the reserve status prevented them from obtaining needed timbers for mine supports and general construction. In many Rocky Mountain mining areas, the major evidence of mining was not so much the shafts that were dug as the once forested slopes that were left denuded.[38] The second problem with the reserves for the mining companies was the question of access. Strictly interpreted, reserves were areas of no trespassing, and they were off-limits for mineral development. Many areas of the new forest reserves had never been explored for their mineral potential, and the mining industry saw the dwindling of their resource base just as the lumbermen did. This same issue would rise again in the 1960s with regard to wilderness lands, carved mostly from the areas of the original forest reserves and eliminated from mineral development.

Grazing interests felt the pinch too.[39] As private land became more common and farmers settled on increasingly large tracts of land, available public land became more valuable to the owners of cattle, sheep, and horses. Public lands provided grass without charge and the elimination of large tracts in forest reserves further restricted the grazing resource. The competition among ranchers for use of remaining public land was fierce, and range wars were common in the disorganized grazing community from the 1870s well into the twentieth century.

Opponents of the reserves frequently trumpeted their concern for the "little man," that intrepid independent settler who was so common in rhetoric, but so scarce in reality. The independent settler may not have been a viable political force in his own right, but his plight was nonetheless real. In the wood-scarce West, the only viable local source for lumber was in national forest reserves. Without forestry policies, the timber could not be released by the federal owners, and because there was no provision for selling reserve wood, the timber that grew there was legally unreachable for the private citizen. "Timber trespass" (that is, theft) became an accepted western solution that was widely practiced and seldom condemned. For the western resident there was no alternative, and it seemed foolish to pay high prices for imported lumber when some of the best materials were on his doorstep. While eastern residents and administrators bemoaned the sinful ways of their western cousins, they had no management policies to alleviate the problem.

Part of the western opposition to the forest reserves was also rooted in an advocacy of states' rights.[40] Western citizens did not condone the control of local resources by a distant federal government.[41] New Mexico, Arizona, and Nevada had significant numbers of settlers descended from Confederate sympathizers who may have been reconstructed but who also saw a diminished role for federal government as appropriate. Mormon citizens of Utah, smarting from federal efforts to undo their theocracy and eliminate polygamy, seized every opportunity to oppose expansion of federal authority, especially in matters related to their hard-won land.

In the case of the forest reserves, the West failed to present a solid block of opposition as it had in the first rebellion. In some areas, lumber companies secretly favored the reserves, because the privately held timber then became more valuable when reserved timber was no longer available. Some of the larger companies with huge tracts of timberlands in their control saw in the reserves a way to eliminate smaller operators who undercut market prices using illegal means to obtain the public timber. The farmers who depended on irrigation waters from forested mountain watersheds were consistent and vigorous supporters of the reservation policy. They were always joined by urban dwellers who depended on similar water supplies and who feared catastrophic floods from steep, logged-over slopes. Both farmers and urbanites opposed intensive grazing by cattle and sheep in the watersheds on water management grounds, and to further complicate the issue, sheep grazers and cattlemen were frequently at odds with each other. Therefore, throughout the second sagebrush rebellion over the forest lands, the West was unable to present a politically united front, and debates among western interest groups were as lively as the interregional conflicts.

The two most obvious issues for western forestry after the 1891 General Revision Act were the lack of legally available timber and the lack of management for the reserves. Senator Sanders (Montana) wasted little time and spared no efforts in obtaining passage of the Permit Act of 1891, which made available free timber on all public lands in the Rocky Mountain states.[42] He pointed out that the bill was needed because of the rapid development taking place in the interior west with the only alternative sources in distant Oregon or Minnesota. He also argued that if the western timber were not logged, it would burn in the stands anyway because of the lack of adequate fire protection. In a peculiar twist of logic he then argued that the western residents had earned the right to the timber because of their fire suppression efforts.

Sanders's arguments carried the bill through the Senate with only three dissenting votes, those of Edmunds (Vermont), Quay (Pennsylvania), and Spooner (Wisconsin). The opponents viewed the bill as a massive uncontrolled giveaway, and after the House approved the bill, President Harrison was sensitive to their fears. He signed the bill, but only after it was changed to allow regulation of the cutting by the secretary of interior.[43]

While it was fairly easy to give away the public resources, it proved to be a much more difficult task to manage the resources retained by the preserves. Several bills were proposed in both houses of Congress to arrange a management system, but House Resolution 119, introduced by Congressman McRae, was the most fully developed.[44] His bill provided for the regulated sale of timber from the reserves, use of sales receipts for fire protection and other management efforts, enforcement of antitheft provisions by troops, and the release of agricultural lands in the reserves for entry by settlers. The bill received strong endorsements and lobby support from the ever-present American Forestry Association, the secretary of interior, and the General Land Office. In addition to supporting the much needed management, the latter two agencies recognized a clear opportunity to elevate their roles in resource management to the level of lumber czars.

Opposition to H.R. 119 was from western representatives piously railing against the "infamous proposition" in a bill "to denude the public forest reservations." They warned that the bill would benefit only the popular boogey-men of the time, "syndicates and great land owners, or speculators and monopolists." Such concern for public resources appeared strange, coming as it did from those opposed to the reserves: Congressmen Pickler (South Dakota), Herman (Oregon), Doolittle (Washington), Coffeen (Wyoming), Rawlins (Utah), Hartman (Montana), and Bell (Colorado). The true stimulus for these public spirited statements appeared when McRae amended his bill to permit mining in the forest reserves. Thereafter, western opposition to the bill in the House disappeared.[45]

Meanwhile, in the Senate, Teller (Colorado) proposed an alternative to H.R. 119, which included the provision that reserves would be set aside only for water or timber preservation, framework for timber-cutting leases, access to the lands for mining, and free timber for settlers and miners. In the end, neither H.R. 119 nor Teller's alternative could negotiate the uncharted wilderness of congressional debate, and the issues remained unresolved. Several years later, legislation did become law, and it was based in large part on the provisions of H.R. 119.

While the issue of what to do with the reserves continued in the early 1890s, Presidents Harrison and Cleveland continued to enlarge the reserve system. Harrison set aside only one reserve the year he gained the authority to do so: the Yellowstone Forest Reserve of 1.2 million acres. The following year he set aside a modest addition of about 2 million acres. When reformer Grover Cleveland became president in 1893, however, he vigorously supported the reservation system, and within two years he added over 15 million acres. Because Cleveland recognized the futility of reserving the forests without providing for their care and management, he refused more reservations until some management capability was provided by Congress. In his second Annual Message to Congress, Cleveland strongly endorsed recommendations for a comprehensive federal forestry system, but he was unable to force action.[46]

Activity in the scientific community led to the initiation of the reserves, and when it became clear that meaningful management of the reserves was stalled, natural scientists again took an active role. In 1895, Charles S. Sargent, a Harvard botanist and director of the Arnold Arboretum at Harvard, called a meeting of two of his close friends at his Boston home for the purpose of devising a way to bypass a recalcitrant Congress.[47] He invited a little-known, but enthusiastic young forester, Gifford Pinchot, as an idea man, and Walcott Gibbs, president of the National Academy of Sciences, as one who could accomplish the task. They agreed that public attention had to be combined with expert advice in order to move the political situation off dead center.

They settled on a complex, circuitous plan of action that began with Pinchot and R. U. Johnson convincing the American Forestry Association to pass a resolution favoring the formation of a forestry commission in the national government.[48] Although Chief Forester Fernow resisted (he favored the more direct legislation), he finally gave in and supported the resolution. Then Pinchot and Fernow responded to their own resolution in their official federal capacities by drafting letters to be signed by the secretary of agriculture requesting the National Academy of Sciences to investigate problems in the reserve system and to recommend solutions.

By February 1896, the scheme had come full circle back to its origin: the secretary signed the letter of request to Gibbs, and the six-member Commission on Forestry was formed by the National Academy of Sciences with Sargent as chair and Pinchot as secretary.[49] The other committee members were Alexander Agassiz (Harvard naturalist), Henry L. Abbott (army engineer), William H. Brewer (Yale botanist), and Arnold Hague (U.S. Geological Survey geologist). Gibbs assured the committee's recommendations by appointing only eastern specialists known to favor forest preservation. A notable absentee from the appointments was Bernhard Fernow, the chief forester, whose philosophy of slow forest acquisition did not square with the views of other commission members. Fernow's views were echoed by the American Forestry Association, which wanted to avoid clashes with western development interests.[50]

In May 1897, the commission issued its full report, which was a ringing call for active federal management of the reserves and a radical expansion of the reserve system.[51] The report advocated the addition of thirteen new units containing 21 million acres. Secretary of Interior David R. Francis agreed with the aggressive no-compromise stance of the report, and used it to urge President Cleveland to take the offensive in the forest issue. Cleveland, about to leave office without having achieved his goal of a management system, reversed his earlier policy of trying to force Congress into establishing a management system. On Washington's birthday, 1897, he set aside the recommended reserves in the largest single expansion of the reserves to date; overnight, the forest reserve system nearly doubled in size.

Western reaction was speedy and predictable. Most of the western states were dominated by Republicans, and their dissatisfaction with the reform Democrat Cleveland had been brewing over other issues.

The expansion of the forest reserve system set the second sagebrush rebellion into full motion. Congress was inundated with letters and memorials of protest with the now familiar arguments that eastern know-nothings were imposing ridiculous constraints on western development. More aggravating to the westerners was their sense of colonial status: one chamber of commerce wrote that "King George had never attempted so high-handed an invasion upon the rights of Americans."[52]

At first, the debate seemed to be strictly on an East versus West basis. "The reserves came from some senator away off in Massachusetts," declared Senator Rawlins (Utah). "The speech of the senator from Delaware is to the effect that he has great concern for the preservation of the forests of the distant state of Washington 5000 miles from the place where he lives. Yet neither he nor the people who may live in the state where he now resides can by any conceivable possibility be affected one way or the other."[53] Senator Wilson (Washington) pleaded that the people west of the Mississippi River be allowed to develop their resources in their own way.[54] He concluded by asking, "Why should we be everlastingly and eternally harassed and annoyed and bedeviled by these scientific gentlemen from Harvard College?"

Many westerners were incensed that the position of the secretary of interior inevitably went to an eastern political appointee. In a popular story, a fictional Irish American told his newly arrived countrymen, in explaining the American government, that the secretary of interior should come from Maine or Florida or any Atlantic seacoast town.[55] If he gets the idea that there are any white people (as opposed to lowly Indians) in Ann Arbor or Columbus, he loses his job, according to the storyteller. Half a century later eastern representatives were complaining that the Department of Interior secretary had become an "Ambassador of Western Affairs" because so many westerners had been appointed to the position.

The initial East-West split was an oversimplification that quickly dissolved into a confused reality that pitted interest groups against each other. Within the West there was great divergence of opinion about the reserves. The same reserve might engender a variety of responses from the local population. Loggers might support the reserve if they owned enough timber of their own, or they might oppose it if they worked mostly on public lands. Cattlemen might support the reserve if they perceived it as a means to drive the sheepmen out of business, or they might oppose it if grazing lands were in short supply. Townspeople and irrigation farmers viewed the reserves as the only way to protect watershed, while miners saw them only as impediments to their rightful efforts at resource development. For every editorial calling for Cleveland's political scalp, there were others that supported the reserves. This splintering of the western interest groups reduced the effectiveness of the second sagebrush rebellion.[56]

Political control was equally fragmented in the West of the 1890s.[57] Lumber interests controlled the states of Washington and Oregon, copper companies held sway in Arizona and Montana, stockmen were in authority in Wyoming and New Mexico, and coal and iron developers controlled Colorado. In California, the Southern Pacific Railroad ran the state legislature and the trains with equal precision.

In the now traditional sagebrush rebellion statement, Governor William A. Richards of Wyoming advocated the cessation of all federal lands to the states. He observed that the reserves were administered by absentee overlords and that reserves were selected by "college professors and landscape gardeners."[58] Governor Frank Steunenburg of Idaho preferred short-range goals more easily attained under state ownership: "the pretext that our lands and forests are the just inheritance of posterity is not only hackneyed, but illegal and overdrawn."[59]

In a rush to respond to the furor stirred by Cleveland's Washington Birthday reserves, Congress attached an amendment to the Sundry Civil Bill to restore the entire area of the new reserves to the public domain, voiding Cleveland's proclamations.[60] Cleveland could not keep both the government and his new reserves funded, but he did not back down. In his last cabinet meeting he made known his choice: "I'll be damned if I'll sign the bill!"[61] When Cleveland left office the government had no operating funds, but it did have over 20 million acres of forest reserves.

The new president, McKinley, called a special session to appropriate a budget, but with that accomplished Congress also presented him with another amendment.[62] The amendment was sponsored by powerful Senator Richard Pettigrew (South Dakota) and was amazingly similar in some respects to the old House Rule 119 that had failed six years before. Charles D. Walcott, a major force in establishing a sound forestry policy, director of the U.S. Geological Survey, and member of the Forestry Commission, lobbied Pettigrew to do the necessary work.[63] In Pettigrew's version, forests could be set aside only for timber and water resources, mineral and agricultural lands were open for use, and settlers could have free timber from the reserves. These progressive provisions were accepted by other western representatives because of a final provision that suspended the Washington Birthday reserves for nine months, during which time President McKinley was authorized to nullify them and to return the lands to the public domain. The western Republicans felt secure that their man in the White House would serve them well.

The plan hatched by Sargent, Gibbs, and Pinchot to enlarge the forest reserves and to establish a national policy for them had been successful, but reaction had been so powerful that they were now in danger of losing all that they had accomplished. Congress had given McKinley the power to undo all of Cleveland's Washington Birthday work. In one final lobbying attempt, the National Academy of Science committee had a meeting with the new president. Led by the Harvard botanist Charles S. Sargent, the committee made its strongest possible arguments for the reserves. After the full committee meeting, Sargent returned to talk to McKinley for almost an hour. When he had finished, the administration paused in its headlong rush to disband the new reserves.

Rather than make a rapid decision, McKinley adopted a slower, surer process that protected him against a charge of sellout by his proindustry supporters. He asked for recommendations from his new secretary of interior, Cornelius M. Bliss. John Muir summed up the conservationist's view of the secretary: "There seems to be little arboreal bliss in Bliss—dry goods and mostly dry rot."[64] Dry goods or not, Bliss asked Pinchot to conduct an investigation of the new reserves, thus closing the political circle that had begun with Sargent. Now Pinchot was being asked to evaluate actions he himself had advocated and had worked to bring about. Pinchot, not willing to become too visible in the highly charged political atmosphere, agreed to take on the task only as a special agent, but the role gave the Sargent/Gibbs/Pinchot group the opening into the bureaucracy that they needed.[65]

In writing his report, the young forester took every advantage of his unique position. He began by making a strong case for retaining the reserves despite what he considered short-sighted opposition.[66] Then he urged the adoption of several basic principles that had been suggested by Fernow a dozen years before and that had been high on the list of priorities for the American Forestry Association: direct forestry management by the federal government, the employment of trained foresters, and an organized forest service. To avoid unnecessary conflicts, he urged that lands valuable for minerals and agricultural production be eliminated from the existing reserves.

Pinchot's report was accepted and his recommendations were followed with only minor exceptions. His success in forcefully stating the need for management of the nation's forests, while at the same time treading dangerous political ground, was rewarded three months later. He was appointed to replace the aging Bernhard Fernow as chief forester in the Department of Agriculture.

In addition to Pinchot's special investigation, Secretary of Interior Bliss also directed Charles Walcott's Geological Survey to investigate the entire forest resources of the nation. With funds from the 1897 Sundry Civil Appropriations, the Survey established a new Geographical and Forestry Division with geographer Henry Gannett in charge. Although Gannett was not an ardent supporter of the reserve concept, his professional evaluation of the resources resulted in a successful three-volume atlas in 1899.[67] The atlas provided a much-needed technical base from which meaningful management and policy decisions could be made. The data showed that the nation's forests were mostly concentrated in the South, Northeast, and Northwest (Figure 7), and that the forest reserves were not locking up even a majority of the nation's timber resources.

Convinced by Sargent and armed with reports from Pinchot and Gannett, McKinley let the new reserves stand. His failure to disband them proved not to be costly in a political sense, and he was reelected in 1900. On September 6, 1901, at a public reception at the Pan American Exposition in Buffalo, New York, he was shot by an anarchist with a pistol concealed in a bandaged hand. On September 14, McKinley died and the presidency passed to an aggressive reformer, Theodore Roosevelt. Roosevelt had a direct personal interest in conservation, and through his friendship with Giffort Pinchot he had access to expertise to translate his policies into active management. The new century suddenly witnessed a new direction in American land and environmental policy as, for the first time, power rested with conservation-minded leaders who pressed the issues rather than reacting defensively against the protests of changes. Roosevelt, "that damned cowboy" according to Mark Hanna,[68] a New York conservative, was more than a match for the sagebrush rebels.

When Roosevelt entered office, the second sagebrush rebellion had been stymied. Over 46 million acres of forest and watershed lands were in reserves, and attempts to destroy the reserve system had been turned back.[69] Scientifically based management was in its beginning phases, and the rebels had been unable to maintain a solid political base. The rebellion was far from over, however. Pinchot and Roosevelt were not about to rest after the initial victories. Sensing that the opposition was in a weak position, and that the American public was

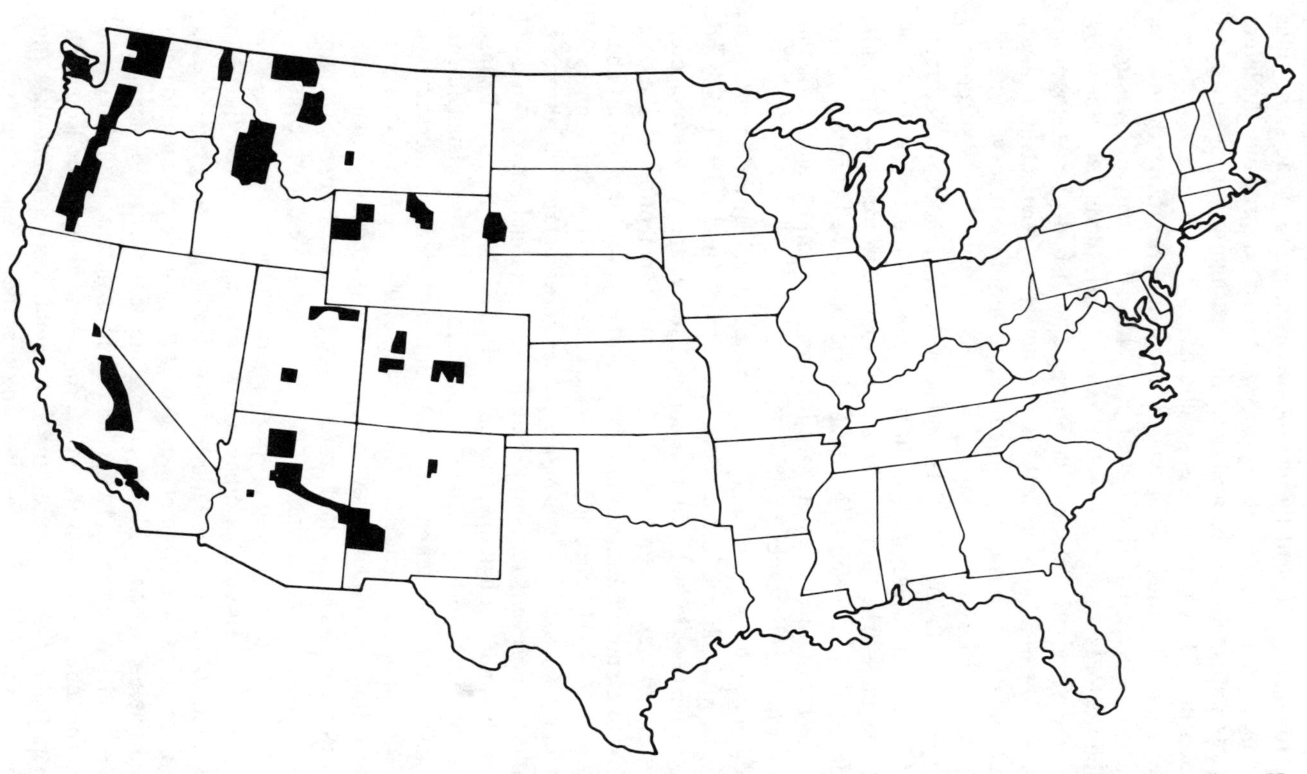

Figure 7. The distribution of the national forest reserve system in 1899 showing concentrations in the West (data from H. Gannett).

ripe for a major education effort, they detected that the still unnamed conservation movement was a possible new force in the new century. Vastly outnumbered, these two individuals embarked on a campaign to reverse the frontier psychology of an entire nation and to infuse a new conservation ethic into the national conscience.

Chapter 6

The Cowboy and the Dude in Washington

The "Cowboy" Roosevelt and his close friend the "Dude" (as Pinchot was not too affectionately known by western detractors) wasted little time in pressing their concepts of federal resource management. Recognizing that their positions of authority gave them the capability to reverse the trends of public land abuse, Roosevelt signaled his intentions without reservation in his first message to Congress.[1] After pointing out the hazards of unregulated "trusts" (or large corporations) and advocating public disclosure of corporate financing and dealings, he turned his attention to natural resources. "The preservation of our forests is an imperative business necessity," he declared. "Additions should be made (to the forest reserves) whenever practicable, and their usefulness should be increased by a thoroughly business-like management." For the first time, a president was outlining his intentions to impose a conservation-oriented approach to resources, not just for the sake of profit, but with a larger vision: "the forest and water problems are perhaps the most vital internal questions of the United States . . . the forest reserves should be set apart forever for the use and benefit of our people as a whole and not sacrificed to the shortsighted greed of a few." Western Republicans began to wonder if this new resident of the White House was really a Republican after all.

Roosevelt and Pinchot had clearly determined that the problems surrounding the forest reserves should be attacked just like Roosevelt's approach to life: head-on. In his congressional message, Roosevelt specifically requested the consolidation of reserve management authority (along with the reserves themselves) in the Department of Agriculture, where Pinchot as chief forester would be in charge. The situation had become increasingly confusing because the reserve lands remained under the jurisdiction of the General Land Office in the Department of Interior, while management expertise was housed in the Department of Agriculture's Division of Forestry. This peculiar arrangement came about because the lands in the reserves were in the public domain, and thus were Interior's responsibility. The Division of Forestry was in Agriculture because of its obscure legislative beginnings in agricultural appropriations.

Pinchot, the professional forester, viewed the Land Office administrators as totally incompetent in forestry matters. He was convinced that the long-term success of the reserves depended on their transfer to his own worthy agency. He reported to Roosevelt that "the Department of Interior never has been and never will be a real Department of Conservation. It has been a government real estate agent, bent always on getting the public lands into private ownership."[2] It was increasingly obvious that the first skirmish in the new century in the ongoing controversy over western forest lands was going to be over which agency would control the lands and the users.

Most loggers and graziers preferred that the reserves remain in the Department of Interior instead of in Agriculture where the crazy "Dude" was likely to pursue aggressively restrictive policies. Interior was also unwilling to lose a substantial part of its bureaucratic and real estate empire, and responded to the growing signs of Agriculture's imperialism by establishing its own Forestry Division. Known as "Division R," the division was to manage the reserves and was staffed by five of Pinchot's men.[3] This arrangement was not as strange as it might seem, because although the men were nominally under control of the Department of Interior, they actually were responsive to Pinchot's direction. Until he could secure official authority, Pinchot saw his men in Division R as a means of indirect management with the added advantage of not being officially accountable. He even wrote direct orders for office and field representatives, but always for the signature of the secretary of interior.[4] Within a few months, Division R of Interior was hardly distinguishable from the Division of Forestry in Agriculture, and there was little question of who was in charge.

The arrangement with Division R was unsatisfactory for Pinchot who considered the Department of Interior hopelessly corrupt.[5] His men there were not promoted as rapidly as those in Agriculture, and their progressive policies ran counter to the prevailing "business as usual" methods in the raid on public resources. Although he aided in the establishment of Division R, Pinchot saw it only as a temporary inconvenience, soon to be remedied.

The first cabinet-level discussions about the proposed shift of the reserves from Interior to Agriculture were in 1899, but no action was taken.[6] Significant influence trading was in evidence even at that early date, however. Elihu Root, secretary of state, supported the shift as a reflection of his long personal friendship with Pinchot. James Wilson, secretary of agriculture, saw a chance to expand his own influence in the shift and also was supportive. E. A. Hitchcock, secretary of interior, was mildly supportive, concerned at the time that the reserves would produce more political trouble than they were worth. President McKinley, retaining perhaps a bit of advice from the aging botanist Charles Sargent, also approved. Action was blocked, however, by Binger Hermann, the commissioner of the General Land Office, who stood to lose the most. The executives were also wary of further antagonizing western interests who feared strict grazing policies from Agriculture and who wanted to retain the less professional but more easily manipulated arrangements in Interior. Before Roosevelt became president, Pinchot was content to work quietly through Division R on the advice of two geologist friends, Arnold Hague and Charles D. Walcott, who told him that westerners would vigorously oppose an all-Agriculture agency. With Roosevelt as president and progressive politics the first bandwagon of the new century, Pinchot and his legions came out of the bureaucratic closet and worked openly for the transfer.

Responding to Roosevelt's request and Pinchot's lobbying, Congressman John F. Lacey (Iowa) introduced bills in the House in 1901 and 1902 to initiate the transfer, as well as to appoint local forest supervisors and rangers and use grazing fees for forest protection. The effort collapsed under the combined weight of opposing western representatives Mondell (Wyoming), Jones (Washington), and Shaforth (Colorado).[7] "Uncle Joe" Cannon (Illinois), the coarse, cigar-smoking Speaker of the House, maintained his typical conservative position that no change was the best change. He was also not partial to Pinchot and the professional foresters who he derisively described as "college professors, students, wise men." He succeeded in gutting the bill that emerged from the committee by removing the enacting clause and cutting the bill "right close up behind the ears."[8] The sagebrush rebels may have lost the fight over Cleveland's Washington Birthday reserves, but they were not giving up yet.

In 1902, the western interests scored a major victory in the passage of the Reclamation Act (also known as the Newlands Act after the sponsor, Francis Newlands of Nevada).[9] Shepherded through Congress and enthusiastically signed by Roosevelt, the act established the Reclamation Service (housed in the docile Department of Interior, of course) and committed the federal government to a massive effort still proceeding eighty years later to "reclaim" arid western regions for agriculture. Almost a quarter of a century after his *Arid Lands Report* and a decade after the destruction of his Irrigation Survey, nearly all of John Wesley Powell's water resource proposals had been adopted by his opponents. Federal investment in major dams and water distribution systems and a framework for cooperative development and management were important parts of the legislation. Roosevelt gave his support to the measure in hopes that the act would benefit individual irrigation farmers because a provision limited participants to landowners holding 160 acres or less. Large corporate farmers easily circumvented the restriction, though, and the bill actually became a method for huge subsidy to western agriculture. By 1980, crops grown on federally reclaimed land were worth $78 billion per year, with agribusiness repaying only about 7 percent of the cost of the projects providing the water.[10]

Winning passage of the Newlands Act compensated the sagebrush rebels in part for their losses in regard to the forest reserves, but time and events were running against them in their efforts to prevent the shift of the reserves to the Department of Agriculture. In addition to the personalities of Roosevelt and Pinchot, three issues played important roles in the process: a major land scandal in the Department of Interior, the development of specialists in range management in the Department of Agriculture, and the activities of a Second Public Lands Commission.

Fraud in the General Land Office was an acknowledged evil at the turn of the century. Commissioner Binger Hermann had made some efforts to rid the office of the problems, but he met with little success. Easterners were suspicious of him because of his known associations with Washington State lumber interests, and westerners were unhappy with his ruling that the federal government could rightfully regulate grazing on public lands by means of a permit system. He appointed his brother, A. B. Hermann, supervisor of the Grand Canyon Forest Reserve in the land-office tradition of nepotism and political favoritism.[11]

Land Office operations were also questionable. When the reserves were established, they frequently contained isolated sections of land held by individuals or more commonly by railroads as part of their land grants. The holders of these parcels were permitted to exchange these reserved parcels for other "in lieu" lands from the public domain. Exchanges were authorized by the forest legislation of 1897, but in many cases nearly valueless lands in reserves were exchanged for timber, mineral, or agricultural lands elsewhere worth several times more. Speculators made huge profits at government expense in the exchanges that were approved by corrupt officials.[12]

When Roosevelt and his Secretary of Interior Ethan Allen Hitchcock began to detect the extent of the frauds in the Oregon timber country, they dismissed Hermann from his position as Land Office commissioner as part of a house-cleaning effort. Before the scandal was allowed to retire from the public eye, illegal activities involving millions of dollars, hundred of thousands of acres in the public domain, and indictments of three of the four congressional representatives of Oregon and Washington came to light. A major component of the schemes involved Department of Interior personnel who notified cronies in the west of impending reserves.[13] The contacts would then buy cheap lands, see them included in the reserves, trade them for vastly more valuable parcels, and then sell them off at large profits. Pinchot's caustic remarks about corruption at Interior appear not to have been far off the mark. Hermann was deeply implicated,[14] but within six months of his dismissal, the voters of Oregon returned him to the nation's capital as a congressional representative.

The significance of the scandals and Hermann's dismissal was that the Department of Interior was so tainted with corruption in the public eye that it became politically dangerous for congressmen to express support for the agency. Muckraking journalists had a field day with the scandals, and most congressmen preferred not to appear eager in their support of retaining the forest reserves in an obviously troubled agency. Eastern and reform politicians were constantly on the prowl for western representatives known to be associated with the ominous "trusts." Congressman Lacey (Iowa) dourly noted that he was "not surprised at western opposition [to shifting the reserves to the Department of Agriculture] coming from sources not far from the headquarters of the great mines."[15] Western representatives also had problems at home where the railroads, primary agents in many land dealings, were viewed as the enemy of the common man. Congressmen too supportive of policies favorable to the railroads were likely to be out of their jobs after the next election.

The most vociferous opponents of the shift of reserves from Interior to Agriculture were the grazing interests. By the turn of the century, grazing areas where free public grass was available were objects of intense competition. The growth of the western livestock industry was as phenomenal as it was catastrophic.[16] For example, in Montana the number of sheep exploded from about one-quarter million in 1881 to over 6 million by 1900. In 1900, there were more than 20 million sheep in the entire sagebrush region. By comparison, in 1980 the sheep population of the entire United States was only 11 million.

Violent confrontations on the public domain were common. In one of the best known, the Johnson County (Wyoming) War of 1892, ranchers from the famous Cheyenne Club hired gunmen from other states to drive off farmer-settlers who had encroached onto former ranges. After several deaths and a visit from the federal cavalry, the situation remained unresolved. Wyoming cattlemen were therefore highly protective of their remaining ranges, and when forest reserves further reduced their available range, they were not about to yield without a political version of the Johnson County War.

Cattlemen also felt restricted by the influx of sheep herders. Generally the open range was divided among cattlemen on a first occupancy basis, so that ranges once occupied by cattlemen were considered by them to be their own irrespective of actual public ownership. When sheep herds were moved in by newcomers, the tenuous divisions of ranges among cattlemen were disrupted, and "sheep shooting" became a widely practiced sport in the 1880s and 1890s. By the turn of the century, however, many operators ran both sheep and cattle, or switched from one type of stock to the other, depending on market conditions.

The sheepmen played important roles in the debates over reserves in Arizona and New Mexico, in part because in these states the sheep industry was a century-long feature of the economy. In Arizona, forest reserves to protect water supplies for the Salt River Valley (the area surrounding Phoenix) removed large areas of sheep range. Department of Interior regulations permitted sheep grazing only on the reserves of Washington and Oregon where precipitation was assumed to be sufficient to allow the replacement of the grazed forage. When the Arizona reserves were proclaimed, E. S. Gosney, a Flagstaff lawyer and wool grower, led a massive public campaign to prevent the designation of new reserves and to open the established ones for grazing.[17] He contended that antisheep propaganda "came to Arizona and elsewhere via magazines and newspaper articles by 'armchair' scientists and by engaging and plausible speakers, particularly in the Salt River Valley, where their frequent orations were as wild and unfounded as they were eloquent and convincing.'' Mass meetings were held, and Gosney went to Washington with a rancher friend, Albert Potter, with the blessings of Governor Murphy, to try to gain grazing access to the reserves.

The division between irrigators and graziers was especially pronounced in the issue of the reserves, and the split prevented western representatives from being able to represent a solid block of constituents. In Idaho, Utah, and Arizona, many town councils petitioned the federal government to create more local reserves, while cattlemen had difficulty rationalizing the establishment of ''forest'' reserves for the sake of watershed protection where there were no forests. At a meeting in Nogales, Arizona, one disgruntled rancher commented, ''Looks like every time one of the government scouts finds a tree in the West he wires Pinchot and Pinchot gallops into Teddy's office and says 'Oh Teddy, we've found a tree in the What-the-Hell Mountains. Let's create a new forest!' ''[18]

The opposition of grazing interests slowly changed in the few years after 1900. Ranchers recognized the problem that Garrett Hardin in later years would call the tragedy of the commons.[19] Each user of the range maintained his herd with as many animals as possible—if he did not, another rancher would use the range and gain the profits. Without regulation there was no way a rational cattleman could conserve the grass resource even if he wanted to. Recognizing that overgrazing was destroying the foundation of the industry, most cattlemen sought some form of loose government regulation as a means of ending the chaos. Even the bylaws of the newly formed Arizona Wool Growers Association supported the concept of reserves. The complete elimination of sheep from the reserves was too extreme, however, and ranchers pressed the issue in Congress. At one point in 1901, the Senate attached a rider to the Sundry Civil Bill to open the reserves for unrestricted grazing. The rider, a product of the efforts of Senators ''Slippery Tom'' Carter (Montana), Teller (Colorado), Rawlins (Utah), Shoup (Idaho), and Hatfield (Idaho) failed in the House.[20]

In 1900, Pinchot recognized that there would have to be some accommodation with the grazing interests. In response to the problems in Arizona, he agreed to a field inspection, partly to learn more about the situation and partly as a propaganda move. In May the field party assembled at Winslow.[21] In addition to Pinchot and his assistant, the party included leading Arizona rancher Albert Potter and Professor E. C. Bunch representing the Salt River Valley Water Users Association.[22] Local stockmen spent a few days with the group as it traveled through the rangelands on horseback.[23]

The western cattlemen expected Pinchot to be a frail eastern bureaucrat ready to fold his tent and return to the comforts of the city. Instead they encountered an outdoorsman accustomed to horseback riding and wise to the ways of the range. Early in their six-week field investigation the party stopped at a water hole for a much-needed drink. As usual on the open range, the standing water was well populated by algae, scum, and insects of a wide variety and great number. One perhaps excessive account also included a dead cow in the aquatic collection. In any event, the thirsty cattlemen held back for a moment, waiting to see what the "Dude" would do. Pinchot, recognizing a chance when presented one, drank his fill, insects and scum included. From that time on he made rapid progress in winning new friends among the cattlemen.[24]

The western trip was an education for Pinchot who returned to Washington with two things. First, he took back a determination to open the forests to limited, controlled grazing.[25] He felt that if properly managed, sheep would not harm timber or water resources, but he agreed with Roosevelt, who was known on occasion to thump the arm of his chair with his fist and declare, "Gentlemen, sheep are dee-structive!"[26] The second thing Pinchot took back to Washington was Albert Potter, the Arizona rancher. Potter and Pinchot were impressed with each other's honesty and forthright manner.[27] Pinchot wanted Potter for the Division of Forestry as a grazing expert to provide experience and advice. He also wanted the goodwill likely to accrue to the federal agency, which included an outspoken western representative.

Potter's new job was not easy. Most of Pinchot's foresters had been trained in Europe or in the eastern United States where cows grazed in fields and not forests. Potter preached continuously that grazing was ingrained in the western life-style, and that if grazing was accommodated in protective management schemes, the westerners would eventually accept the reserves. Potter's education efforts in Washington were twofold, one directed to the specifics of grazing management and the other directed to the generalities of western attitudes. His success was reflected by the decline in grazing-forestry clashes by 1905. Sheep and cattle grazing associations developed in many states, and Pinchot delegated the authority to distribute grazing allotments to them.[28] The modest controls imposed on the reserves brought order to the range wars, and when Potter sent a questionnaire to western stockmen in

1903, the 1,400 who responded favored federal regulation by a five-to-one margin.[29] With such quick success behind him, Potter became one of Pinchot's most trusted assistants, and he was instrumental in building working relationships between managers and users.

In the effort to transfer the reserves from Interior to Agriculture, two of the three major issues had been resolved by late 1903. First, the Hermann scandal gave Interior a public black eye and weakened its position to retain the reserves. Second, Pinchot and his division in Agriculture strengthened their positions by reaching accommodation with the grazing interests. The final issue was strictly a political one: how could Pinchot and Roosevelt best accomplish the transfer? Two routes were available, one through legislation and one by means of executive order. Legislation was a slow and difficult process but one that produced a durable result because of the compromises needed for passage. Executive orders could be quickly consummated but might produce results that would be overturned by Congress, later, as almost had happened with Cleveland's Washington Birthday reserves. In order to test the political climate, Roosevelt appointed a Public Lands Commission in October, 1903, to report on the general situation.[30]

Officially the Public Lands Commission was charged by the president with the task of reporting on the general condition of the public lands, the status of the nation's land laws, and the outline of potential changes in the federal approach to the issue of resource management. He appointed W. A. Richards (Land Office commissioner), F. H. Newell (Geological Survey and the Reclamation Service), and the ubiquitous Pinchot. After several weeks in Washington, the commission undertook an investigative tour of western states with large amounts of public lands. Their first stop was in Cheyenne, Wyoming, where they met the "enemy" face to face in a statewide convention that had been called by the governor, Fenimore Chatterton.[31] A public debate, or perhaps better styled an argument, ensued between Chatterton and Pinchot. Pinchot's performance was ineffective, and the conference later voted to oppose reform of land laws and to support the western resolution to land management problems: they demanded that all public lands be ceded to the states.

Elsewhere on their western tour, the commission, represented by Newell and Pinchot, found vague and divided opinions about the reserves. They found that the fractious economic interests were not likely to unite solidly either for or against expansion of the reserves, and that Roosevelt's critics could not mobilize enough influence to derail reform generally or to stop the formation of specific reserves.

These findings were not part of the official report of the commission of course, but they were probably relayed privately to the president and they confirmed his belief that the stubborn Congress could be avoided safely by decisive executive action.

The official preliminary report in 1904 and the final report in 1905 recommended the concentration of the reserve activities in the Department of Agriculture and called for sweeping reform of the land laws, especially those that permitted "in lieu" trades of forest lands.[32] The report also contained the most complete statistical summary of the public domain up to that time, and in this respect it served a data collection function just as did the report of the first Public Lands Commission of 1879. But like the first commission's report more than two decades earlier, Roosevelt's commission had little impact on legislation and its efforts went largely unnoticed.[33]

Reorganization of the executive branch in general and of the Departments of Interior and Agriculture in particular were the subjects of still another committee appointed by Roosevelt in the same year as the Public Lands Commission. The second committee (called the Committee on Organization of Scientific Work in Government) was directed by Charles D. Walcott, and was concerned with the evolution of federal science.[34] It recommended that most environmental-related research, including the work related to the forest reserves, should be transferred from Interior to Agriculture. Interior's reputation as a den of bureaucratic iniquity and Agriculture's consistent progress in scientific research were probably major influences in the committee's decisions. The report even recommended shifting the Geological Survey and the national parks to Agriculture jurisdiction.

The report was useful to Roosevelt and Pinchot as one more piece of evidence that their plans were rational and widely accepted. By 1902, they had succeeded in orchestrating a massive campaign to consolidate approval from widely divergent sources for their plans, and by 1905, they had overcome their most strident opponents. With support mounting in Congress, they decided against the high-handed and impermanent executive order and opted for the legislative approach. Congressional opposition centered around Senator Alfred B. Kitteridge of South Dakota. Kitteridge was concerned that the timber on the Black Hills Reserve would be unavailable to his major political supporter, the Homestake Mining Company.[35] Pinchot privately pledged that he would personally assure that the timber would not be sold out-of-state and that it therefore would be available to Homestake if Kitteridge approved the transfer of the reserves to Agriculture.

Kitteridge agreed, and the deal was kept by both sides. On another front, California's congressional delegation was bombarded by letters from a fledgling group calling itself the Sierra Club.

On February 1, 1905, the control of the reserves was transferred to the Department of Agriculture's Division of Forestry where Gifford Pinchot was now king of the woods. Pinchot's education by Albert Potter on forests and grazing had been successful, and his more flexible attitude resulted in a surprise sponsor of the 1905 transfer bill: Wyoming's rancher-Senator Frank Mondell shepherded the bill through to approval. Pinchot had won at least a part of the West.

Executive power in control of the forest reserves was now concentrated in the hands of two men determined to radically expand the system. Pinchot and Roosevelt set out to systematically enlarge the reserves to encompass significant stands of timber and areas of critical watersheds. In 1904 the reserves contained 62.7 million acres, while in 1905 the total climbed to 85.8 million, in 1906 to 107 million, and in 1907 to almost 151 million.[36] It was a remarkable period for resource conservation as the federal government became an increasingly influential land owner and manager. The tireless Pinchot generated management guidelines, administered the burgeoning Division of Forestry, and provided a continuous stream of suggested reserves for Roosevelt's approval. Roosevelt's second administration from 1904 to 1908 signaled the true shift of the government's land policies from disposal to active management. The "Cowboy" and the "Dude" had succeeded in convincing an entire nation of the utility of a new ethic, and in the process even gave the ethic a name: conservation.

Change and growth was occurring rapidly because both Pinchot and Roosevelt perceived themselves as locked in a furious race with lumber companies to find, survey, and claim prime timber on the public domain. The mass creation of several reserves in the sagebrush region was typical. During Roosevelt's second administration, Pinchot was invited to receive an honorary degree from the Michigan State Agricultural College in East Lansing. Despite his pressing duties in Washington, he agreed to provide a speech at the ceremonies. While he was in Lansing, however, boundary surveys were completed on a number of potential reserves.[37] Not wishing to wait even the few days until he returned to Washington, Pinchot ordered Arthur Ringland (chief of Boundaries) to bring the boundary notes and maps to Michigan for final analysis. The night before Pinchot's address Ringland appeared at Pinchot's room at the Downey Hotel carrying a golf bag stuffed with oversized maps. He and Pinchot pushed the furniture against the walls of the hotel room and covered the floor with maps. While Ringland read from the notes, Pinchot crawled around the room, penciling in the blue lines that eventually would be the boundaries of future national reserves. The next day while Pinchot gave his address, Ringland returned to Washington with 17 proclamations for Roosevelt's signature to create new reserves in Arizona, Utah, California, New Mexico, and Nevada.

The process of reserve formation and designation proceeded so rapidly because few people and almost no bureaucracy were involved. A particular area possibly qualifying as a reserve was identified either by local requests for investigation or by initiatives of federal officials. Boundary surveyors then investigated the area. These individuals rode alone on horseback through the country to be investigated, mapping forests and drainage basins as they went. For three, four, or five months at a time, depending on the length of good weather in the summer, they camped in the open at night and rode and mapped by daylight. One boundary surveyor claimed to have mapped 3 million acres in one season.[38] Pinchot was especially proud of his "boundary boys" who he contended were competing with "as competent a body of land thieves as ever the sun shone on."[39]

The surveyors then forwarded their notes to Washington where Pinchot personally decided on the suitability of the lands for inclusion in the system. He drafted the proclamations officially creating the reserves, and sent them down the street to the White House for Roosevelt's approval. That approval was almost always forthcoming without delay. The streamlined process completely excluded Congress and was so efficient that by the end of Roosevelt's term of office the forest reserve system was four times larger than it was when he entered office.

An example of how this rapid growth was not always successful in forest protection is provided by an experience in northern Washington State.[40] In late 1906, the General Land Office announced that on February 6, 1907, it would open for public entry large tracts of timber near the Canadian border. Arthur Ringland was sent to survey the area, and after a quick reconnaissance, Ringland wired his report back to his boss. In the usual fashion, without prior public notice, Pinchot prepared the proper descriptions and Roosevelt signed the proclamation declaring a new forest reserve in the area of question on January 7, 1907. The entire state of Washington was incensed, and its congressional delegation discussed legislation to restore the area to the public domain where its timber would be accessible to harvest. Not wishing a fight on the issue that was so strongly supported by the local population, Roosevelt wired Albert E. Mead, Washington's governor, that the order was being rescinded to correct a "clerical error."

The Washington reserve incident and the general explosive expansion of the system began to generate a reaction among western representatives and developers. The most vocal of the opponents was Senator Charles W. Fulton of Oregon, one of those implicated in the Oregon forest frauds four years earlier. Fulton organized a congressional attack on the administration and particularly, on the Forestry Division.[41] Along with Senators Heyburn (Idaho), Tawney (Minnesota), and Mondell (Wyoming), he attacked every aspect of the reserve system from the methods of marketing timber to Pinchot's salary. Heyburn agitated for an amendment to the appropriations bill to eliminate Roosevelt's declared reserves, but the conservation ethic was too strongly entrenched for such a reversal of policy. The reserves and the Forestry Division were protected by Senators Beveridge (Indiana), Proctor (Vermont), Spooner (Wisconsin), and Dolliver (Iowa), who countered charges of dereliction of duty by the forester, saying that the federal conservationists in the person of Agriculture Secretary Hitchcock had "piled up a mountainous presumption of plain honesty."[42]

The sagebrush rebels saw that they would be unable to undo the work of Roosevelt and Pinchot in creating the reserves, so they determined at least to curb the power of the executive in the reservation process, and to move that authority to Congress. The rebels were able to forge an amendment that prohibited the creation of any new reserves in six states (Oregon, Washington, Idaho, Montana, Colorado, and Wyoming), abolished the forestry fund built from lease receipts, and returned 10 percent of all receipts from the reserves to the states where they were located.[43]

Conservationists were powerful enough to prevent passage of the amendment, but not powerful enough to kill it as well as pass the appropriations bill. They were forced into a compromise. As the price of their acceptance of the rebels' restrictions, the conservationists inserted into the amendment important provisions. The Forestry Division was renamed the Forest Service (putting it on par with the Reclamation Service); the reserves were named national forests; the agency's budget was increased; Pinchot's salary was increased; and timber from the national forests was permitted to be sold outside the state where it was cut. Fighting to the bitter end for his benefactor at

Homestake Mining Company, Senator Kitteridge secured an exemption for South Dakota from the final provision.[44]

Pinchot and Roosevelt were not altogether pleased with the final bill, but they saw that if the president refused to sign the measure, Congress could attempt to eliminate the new reserves in the same way they attempted to undo Grover Cleveland's work in the Washington Birthday reserves. Roosevelt was approaching the end of his final term and had little control over future events. It appeared that the remaining forests in the public domain of six major states had been saved for the developers.

In characteristic fashion, however, Roosevelt snatched victory from seemingly certain defeat.[45] For several months, he had been signing proclamations declaring new reserves and then storing them secretly in the State Department where they were safe from the prying eyes of agents in the General Land Office. Pinchot and his staff rushed to complete surveys on additional reserves. Roosevelt signed the bill that removed his power to create national forests, but just before he signed the fateful documents he proclaimed twenty-one new reserves in the six affected states.[45] His contributions to the system were complete.

Years later Roosevelt looked upon his national forest victory with the smug pride of a reformer who literally changed the map of the nation. "When friends of the special interests in the Senate got their amendment through and woke up," he wrote, "they discovered that sixteen million acres of timberland had been saved for the people by putting them in the national forests before the land grabbers could get them. The opponents of the Forest Service turned handsprings in their wrath, and dire were the threats against the Executive; but the threats could not be carried out and really were only a tribute to the efficiency of our action."[47] With the forest lands secured, however, the outcome of the second sagebrush rebellion was not yet decided. In the West, organizers began a long-running campaign to wrest control of the national forests from the federal government. The "Cowboy" and the "Dude" had won, but the unanswered question was whether or not they could ride out the coming storm of western reaction. They did not have long to wait for an answer.

Chapter 7

Conformers and Reformers

The second sagebrush rebellion with its political pyrotechnics, long-standing disputes, and potential for sweeping fundamental changes in the relationship between Americans and their land can be understood by exploring its central issues. To end exploration there, however, would be to miss an important dimension of the events and their consequences—the roles of individuals. The rebellion was a product of clashes of personalities and competition between firmly held beliefs that had developed from the educations and experiences of the participants. Just as the individual characteristics of Powell, Teller, and Stewart were driving forces behind events and resolutions of issues in the irrigation lands controversy, the individual characteristics of Pinchot, Roosevelt, and their supporting cast in the second rebellion were significant in explaining events.

The following pages present a review of the people who influenced the course of the second rebellion. Some are well known and have left their names in many places in the American landscape. Some, like the scientists and congressmen, are known only to a few in more modern times. Others, like the western ranchers, the timber cutters, and the forest service employees, played important roles in the dispute but did so only as faceless members of powerful groups. Finally, there was a small group of naturalist writers and conservation leaders who were quietly beginning to shape the way Americans viewed the resources of their land.

Of the various people influencing issues and events in federal land management as the twentieth century dawned, none were more central to the second sagebrush rebellion than the ranchers, miners, and timber men advocating free access to all public lands. Timber companies in particular played for high financial stakes, and they occupied pivotal positions in orchestrating the rebellion.

Before 1900, most of the timber companies operating in the West were locally owned and openly hostile to any interference from Washington. Although timber laws specified that logs from public lands were to be used only in the state of origin, timber company representatives argued against any restrictions on cutting for distant markets, and export was common. The Mexican Central Railroad, for example, openly advertised for ties from public lands in Arizona and New Mexico, and its needs were speedily filled.[1] In 1885, the U.S. attorney general sued the Sierra Lumber Company in California for an illegal cut of 60 million board feet.[2] Elsewhere the attorney general found systematic illegal clear-cutting around Eureka, Nevada, and 9,400 cords of illegal wood stacked along the Montana right-of-way of the Northern Pacific Railroad. Even when caught, many of the western timber thieves escaped, thanks to connections in the General Land Office. In 1883, a special agent for the federal government built a complete case file on a fraudulent claim involving 10,000 acres of timber, a $5,000 bribe offer, signed affidavits of timber company employees, and the names of thirty-six fake entrymen, plus the amounts of their bribes. The General Land Office responded by approving the claim and firing its special agent.[3]

Once the president established forest reserves, another problem arose involving the railroads. Original railroad grants included alternating survey sections that became included in the reserves, so that the 1897 Forest Lieu Lands Act allowed the railroads to trade their parcels inside the reserves for lieu lands elsewhere. The approach proved to be a bonanza for the railroad as they traded worthless lands in the reserves for densely forested tracts. The biggest winner was the Santa Fe Railroad, which exchanged 735,000 acres in Arizona and New Mexico.[4] Pinchot's forest planners tried to combat the lieu lands problem in the new San Francisco Mountain Reserve near Flagstaff, Arizona, by including only alternate sections in the reserve.[5] The result was an unmanageable checkerboard pattern of land holdings that included within the general outline of the reserve 507,000 acres held by Santa Fe, 134,000 by the Aztec Land and Cattle Company, 79,000 by a single private individual, and 40,000 by the Saginaw and Manistee Lumber Company.[6] A congressional act in 1905 suspended the lieu lands provision of earlier acts and ended the chaos.

The western loggers were adept at handling the federal bureaucracy, but they were no match for their unexpected competitors: large, well-financed logging companies from the upper Midwest.[7] As the pine forests of Michigan, Wisconsin, and Minnesota became depleted in the 1880s and 1890s, the nation's largest lumber companies looked to the West for new resources. The first to make the move was Chauncey Griggs, who logged 80,000 acres of Northern Pacific Railroad lands. He was followed by David Whitney who operated his vast lumber empire in the western states from a mansion in Detroit. By 1900 the flood of midwestern companies dominated the western timber industry, driving out the smaller local operators with seemingly endless capital, efficient, mechanized mills, and access to voracious eastern markets. The corporate shift boosted production from western forests from 660 million board feet in 1880 to a staggering 8 billion in 1910, but it also transferred the control of the industry to eastern investors.

The most significant of the new breed of lumbermen was George H. Weyerhaeuser, a German-born immigrant who began his career in a Rock Island, Illinois, sawmill.[8] By judicious investment he built successful companies in the upper Midwest, and through skilled administration he dominated the market. Recognizing the depletion of the midwest timber reserves, Weyerhaeuser was convinced by his friend James J. Hill to move to the West. He bought 900,000 acres of timber from the Northern Pacific Railroad (which Hill controlled) in one of the largest private land transactions in U.S. history. To develop the timber he formed a new investment group with fifteen others who insisted that the new corporation bear a name symbolic of sound management and good judgment: The Weyerhaeuser Timber Company. In 1905, the company administered 1.5 million acres and dominated the industry. Seventy years later, the situation was roughly the same with Weyerhaeuser owning 5.7 million acres and controlling timber rights to another 10.7 million acres.

Weyerhaeuser himself was a new kind of business executive: he was an administrator, not a tycoon. He was quiet, cautious, and well-liked. He lived a modest personal life, went to bed early, attended church regularly, and ascribed his good health to his fondness for buttermilk. He always appeared well-dressed and sported a trimmed beard, but his huge hands belied his background as a laborer. Unlike the small western operators he replaced, he viewed the concepts of conservation as wise, long-term perspectives, and he frequently attended conservation conferences. He organized and saw to completion an industry-wide effort to raise $100,000 to establish a Chair of Forestry at Yale University. Throughout his life he exemplified the best of a new breed of modern businessmen.

Weyerhaeuser's progressive social and political views were evidence of one of the major problems encountered by the second sagebrush rebellion.[9] The rebels drew heavily on the timber industry for support, yet that industry was strongly divided: Western leaders against the interlopers from the East; small operators using public land against large operators using private forests; local operators against national corporations; conservative westerners against progressive easterners. Despite the focus of the controversy on timber, the antifederal timber lobby was fragmented and provided little in the way of united political or monetary support for the rebellion.

A more united antifederal group was the mining lobby. By 1900, the industrialization of the western mining industry was even more advanced than in the lumber industry. Because western capital was scarce, major mining and milling operations for gold, silver, copper, lead, and zinc were possible only with money from eastern financiers. The limited number of companies and individuals involved in directing the minerals industry allowed the formation of an influential group of men who fought bitterly among themselves for financial gain but who stood unified against outside interference with the industry. The shift toward conservation and the development of forest reserves ranked alongside federal "trust-busting" efforts as political obstacles for the mineral lobby. The careers of Senators Hearst, Guggenheim, and Clark represented the means and the ends of mining representatives.

George Hearst began his mining career as a pick-and-shovel artist in California, but, like Weyerhaeuser, he parlayed meager earnings into a financial empire that included mining ventures throughout the sagebrush region. His most striking success was his purchase, along with two coinvestors, of the run-down Homestake Mine in South Dakota for $70,000. Under Hearst's administration the mine became a symbol of success, and between the 1870s and 1930 it produced $233 million in gold (and $58 million in dividends to stockholders).[10] Part of the mine's success was attributable to free access to forests for mine timbers, and as corporate profits went up, down came the public forests of the Black Hills. Hearst protected his investment most efficiently as a U.S. senator, but not without a price. His election to the post by the state legislature of South Dakota required "a cool half million" in bribes.[11]

A second member of the Senate mining force was Simon Guggenheim, a strident anti-Pinchot spokesman. One of seven sons of a Swiss immigrant who made his American fortune in the silver mining district of Leadville, Colorado, Guggenheim was a soft-spoken but deadly political infighter.[12] After a financial donnybrook with Standard Oil's

Amalgamated Copper Trust, he and his brothers gained control of a new group in 1901 that came to dominate western ore processing, the American Smelting and Refining Company (ASARCO). Operating from Colorado, Guggenheim spruced up his public political prospects by donating large sums to the poor, a new building to the Colorado School of Mines, and $50,000 to elect state senators who, in 1907, voted him into the U.S. Senate. Expansion of forest reserves that restricted access to mine support timbers obviously received little support from the new senator from Colorado.

Not all the members of the mining lobby in the Senate were cool, calculating individuals. The flamboyant William Andrews Clark was a constant and highly visible opponent of the forest reserves, always ready with inflammatory quotes for an eager press and a willing participant in organized anticonservation efforts. Trained to rough dealing in the struggles for control in the brawling copper politics of Montana, Clark chaired the state constitutional convention there in 1889. He and his fellow mine owners spent much of their time fighting each other for the mineral wealth of the new state, but when necessary they buried their collective hatchets when threatened with the ultimate weapon: high taxes.[13] Under Clark's direction, the Montana constitution ensured that the mines paid scandalously low state taxes. The mine owners hung together and prevailed against outraged cattlemen who discovered they were to carry an unequal share of the tax burden. Clark owned and operated mine properties throughout the West, including a major producer in the bonanza town of Jerome, Arizona, but his base of operations remained in Montana. His properties ranged from newspapers to sugar plantations in California, and a railroad from Los Angeles to Salt Lake City. His legacy is the Arizona smelter town of Clarksdale.

After intensive efforts to influence the Montana state legislature, Clark was elected to the U.S. Senate in 1900. Although he made numerous quotable statements, the most famous relate to his method of dealing with politicians: "I never bought a man who wasn't for sale,"[14] and his perspective on conservation: "Those who succeed us can take care of themselves."[15]

Ranchers almost always lent their support to the cause of the second sagebrush rebellion, in part because of changes in the grazing industry during the last decade of the nineteenth century. After the harsh winters of the late 1880s, most cattle and sheep operations in the sagebrush country were locally owned. Most successful operations were moderate in size, a few hundred head of cattle or a few thousand head of sheep. This arrangement had three important implications for the course of the rebellion. First, the stockmen were important members of western communities, and they spoke from the position of the grass roots as well as the leadership. Second, they injected a marked conservative bent to local politics. On almost all worldwide frontiers of the era, cattlemen were politically conservative, strongly self-interested economically, and generally cynical.[16] Those on the American frontier were no exception. Finally, they were directly involved in the forest reserve controversy because they wintered their herds in the low-altitude deserts, but they used the high-altitude ranges in summer.[17] Invariably the high summer ranges were forested and likely candidates for inclusion in the reserves. By 1905 the forest reserve system served as grazing area for 700,000 cattle and 1.5 million sheep, owned by almost 8,000 ranchers.[18]

Ranchers were politically powerful in the sagebrush states, especially after the formation of cattle growers' and sheep growers' associations. Because each rancher did not usually have a large enough operation to dominate his local community as the cattle barons did, he had to join associations to influence statewide decision-making processes. In most sagebrush states, the associations were formed between 1890 and 1905 in response to the political turmoil associated with the second rebellion.[19] The associations dispatched especially articulate spokesmen to Washington to pry concessions from the tight management policies of Forest Service. Compromises and some success for the ranchers usually resulted, as when Gosney convinced Pinchot of the need for policy changes for grazing in Arizona forests.[20]

Another example of successful dialogue involved William A. Richards, a Wyoming sheep rancher. Typical of the breed, Richards was a frontier jack-of-all-trades. He had experience in law and engineering, surveyed territorial boundaries, farmed, irrigated, ranched in a variety of places, and of course, played politics. As governor of Wyoming, he played to his local audience by denouncing the reserves as being selected and administered by absentee federal landlords.[21] He advocated the disposal of all federal lands to the states. After his term as governor, he became active in the ranchers' cattle and sheep associations. In a difficult period shortly after 1900, incensed Wyoming sheepmen had burned some of the new forest reserves, and Richardson sent a messenger, A. A. Anderson, to Washington to protest the reserves. President Roosevelt, in a typical maneuver, redrew some boundaries, appointed messenger Anderson as the new forest superintendent, and sent him back to Wyoming.[22]

Richards, seeing the possibility of debate and compromise, also went to Washington on behalf of the Wyoming sheepmens' association. The former governor, outspoken but cordial and unpretentious, found opposing opinions but kindred spirits in Pinchot and Roosevelt. After settling the worst of the sheep-forest disputes, Roosevelt asked Richards to stay in Washington as commissioner of the General Land Office. The Wyoming rancher accepted the offer and was instrumental in avoiding local conflicts in the development of Pinchot's expanded forest reserves and F. H. Newell's irrigation projects by the Bureau of Reclamation.[23]

The timber cutters, miners, and ranchers either went to Congress or were represented there by aggressive anticonservation spokesmen. Two Wyoming congressmen, Francis Warren and Franklin Mondell, were typical. Francis E. Warren directed the booming operation of his Warren Livestock Company to success by using public grazing lands in the early days of the Wyoming Territory. By the 1880s, he became governor and sparred with federal land commissioner William A. J. Sparks over the question of whether graziers could fence public lands. After Wyoming became a state in 1890, he rose to prominence in national politics as the first president of the National Wool Growers Association and then as one of Theodore Roosevelt's major contacts and friends in the U.S. Senate. To his colleagues he was modestly known as the "greatest shepherd since Abraham."[24] His influence with the president was demonstrated when an Interior Department investigation implicated him in illegal fencing of public lands. The report disappeared into the paperwork jungle of the executive branch never to be seen again.[25]

Franklin W. Mondell was also a successful long-term Wyoming frontiersman. An orphan boy from St. Louis who worked his way west as a miner, store clerk, and railroad worker, he made his fortune in coal mining and founded the town of New Castle, Wyoming. He was known locally as an ultraconservative law and order Republican, who immeasurably added to his image by sustaining a wound while leading a posse. He carried the two slugs in his body for the rest of his life. After a successful career as a Washington lobbyist for the Wyoming Wool Growers Association, Mondell served as an aggressive anticonservation representative from 1895 to 1923 (with a two-year break). His antifederal opinions sprang from the heart and from his frontier experiences, and they fueled more than two decades of his congressional activity.

Like the first sagebrush rebellion, the second represented a focused reaction to land management initiatives by the federal government, so that an understanding of the second rebellion, in part, rests on an understanding of those individuals who agitated for change in federal land policies. Three eastern scientists—Franklin Hough, Bernhard Fernow, and Charles S. Sargent—were instrumental in setting the new federal course toward forest reserves that so strongly affected the West. Franklin B. Hough was more than a scientist: he was a physician, historian, naturalist, statistician, and author.[26] Born in Martinsburg, New York, he maintained a lifelong interest in forests despite numerous distractions.[27] He earned his medical degree in 1848 and served as a doctor in the Civil War. Later, his statistical abilities led him to the positions of director of the New York State census and in 1870 as director of the U.S. census. In a period when the census assessed natural as well as human resources, Hough noted with alarm the drastic decline in eastern forest reserves. Spurred by Marsh's book *Man and Nature* and Carl Schurz's dire predictions of a coming timber famine, Hough spent the remainder of his life seeking a role for the federal government in forest protection.[28] Surviving photographs show Hough as a smiling, authoritative figure with friendly eyes, but his government reports reveal a tough, no-nonsense approach to the major environmental problems of his day.[29]

Hough was succeeded as the chief advocate for forestry by Bernhard E. Fernow, who lacked the side interests of his predecessor but who was more of a professional forester. Fernow came to the United States as a licensed German forester to attend a meeting of the American Forestry Association in 1876. He found a job and a wife in Pennsylvania and close friends in the American Forestry Association, and when the position of chief of the Department of Agriculture's Division of Forestry became available a decade later, he had enough friends in high places to secure the appointment. Personally, he was friendly and unflappable; professionally, he was intense, deliberate, and uninterested in politics. He was also a keen judge of character, and he recognized in Gifford Pinchot a successor likely to serve the interests of American forestry despite a vastly different personality. Fernow's major contribution to forestry was his solid economic perspective, wherein he viewed forest reserves as capital, the harvest as interest, and overcutting as wasteful depletion of the capital base.[30]

On a somewhat more intellectual level than the practitioners Hough and Fernow stood Charles S. Sargent. Born in Boston in 1841 the son of a trade merchant, Sargent cultivated an obsessive interest in botany and forests while at Harvard. After serving for the Union in the Civil

War, he returned to academic life and assumed the directorship of Harvard's Arnold Arboretum, where he pursued pioneering work in horticulture and arboriculture and published several classical books. Unlike Hough and Fernow, Sargent viewed the forest reserves in a preservationists's light, to be protected (by the U.S. Army, if necessary) from all cutting.[31] Sargent was one of the most widely recognized scientists of his day, and when he chose to speak out on issues of forest preservation, he carried with him the authority of a distinguished scientific career in botany and the power of his office as director of the Harvard arboretum. One of his major platforms was a semipopular magazine, *Garden and Forest*, of which he was publisher.[32] Because of his national recognition, Sargent could orchestrate efforts of the American Forestry Association and the National Academy of Sciences. When Charles Sargent spent an hour convincing President McKinley to keep the forest reserves instead of returning them to the public domain,[33] he literally spoke for American science. The backgrounds of Hough, Fernow, and Sargent lent credence to the charges of the sagebrush rebels that eastern-bred and trained specialists were influencing the everyday lives of common westerners in the forest reserve controversy.

Scientific arguments convinced Presidents Harrison, Cleveland, and McKinley of the importance of proconservation decisions, but the administration of Theodore Roosevelt brought about a wholesale change in government policy toward conservation. From the highest office in the nation, Roosevelt provided the political protection needed by the embryonic forest reservation efforts, and without him it is doubtful that wilderness areas would have remained half a century later for preservation in a wilderness system.

Roosevelt's affinity for the wild, nonurban West was a product of his life experience.[34] Born and raised in New York City, he led an aristocratic life in his early years, but his extended family imparted to him progressive political opinions that at the time were characterized as "progressive" or "reform." In later decades they could be characterized as decidedly liberal. As a youth, he was physically weak and frequently ill, but at the urging of his father he undertook a rigorous physical training program including boxing, and later he developed into a rugged outdoorsman. He cultivated a lifelong interest in natural science, and as a boy he was constantly adding to his natural history collection, which he referred to as the Roosevelt Museum.

As he grew to manhood, Roosevelt's interests in wild nature took different turns. In the early 1890s he became interested in ranching in

North Dakota, and he eventually operated his own Maltese Cross Ranch.[35] He could ride as long and as hard as many of his hired hands: on one occasion he rode for almost 40 continuous hours, wearing out five horses in the process. The local townspeople were at first skeptical about their new neighbor, but Roosevelt's blustery version of the macho man won them over. Notorious as a good shot with a Sharps .45 caliber rifle, he fit well in the rough and tumble frontier society. When Roosevelt's ranch workers were threatened by a known gunman and local tough, E. G. Paddock, Roosevelt followed the character to his cabin and called him out. "I understand that you have threatened to kill me on sight. I have come over to see when you want to begin," the bespeckled Roosevelt challenged.[36] Paddock replied that he had been misquoted.

In another incident, a local bully trained two pistols on Roosevelt and demanded that he buy a round of drinks at the saloon. Roosevelt, employing his pugilistic training, decked his assailant with a single punch and thus began one of the more oft-told tales of the Dakota frontier.[37]

Roosevelt's western experiences were important during the years of the conservation debate, because they gave him valuable insight into regional politics. He knew the antagonism felt by the average voting citizen for the big lumber companies and railroads, and he knew that the forest reserves could be cast in an antitrust framework appealing to western voters. His ranching experience also demonstrated to him the problems of allotting the grazing resources among powerful stockmen. He knew that in marginal areas the forest reserves could serve as a means of welcome outside regulation for ranchers. Roosevelt returned to the East and to reform politics with an appointment to the U.S. Civil Service Commission in 1889, the New York Police Board in 1895, and assistant secretary of the navy in 1897. In 1898 he assembled the first U.S. Volunteer Cavalry (the Rough Riders), which led to an important victory in the Spanish-American War and further favorable publicity, plus a spot as vice-presidential candidate with William McKinley on the Republican ticket of 1900. When Roosevelt succeeded to the presidency after McKinley's assassination, he brought to the White House a unique combination of eastern liberalism and western experience.

Roosevelt was a popular president, at least during his early years in the White House. He was a stocky, handsome man, equally at ease in the saddle or in high society. He had a ready smile, a decisive manner, a sense of humor, and he always championed the "little man." His blustery manner and aggressive speaking style, with his fine white teeth clipping the end of each word, made Roosevelt the leading exponent of the rapidly maturing nation. His personal interest in the conservation movement set the stage for successful conversion of the national body politic from a mentality of exploitation to one of conservation.

The scientists provided the intellectual background and the initiating spark for the development of a forest reserve system, and Roosevelt provided indispensible political support; but the success of management and public education jobs required the dedication, ambition, and political abilities of Gifford Pinchot. Pinchot was born in Simsbury, Connecticut, and raised in New England as the first son of a wealthy family whose roots extended to Napoleonic France.[38] His family's wealth allowed Pinchot to choose his career based on idealistic service as opposed to monetary need. His father, once a vice president of the American Forestry Association and author of several magazine articles calling attention to the nation's disappearing forests, recommended the profession of forester to his son at a time when there were no professional foresters in the United States. When a fellow student at Yale asked young Pinchot about his postgraduate plans, his confident response was, "I am going to be a forester."

"What's that?"

"That's why I am going to be a forester."[39]

He was so enthusiastic that when he met Bernhard Fernow at Yale in 1889 he offered to work for the federal forester at no salary; his offer was quickly accepted.[40] Then, after a year of study at the National School of Forestry in France, Pinchot returned to work for Fernow, but in a preview of his later practicality, he thought the older man to be overly scientific and politically insensitive. In 1892 he took on a major challenge—the management of several thousand forested acres associated with George Vanderbilt's Biltmore estate in North Carolina. His phenomenal success in the following three years boosted his professional reputation, which was solidifed by service with his close friends Walcott Gibbs and Charles Sargent on the 1896 Forestry Commission of the National Academy of Science.[41]

Given his aristocratic background, it is not surprising that Pinchot became influential at the national level, but he was hardly a classic Ivy League blue blood. He was an accomplished horseman, could place three out of five shots in the bull's eye with a pistol at 50 yards, and was an experienced outdoorsman. During the western swing of the Forestry Commission, he struck up a warm friendship with John Muir.

At the Grand Canyon the two slept out overnight to tell stories and stargaze. They had to sneak back into the hotel where their more staid (and worried) colleagues "told us what they thought of us with clarity and conviction," according to Pinchot.[42]

Personally, Pinchot seemed to engender either fierce loyalty and respect, or antagonistic opposition. Tall, lean, with a full moustache, eaglelike nose, and deep-set blue eyes, he was outspoken and friendly, but he had a strong moralistic streak. Sometimes this moralism manifested itself in trivial ways, as when he admonished his field foresters to avoid hard liquor. When he organized the Society of American Foresters, originally a closely knit group of professionals similar to Powell's "Great Basin Mess" in earlier years, it became known as the "Baked Apple Club" because he served ginger bread and baked apples at meetings in his Washington home.[43] When combined with his obsession for achievement, his moralistic approach fueled his drive for a national forest conservation movement.

Pinchot's hidden side centered on his social activities.[44] Generally comfortable in the circles of high society, his diary nonetheless contained the frequent after-party comment, "made an ass out of myself." While working at the Biltmore estate, he planned to marry Laura Houghtling, but in 1894 his fiancée died. Pinchot grieved for years, feeling that his good days were those when she was close. He finally married Cornelia Bryce in 1913 when he was 49 years old.

Whatever his personal grief, Pinchot was always an extrovert who could converse comfortably with political enemies as well as with allies. For example, when Franklin Mondell (Wyoming's rabid anti-conservationist representative) was chair of the House Committee on Resources, Pinchot attempted to iron out differences of opinion with him over box lunches in the peaceful surroundings of Washington's Rock Creek Park.[45] On the other hand, Pinchot was such a close friend of President Roosevelt that strategy for conservation was frequently the object of discussion while the two went horseback riding or swimming in areas around Washington. In an era of somewhat lesser concern for the president's safety than in later decades, a pistol carried in Pinchot's pocket provided security for the president.

Pinchot's magnetic personality attracted a variety of enthusiastic and highly competent assistants, the most important of whom was Albert F. Potter. Potter, a native of California's Central Valley, moved to Arizona as a cure for incipient tuberculosis.[46] He became an expert cowboy, roper, and horse breaker in northern Arizona. He developed a thriving cattle business in the country south of what later became

Petrified Forest National Park, but the financial panic of 1893–1894 wiped out his business. By 1896, he had invested heavily in sheep, hoping that, if McKinley were elected, an import tariff on wool would increase the value of sheep. After he received news of McKinley's victory, he rode day and night to buy his sheep options. In 1900, he sold his outfit at a fine profit.

It was in 1900 that Potter met Pinchot during the forester's western inspection trip, and he accepted the offer to participate in the Forest Service. Pinchot's general conservation philosophy was acceptable to Potter, and Pinchot was impressed by the rancher's knowledge of the western cattle and sheep businesses and by his many connections. Potter's personal manner was also attractive: he was fearless in debate, fair, emphatic, yet quiet of speech. He was just the right man to oversee the development of the Forest Service's grazing policies and to deal with western ranchers.

Potter had a dramatic impact on the Washington bureaucracy of the Forest Service.[47] He schooled his superiors on western attitudes and acted as a sounding board for new ideas. His suggestion that some Forest Service revenues be returned to western counties initiated an important practice for federal land management agencies. He arranged policies for renewing grazing permits every ten years to encourage interest in conservation practices by lessees, and he emphasized the issue of grazing permits based on carrying capacity. As an ex-rancher his public statements against overgrazing were especially convincing. Every early Forest Service policy and regulation pertaining to grazing bore the stamp of his efforts. Within a few years, Potter assembled his own staff that ensured continuing input from those who were regulated. The staff included W. C. Clos, a Mt. Pleasant, Utah, sheep breeder; Joe Campbell, an inspector from the Arizona Livestock Board; C. H. Adams, a livestock businessman; Will C. Barnes, an Arizona-New Mexico territorial legislator and manager of the famous Esperanza Cattle Company; and Leon F. Kneipp, a forest ranger from Prescott, Arizona.[48] This staff of westerners was an important counterweight to the eastern-trained foresters who dominated the agency. Throughout his career, Potter emphasized the rights and needs of the ''little man,'' and he was one of the few forest officials who enjoyed immense respect in western communities and on the range.

As the Forest Service developed in its early years, Pinchot, Potter, and a few other professionals relied heavily on large numbers of assistants in Washington and in the field. In order to obtain much-needed, inexpensive labor for the agency, Pinchot developed a student assistant program that was tremendously popular with eastern college students.[49] Pinchot hyped the program with public speaking and articles in popular magazines such as the *Saturday Evening Post*. He developed for the service a reasonably well-trained corps of young men who idolized their leader and who were willing to take on difficult jobs at low salaries. The disadvantage was that they lent credence to western complaints that eastern college students were making western resource policy. A popular story in the western lumber camps concerned one of Pinchot's green student assistants recently assigned to monitor a logging operation. The first day on the job the assistant rushed into camp exclaiming, "There's a *bear* on the ridge!" "Yep," replied the timber foreman. "I thought I'd have to tie up that old black sow when you came around."[50]

Not all Forest Service personnel were eastern dandies in cowboy boots and lumberjack shirts. The majority of forest rangers and surveyors were westerners working in their own communities or at least in their own states. Many had failed in other professions, and had previously been cowboys, stockmen, lumberjacks, timber cruisers, miners, engineers, artists, pharmacists, or ministers.[51] Some were outright adventure-seekers. These men were comfortable with the ranchers, timber cutters, and miners with whom they had to deal, and they eased the way for increasingly strict forest regulation because they fit well in western society and could handle problems in locally accepted fashion. When a dispute broke out among shepherds over grazing allotments on the Uinta Reserve in northern Utah, for example, forest ranger Bill Anderson settled the matter in a meeting around the campfire with the principals. The final agreement was based on a stick-drawn map made in a cleared space of ground.

The forest rangers were usually as tough as their fellow rural westerners, and most early rangers carried a gun, sometimes hidden in the bedroll for the sake of peaceful appearances. Charlie Berry, forest ranger at Tres Piedras, New Mexico, had to settle some particularly difficult disputes in 1906 and wound up defending his home of fifteen years with his Winchester rifle.[52] In some areas, the local forest ranger was constantly threatened and never went to town unarmed. Rangers who lived in small communities were considered outsiders by the residents and were literally social outcasts.

Field employees of the forest service had positive experiences with western locals as well. In many remote towns the first telephone service was a by-product of the establishment of a new ranger station, and forest-related road improvements benefited many isolated settlements.[53] As the forest rangers were more accepted, they entered the social lives of the small towns near the reserves. Many local histories indicate that the rangers married local women and were early contributors to out-migration from small western towns, because when the rangers were transferred, they took their new wives with them.[54] The forest service field workers altered the social map of the west as surely as their Washington sponsors altered the political map.

The efforts of highly visible scientific and political leaders bore fruit in the effort to establish forest reserves, because a significantly large segment of the general population had already accepted the general principles of conservation. The way toward acceptance had been prepared by the popular writings of John Muir who, in old age, was widely recognized as the leader of a new American perspective on the natural environment. As a personal friend of Roosevelt and Pinchot, Muir lent their conservation efforts an air of intellectual legitimacy, and they provided him with some semblance of political respectability.

Roosevelt contributed to the public awareness of the value of wilderness experiences primarily through his writings on his hunting experiences. His first book, *Hunting Trips of a Ranchman*, was popular fare in which he criticized the slaughter of wildlife by commercial hunters and presented the case for sport hunting in the wilderness.[55] His style, more journalistic than literary, enjoyed wide appeal and increased public interest in outdoor affairs. He further enhanced his literary reputation with a subsequent widely read work that was a collection of his best magazine articles. *Ranch Life and the Hunting Trail* was published in a deluxe edition with illustrations by the popular western painter Frederic Remington. By the time of the forest reserve controversy, Roosevelt was firmly established as a naturalist.[56]

Writers more skilled than Roosevelt were also preparing the public for the coming development of conservation. George Bird Grinnell edited the successful magazine *Forest and Stream* and provided an outlet for many naturalist writers with reputations for scientific accuracy.[57] Using his editorial position, he helped found and nurture the powerful Audubon Society for the protection of birds and the Boone and Crockett Club to protect big game animals for sport hunting. Grinnell's pen occasionally received valuable support from William Temple Hornaday, director of the New York Zoological Society and author of *American Natural History* and *Our Vanishing Wildlife*.[58] Hornaday began as an avid hunter but later developed a preservation perspective that presaged the preservation movement of half a century later. He frequently argued that wild animals and places were part of the nation's trust and were not a resource to be used and wasted.

Two writers at the turn of the century presaged a change in American attitudes about seemingly worthless wild areas. The arid Southwest, where water development gave impetus to forest preservation, had a forbidding appearance to most nineteenth-century Americans, but to John Van Dyke and Mary Austin the stark landscapes held beauty, charm, and intrinsic value. Van Dyke made his case in his book titled simply *The Desert*, published in 1901.[59] In his writings about landscapes and wildlife in the Mojave and Sonoran deserts, he was one of the first nationally recognized authors to argue from the uncompromising position that "the deserts should never be reclaimed. They are the breathing spaces of the west and should be preserved forever." Van Dyke was born in the East and grew up in the Midwest, but his travels and an interest in art trained his appreciation for desert landscapes.[60] As professor of art history at Rutgers, he brought an intellectual acuity to conservation writings that previously had been lacking.

Mary Austin turned to the desert for solace from a life of tragedy.[61] From infancy, she was rejected by her mother, and when she was ten, her beloved father and sister died. Her schoolmates considered her odd and perverse despite (or perhaps because of) her brilliance in literature and science. After graduation, her family moved to a small town in southern California. She did not enjoy the social company of men in confining turn-of-the-century society. "When we went places they usually came home mad," she later wrote. "Either I had more attention than they did, or I had to tell them to shut up and let me talk."[62] By the time of her socially unacceptable divorce, however, she had achieved national recognition for her writings about the Southwest deserts as places of beauty and personal renewal. Reserving her passion for her work, she wrote her book of influential essays, *The Land of Little Rain*, in only one month, and after its publication in 1903, her works, with their message of preservation and natural appreciation, were part of the parlor accoutrements of nearly every educated family in the country.[63] She and Van Dyke ranked with Muir in terms of literary influence and paved the way for the rise of conservation.

Efforts of individual authors were augmented by increasingly powerful conservation organizations. The Sierra Club was officially formed in 1892 by Henry Senger (a Berkeley socialite), William Armes (an English professor at the University of California), and Warren Olney (a San Francisco attorney).[64] With Muir as president, the club collected 175 members within a few months. By the time of the congressional effort to overturn President Cleveland's "Washington's Birthday Reserves," the club was an active proforest lobby. In 1900, the club began sponsoring wilderness trips to stimulate membership and to increase wilderness awareness. Muir almost always went along and continued to preach his gospel. "I had a long talk with Mr. Muir this afternoon," wrote one participant, "or rather he did all the talking."[65] During the first decade of the twentieth century, the club spent most of its efforts attempting to transfer jurisdiction for Yosemite Park from the state to the federal government and then attempting to keep dam-builders out of the park. As a result the club was less visible than in later controversies. Also, the club's preservation perspective did not parallel the conservation or "wise use" principles of Pinchot and Roosevelt.

Grinnell's Audubon Society succeeded in raising consciousness levels among many easterners with regard to preservation of wildlife, but state organizations in the society were very influential and internal fighting was common. Viewed by outsiders as mere bird-watchers, Audubon members worked to support forest preservation as a means of protecting game habitats, but as a group they were ineffective.

A little-known group, the American Civic Association, worked effectively on a parallel course with Pinchot and Roosevelt through the efforts of its publisher-director, J. Horace McFarland.[66] McFarland's major contribution was his effort to preserve Niagara Falls from development as a hydroelectric site. He motivated bills in Congress to protect the area and fought in the legislative trenches for its passage. Power industry engineers and lawyers from General Electric Company, along with J. P. Morgan's attorney, organized powerful opposition, but with Roosevelt's support, the site was protected in 1906. Part of the success was the result of an aggressive grass roots campaign by McFarland fueled by articles in *Ladies' Home Journal* and *Harper's Weekly*. In one year, administrators received 6,450 letters supporting preservation and only 100 supporting diversion of the river for power production. After several years, diversions were permitted, but only to a limited degree. McFarland had waged a fight that provided a glimpse of the basic wilderness preservation controversy that would come late in the century: amateur versus the monied interests. Of more importance was the fact that McFarland's efforts brought the preservation/conservation concepts to the very backyard of the eastern population centers and attracted attention to the growing conservation ethic that was not limited to western environments.

The individuals who participated in the forest reservation movement and the response represented by the second sagebrush rebellion reflected a developing pattern. As in the first rebellion over the irrigation lands, the individuals favoring exploitation were those who would benefit monetarily from development. Many believed that development was the destiny of the nation and saw no value in preservation or conservation not related to their own financial interests. Unfortunately for them, the public and the press could not separate philosophical from financial motives for the rebels. The conservationists, on the other hand, had little to gain financially from their position. Pinchot was independently wealthy and adopted any position he chose without charges of conflict of interest. Roosevelt rode a wave of popularity and took progressive positions contrary to trends in his own political party. Although small in numbers, the authors, scientists, and organization people promoted proconservation perspectives with a zeal that sustained their arguments and that allowed them to literally change the course of American history and to alter the geography of the North American continent.

Chapter 8

The Hole in the Donut

During the period of the national forest controversy, the interior western United States was a unique region, with economic, political, social, cultural, and natural characteristics that made it very much unlike any other region of the continent. But like the nation of which it was a part, the sagebrush region was changing rapidly, and the changes were inextricably bound up with the forest movement and later, the wilderness movement. By the 1970s, the region had become what one Utah official referred to as "the hole in the donut."[1] He noted that this "empty quarter" was virtually surrounded by economic growth that it did not share: California to the west, Seattle to the northwest, Wyoming to the northeast (with its oil and coal booms), and the Denver to Pueblo urban corridor along the Colorado Front Range to the east.[2] The hole in the economic donut began to appear clearly as the twentieth century dawned, and the trends that produced it were as strongly related to the second sagebrush rebellion as they were to the endemic characteristics of the region. In the following pages is a review of the characteristics of the region that ensured its continued conflict with a distant federal government over land management policies: land ownership held mostly by outsiders, ineffective state political structures, economic colonial status, and above all an environment that was more unstable and difficult to manage than anyone imagined.

Throughout the second sagebrush rebellion, land ownership formed the core of the controversy. As long as lands remained in the public domain without meaningful management and without on-site administrators, western residents could use the land as they saw fit and answer to no authority. The transfer to national forest status of large amounts of potentially productive land drastically reduced available options for public land users, especially ranchers. At the same time, remaining public lands available for settlement dwindled as the number of new entries for homesteading (real or otherwise) remained consistently high throughout the period of the controversy.[3] In addition to these depletions of available land, a third highly visible process of reservation was also occurring in the setting aside of national parks.[4] During the period 1891–1911, fourteen outstanding scenic areas were designated as scenic national parks, while an additional fifteen were preserved as national monuments. The majority were reserved during the Roosevelt administration, and all were west of the Mississippi.

During the second sagebrush rebellion the great disparity between the East and the interior West in terms of land ownership became firmly embedded in the national map. Despite continued entries and claims, it became clear that the western states were bound to be public land states: settlement of the western states to the degree that had occurred east of the Rocky Mountains would never take place. The East also had no national forests, no national parks, and almost no wilderness.

If population density were to be used as a guide and wilderness areas defined as those with population density of less than two persons per square mile, the only qualifying areas in 1900 in the East were the marsh areas of southern Florida and the deep woods of northern Maine (Figure 8).[5] The West, on the other hand, had huge areas with no population, and wilderness areas by even strict definitions were common. Arizona, Nevada, and Utah, in particular, had large areas with no evidence of human activities. Even though major railroads crossed the region and provided some access, branch lines and connecting roads were few and far between. Except for widely scattered mining camps focusing on copper, gold, and silver deposits, rugged mountains remained unspoiled, and remote plateau/canyon country was preserved by inaccessibility.

Government geologic reports on many areas, even into the 1930s, included explanations on wilderness travel and advice on survival.[6] Designation of some wild areas as national forests during the second sagebrush rebellion ensured that wasteful logging operations in marginal, remote sections were not allowed, thus saving the areas for eventual permanent preservation. The influence of the forest reserves on wilderness preservation was confirmed in 1964 when the first fifty-four areas preserved in the National Wilderness System were drawn from national forests.

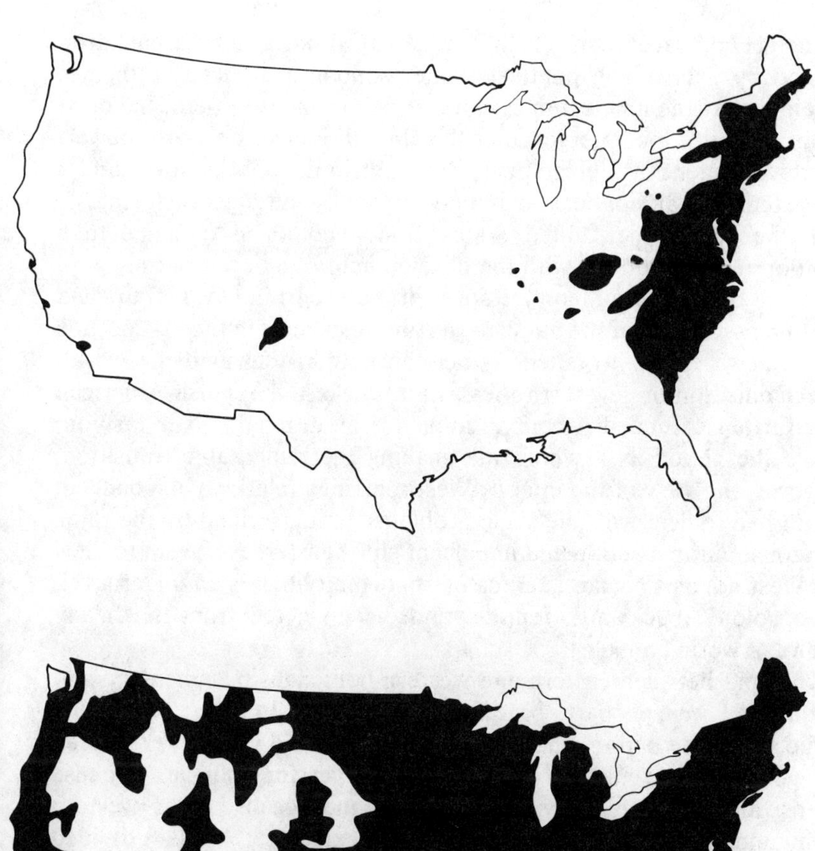

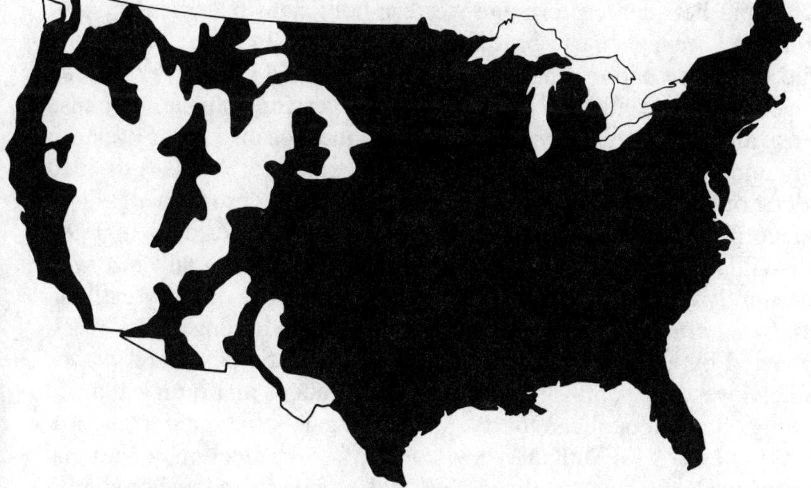

Figure 8. The distribution of the settled areas defined by population density of two or more persons per square mile (shown in black) in 1790 (above) and 1980 (below) (Data from R. Sale and E. Karn).

The second sagebrush rebellion was played out against a backdrop formed by a changing political scene. Among a number of themes developing as the nineteenth century turned to the twentieth, the most important were the emergence of the United States as a world power, the development of a clear East/West split in national politics, and a consistently weak political position occupied by the sagebrush states.

In the 1890s, the United States completed its ascendance to a position of world power with the development of an eastern industrial base and of a powerful navy.[7] Roosevelt, an expert in naval affairs and one-time secretary of the navy, used American fleets in the first decade of the new century to extend American authority around the world. The acquisition of new territories as a result of the Spanish-American War further extended American influence in an imperialistic fashion. While the nation as a whole was making the remarkable transition, however, the view of the interior West remained relatively unchanged. It still had to deal with the same problems circumscribed by the same horizons, but in a subtle and important shift, easterners began to view the West not as a romantic arena of opportunity but as an underdeveloped colony, backward, remote, and disconnected from the heady affairs of world conquest.

The split between eastern and western politicians became deep and lasting and was, in part, based on economics.[8] In 1893, a financial panic swept the nation when gold reserves fell and Congress repealed the government's authority to purchase silver for coinage. Because almost all silver mining was in the West and because some western communities depended on silver mining for existence, the issue divided leaders on a sharply defined regional basis. The problem of silver coinage produced regionally based candidates for president in 1896, with William Jennings Bryan advocating free coinage of silver. Bryan consistently made the case for the "hardy pioneers" of the West who were being crucified on a "Cross of Gold."[9] This darling of the West, supported by a splinter group from the Democratic party and by the strongly western Populist party, was more adept at oratory than at winning. The Republican party, advocating a strict gold standard, swept to power with McKinley's victory. After the election, a National Monetary Congress formalized the gold standard and ushered in a decade of financial stability and national prosperity. The West remained in a subservient position, and few westerners were visible on the national political scene, increasing the sense of political isolation and setting the stage for another rebellion based on sectionalism.[10]

Because of the larger population in the East, the House of Representatives (apportioned by population) tended to favor the eastern and federal positions in the forest reserve controversy. In the Senate, the West was on a more equal footing and received more support. In reality, the states of the sagebrush region were poorly developed politically. They lacked the strong internal organization that might have allowed them to carry out the rebellion more successfully, a weakness that resulted from the recency of their formations and from domination by commercial interests. During the entire period of the rebellion (about 1891–1911), Arizona and New Mexico were only territories and thus did not even elect their own governors. Fearful that the two territories would elect Democratic candidates, the Republican-controlled House, Senate, and White House delayed their transition to statehood, adding to resentments in the region. Republican fears were well founded. After Arizona was granted statehood in 1912, Democrats won the governorship in two out of every three elections through 1982 and always sent at least one of their party to the Senate, and held the only House of Representatives seat for thirty years.[11]

The case of Utah was different from that of Arizona and New Mexico, but the antagonism between the state's general population and the federal government was the same. Settled primarily by members of the Church of Jesus Christ of Latter-day Saints (Mormons), Utah, a cultural anomaly in the United States dominated as it was by a theocracy, was viewed with suspicion by outsiders.[12] As a territory, it had been occupied by federal troops in 1859 to prevent secession. The territory tried repeatedly to achieve statehood in the 1880s, but anti-Mormon groups in the East successfully kept Utah out of the Union. "The Mormon Question" revolved around the open practice of polygamy and the dominant position of the church in controlling business affairs. The territory reached a tacit accommodation with the federal government by the early 1890s by divestiture of major church investments and cessation of polygamy in return for statehood, granted in 1896.

However, statehood did not end the differences between Utah and its eastern cousins. When Brigham H. Roberts, a high-ranking church official, was elected to Congress in 1898, the House of Representatives refused him his position by a vote of 268 to 50 on the grounds of his church association.[13] In 1903, Utah's Senator Reed Smoot, also a church official, was forced to defend his election during a three-year investigation by the Senate Committee on Privileges and Elections.[14] The investigation was fueled by an anti-Mormon petition of more than a million signatures. By 1906, the Mormon church had relented under charges of trust-building, monopoly, and anti-Americanism and sold almost all of its major business interests.[15] The purchasers were large national corporations curiously similar to the church in the scope of their activities, but acceptable to the general U.S. population. Although the church eventually reinvested in the business community, resentment over these difficulties served to establish strong local support for the second sagebrush rebellion, which was seen as a simple extension of antagonism between Utah and the federal government.

Politics in the other two sagebrush states, Colorado and Nevada, were controlled by oligarchies, but they were corporate rather than ecclesiastic. Both states entered the Union decades before the forest reserve issue surfaced, both owed their creation and economic vitality to the mining of precious metals,[16] and both were controlled by the companies that managed the mines and the railroads that provided links with the outside world.[17] It might at first appear that this monolithic political structure would aid the sagebrush rebels, and in many cases ardent defenders of the proposition of decreased federal control were from these two states. The fatal flaws for the rebels in this arrangement were the rising tide of antimonopoly and "trust-busting" opinions of the general population and the skill of President Roosevelt in exploiting the resulting political divisions in western states. Many citizens in Nevada and Colorado bitterly resented the economic domination of their economies by large corporations controlled by Bostonians, New Yorkers, and San Franciscans, and opposition to "the company" became fashionable.[18]

Reform candidates at state and local levels achieved increasing success as the second rebellion wore on, so that by Roosevelt's second term beginning in 1904, his opponents could not count on solid political fronts in Nevada and Colorado. Behind his bombastic oratory, Roosevelt was an astute politician, and he correctly sensed that though his massive forest reserves might stir significant vocal opposition, the opposition was fragmented and not necessarily shared by the general population. He counted on his personal popularity and the preference of the average voter for the federal government as opposed to special economic interests. Without broad support in states like Nevada and Colorado, the rebellion survived on borrowed time.

The economic conditions in the sagebrush region during the 1890s also set the stage for discontentment and regional rebellion. In the cases of extracted resources, the interior West was the victim of market conditions and increasingly capital-intensive technology, but in the case of the railroads the region's disadvantages were the products of deliberate discrimination.[19] For the major commodities of minerals, lumber, beef, and grain, the railroads provided the only route to eastern markets, a factor that they ruthlessly exploited with rate structures biased against western producers. The Burlington Railroad (Chicago, Burlington, and Quincy Railroad) rates provide an example: east of the Missouri River, charges to transport one ton of freight one mile were $1.32, while the same ton/mile charge on the same railroad west of the Missouri was $4.80. As the twentieth century approached, the regional variation decreased, but even by 1900, charges in the West were 50 to 100 percent higher than in the East, and in terms of real value, freight rates increased.[20] The railroads claimed that the western lines were simply more expensive to operate, but western producers contended that the rate structure was obviously a product of eastern monopolies exploiting the West. In the eyes of many westerners, the railroads' villainy was enhanced by the huge amounts of land controlled by the companies. When the second rebellion became a public issue, many citizens sided with the federal government because they feared state or private ownership of the forests would eventually channel the resources to the ownership of unscrupulous businesses.

As recounted earlier, the western silver mining industry suffered irreparable damage in the monetary crisis of 1893–1896.[21] The collapse of the industry caused economic ripples that touched all aspects of the economy of the sagebrush region, especially in Nevada and Colorado.[22] As the mines closed, their workers departed, decreasing demand for everything from beef to lumber for mine timbers. Mining for other minerals, particularly copper, was important in Utah, Arizona, and New Mexico, which also felt the impact of the 1893 financial panic, but western workers bore the brunt of the hardships.[23] Eastern investers with their diverse holdings simply deemphasized their mining properties for a time and waited for a return to prosperity in relative comfort. Because the mines were becoming increasingly mechanized, eastern capitalists controlled the mining industry and the western states became incidental to the decision-making process. They were where the mines happened to be located, and their communities were at the mercy of corporate decisions made 2,000 miles away. This arrangement weakened the case of the sagebrush rebels because the companies representing western properties did not represent western people.

Lumber companies had a major stake in the forest reserve controversy, but they were unable to develop a solid front on the debate.[24] A wide separation existed in spatial as well as in philosophical terms between owners based in the East or in San Francisco, and the local population in the interior West. During the controversy, most western lumber production was from California, Oregon, and Washington, so that those states (especially the latter two) aggressively supported the rebellion on grounds of free access to the timber resource. The vise-like control of state politics in the Pacific Northwest by lumber barons rivaled the degree of control by mining magnates in the interior states. However, this control did not extend to increasingly urbanized populations who were concerned about water resources and trust-busting. In the first few years of the twentieth century, unionism was at its strongest in the lumber camps and mills, diverting some of the companies from the business of the rebellion.[25]

Throughout the West the lumber market was in disarray.[26] During the 1890s, real lumber prices declined as more timber was harvested. In order to maintain profits, companies cut increased amounts to make up in volume what was lacking in unit price. This strategy further depressed prices by glutting the market so that wasteful cutting practices and anticonservation positions made short-term economic sense. No operator could afford to wait until tomorrow to cut what he could today. Pinchot had a clear understanding of these market processes, and he convinced many timber operators that the forest preserves would be a market stabilizer. By 1900, the vicious market conditions had eliminated almost all of the small operators, and the big timber companies had forests of their own. Most of the large companies eventually agreed with Pinchot's position because the reserves kept competing timber off the market while they could log their own private forests without nagging competition from small companies dependent on public lands.

The grazing industry during the forest reserve controversy was the only major industry that enjoyed substantial western ownership, but it was weakened by internal disagreements. Ranchers were unable to settle the question of dividing the public grass among themselves, and when the federal government began to exert a stronger influence in the process through fees and allotments, many ranchers (especially those with smaller operations) welcomed the change. Beef and wool prices fluctuated substantially from one year to the next, and graziers without assured range resources could not survive the low-price periods. The internal bickering among ranchers in public land states gave Pinchot and Roosevelt an opening that they exploited to the fullest by touting their support of the "little man." The ranchers were a potent force in the second rebellion, but the fragmented nature of their lobby proved to be their undoing.

These reviews of the transportation, mining, timber, and grazing industries (summarized in Figure 9) show a West that was divided against itself, with corporate and political leaders, and general citizens splintered on the rebellion's fundamental economic issues. Their divergence of political philosophies was increased by unstable economic conditions in the 1890s and the economic imperialism of the East that rapidly reduced the West to a colonial position. The sagebrush region had little choice, for no capital was available in the West, and if development was to take place it would have to be funded by eastern dollars. In exchange for these dollars, however, the interior West had to relinquish control over its own political and economic destiny. There was little incentive for the eastern investors to develop the interior West beyond the minimum required for resource exploitation, and the logical outcome was realized in the twentieth century: a region that lagged behind its surrounding areas in terms of economic development.

In the sagebrush region, economic instability was complemented by

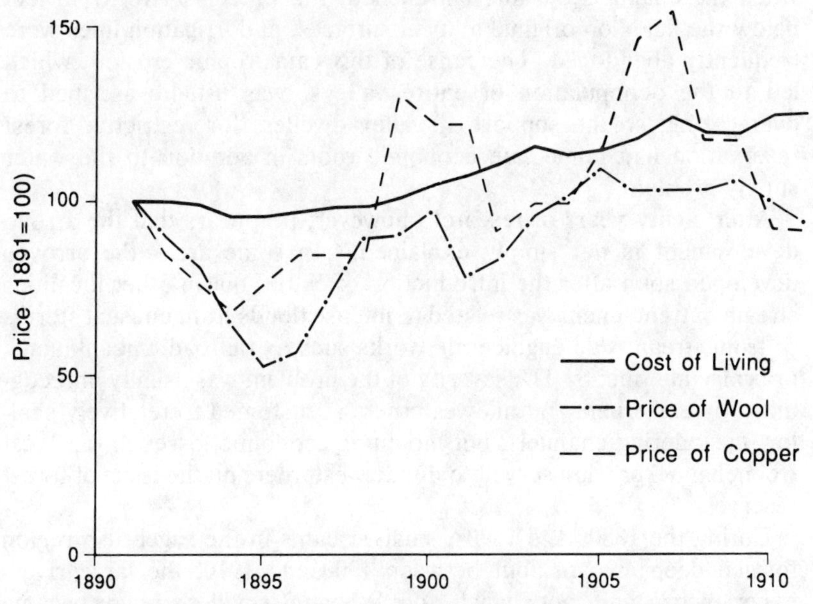

Figure 9. Trends in cost of living, price of wool, and price of copper during the debate concerning the national forest system, showing the steady increase in cost of living, but uncertain trends in agriculture and mining that impacted the West (U.S. Department of Commerce data).

environmental instability hinged on the resource second in importance only to land: water. As in most arid and semiarid parts of the world, the sagebrush region experiences extreme variability in its rainfall patterns so that from the human perspective it appears that there is always too much or too little. This problem eventually leads to the development of dams that are economically justified on two points: the dams store water for use in dry seasons and droughts, and they aid in flood control during wet seasons. Before 1912, however, no large dams existed on the region's major rivers, and nearby residents were at the complete mercy of changing hydrologic conditions.

One of the most important hydrologic-related changes was the development, throughout the sagebrush region, of arroyos—stream channels that excavated deep trenches in the rich alluvial fill of the valleys.[27] The arroyos represented the loss of valuable soils that were suitable for crops: they also disrupted transportation lines by lateral erosion, and irrigation works by vertical trenching. Before extensive erosion, when the channels flowed on top of the alluvial fill, removal of water for irrigation required only a simple brush diversion dam, but when the channels became entrenched, the water was 10 to 50 feet below the fields on original alluvial surfaces, and irrigation lines were frequently abandoned. The cause of this catastrophic erosion, which led to the depopulation of entire valleys, was usually ascribed to overgrazing, so the support of valley dwellers for restrictive forest reservation had immediate economic roots in addition to the water supply problem.

After eighty years of research, however, it appears that the arroyo development is not simply explained.[28] In some areas the arroyos developed soon after the introduction of cattle, but in other localities stream entrenchment was related to intense floods from unusual storms or from stream-side engineering works such as railroad embankments or bridge abutments. The severity of the problem was usually unrecognized by easterners and midwesterners accustomed to relatively shallow meandering channels, but mounting economic losses in the West from channel erosion served to divide westerners on the issue of forest reserves.

During the 1880s and 1890s, small streams in the sagebrush region formed deep arroyos, but between 1900 and 1910, the large rivers experienced significant changes too. When the smaller streams became entrenched, much of the sediment traveled into the channels of the major rivers where it restricted channel capacities.[29] From 1900 to 1905 the region suffered a severe drought, and cattle grazed the relatively lush vegetation along river banks.[30] By removing the vegetation and trampling the soil, they weakened the resistance of the banks to erosion.[31] When massive flooding occurred in 1905, rivers overflowed, destroying fields and towns, and the channels experienced drastic changes. In Utah the Fremont River cut its channel downward almost 20 feet and widened itself from 120 feet to a quarter of a mile. The San Juan River became a milewide ribbon of sand, destroying irrigation works along its banks.[32] In New Mexico and Arizona the Gila River changed from a narrow meandering channel to a braided sand river more than a mile wide.[33] The Colorado River burst through an irrigation inlet and changed its course completely, forsaking its original route to the Gulf of Baja California for a new course into the Salton Sink of southern California.[34]

The floods played a consequential role in the sagebrush rebellion because of the possibility that their cause was poor land management. When mountain forested areas were logged, runoff from major streams collected more rapidly in stream channels than in inhabited lowland areas. The branch and leaf canopy of the forest intercepts raindrops, while the forest floor litter of downed materials inhibits overland flow. Eventually the precipitation finds its way to the channel, but at a much slower rate than if grass-covered or barren slopes are encountered. Farmers and urbanites who made up increasingly large proportions of the western population saw the cut and run loggers as a major hazard to their livelihoods, and so they aggressively supported reservation of forested mountain slopes upstream from their communities and farms.[35] The *Grand Valley Times,* a southeastern Utah newspaper, noted that the formation of a new national forest "will be of great benefit in the long run, though it may immediately affect a few adversely."[36] The first forest rangers in the area carried revolvers to deter those "few adversely" affected individuals.

Maintenance of forest cover not only retarded and reduced floods, it also produced a more dependable discharge for irrigation. Channel flow consists of two parts: runoff and base flow.[37] The runoff comes from surface water and is greatest in the interior West in the spring as melting snow feeds the channels. Base flow is water that moves laterally as ground water: it represents precipitation that percolates into the ground, moves to a channel, and then surfaces. In the ephemeral mountain streams of the West, the base flow is relatively small compared to runoff, but it is vital because it causes the streams to flow well into the summer dry period when the water is needed for irrigation. Forest cover promotes base flow by slowing the runoff process, allowing water to percolate into the soil instead of quickly filling the channel for only a short period followed by no flow.[38] Even with the moderating influence of naturally occurring forests, the water supply was highly variable, but overcutting of the forests led to disastrous results in some areas.

By 1900, popular and scientific thinking about precipitation had graduated beyond the "rain follows the plow" concepts of earlier decades, but some important misconceptions persisted. It was widely believed that forests stimulated rainfall because the trees expired large amounts of moisture, which then condensed and fell as rain.[39] To the casual observer in the interior West, there was convincing evidence to support this notion because clouds formed most frequently over the mountains where the forests were located, and those streams draining forests carry more water for longer periods of time during the year. Subsequent research has shown that base flow explains the longer-flowing streams rather than increased precipitation, and that more clouds and rain occur on the mountains because of their elevation rather than their forest cover.

When air flows over a mountain mass and is elevated in what is termed an "orographic effect," its pressure and temperature decrease, resulting in condensation of moisture and cloud formation.[40] Although this concept was published in the 1800s, wide understanding was slow in developing.[41] Misunderstandings about precipitation mechanisms were not serious because the end products were the same, but they represented an additional complexity in an already complex situation.

An important by-product of the orographic effect in the interior West is a particular distribution of vegetation communities.[42] Forests grow almost exclusively on mountains and high plateaus because those are the only areas with sufficient precipitation to support treelike growth (Figure 10). At the other extreme, lowland areas generate no orographic precipitation and thus are the locations of desertlike growth. Between these extremes are a variety of communities that have increasing amounts of vegetation as elevation and concomitant moisture increases. This altitudinal zonation, first elucidated by botanist C. Hart Merriam (G. K. Gilbert's host in Washington, D.C.) gave rise to a grazing practice peculiar to the West whereby animals grazed grasslands in the warmer valleys in winter and in the cool, moist upland areas in summer.[43] The formation of forest reserves and elimination of forest grazing truncated the summer half of the process, something not always appreciated by a forestry graduate from a school in far-off Massachusetts or Rhode Island.

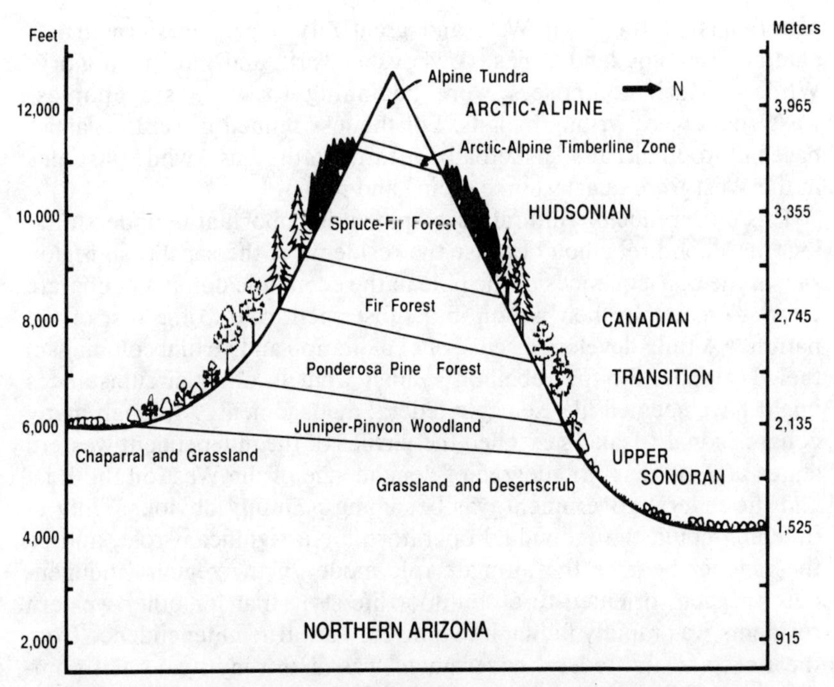

Figure 10. Schematic diagram showing the distribution of vegetation in northern Arizona, a typical example of areas included in the early national forest system. This distribution of vegetation communities, combined with the mountain and valley terrain of the western states, largely explained the resulting distribution of national forest units. (Redrawn from work by D. Brown and C. Lowe, U.S. Forest Service).

Along with the characteristics of land ownership, politics, economics, and an unstable natural environment that gave the interior West a unique sense of place at the turn of the twentieth century, a unique culture had also developed that set the region apart. Just as easterners were comparing themselves with and competing with Europeans in a cultural sense, so the West tried to develop itself culturally to levels of sophistication found in San Francisco or New York.[44] That the effort met with little success outside the older established cities of Denver and Salt Lake City was less a testament to the backwardness of the people than it was a logical outcome of sparse population and lack of money available for cultural refinements. For the wealthy mine, railroad, and sawmill owners, the West was the workplace, while the East was a place where donations to museums, art collections, and the performing arts were made. While eastern painters created portraits, panoramas of the Civil War, and great cityscapes, western artists painted sweeping landscapes, cowboys at work, and Indians in glory. While eastern composers were beginning to write symphonies, westerners were writing ballads. For the less refined but enthusiastic, baseball, football, and basketball appeared in the East, while pastimes in the West were quarterhorse racing and rodeo.[45]

This divergence of cultural development is important in understanding the second rebellion because the residents of the sagebrush region saw in the consequences of the hole in the economic donut a deliberate effort to maintain their region in a subservient role to the rest of the nation.[46] A fully developed sense of exploitation and virtual colonialism fueled support for the rebellion against what in other circumstances might have seemed like sensible federal management. Although many congressional speakers extolled the virtues of the independent western states and citizens, the degree of dependence of the West on the East and the federal government was becoming painfully obvious. Only in ranching could the individual operator play a significant role, and so the rancher became the ultimate role model of the region—independent, rugged, oriented to an outdoor life-style that for other western residents was rapidly fading into a myth. That true independence from the East and the federal government was also a myth was not commonly recognized, but it was ensured by the lines of steel rails, credit, and bureaucratic authority that penetrated the region and bound it to the East. When Roosevelt reserved huge amounts of lands for forests, he did so in the certain knowledge that the objections would be long and loud, but that the western colonies, divided and dependent, would ultimately acquiesce.

Chapter 9

Resolution . . . Again

When it became clear that Roosevelt's midnight reserves were not likely to be dissolved, the defeat of the western development interests in the question of forest preservation was only a matter of time. The new ethic of conservation, established in just a few years by Roosevelt and Pinchot, had become so powerful in the political arena by 1907 that even developers embraced the label if not the idea. Lumbermen,[1] miners,[2] and graziers[3] all characterized themselves as "conservationists" in what they perceived to be wise management of natural resources.[4] They saw federal regulations and use fees proliferate, especially in grazing matters, and they began to feel a loss of control. In earlier times, disputes over western land management had been solved on authority of the gun or at best a showdown meeting of local interests. With the failure of these approaches and the advent of extensive national forests, the decisions affecting the everyday lives of western residents shifted to Washington. With the shift, the second sagebrush rebellion was not about to fade away even with the extensive victories of the conservationists. The first step in the Western counterattack was the most simple but also the most difficult: to get organized. The place to start was the Denver Public Lands Conference.

Colorado was the most likely state for an anticonservation convention.[5] Roosevelt's withdrawals included almost half the remaining unappropriated public land in the state, so there was no shortage of antifederal sentiments among the general population. The state legislature included an assortment of ranchers, real estate developers, and mining interests, all spoiling for a fight. Last, but not least, Colorado was the only western state not controlled by the Republican party. The many rebels, perhaps a majority of whom were Republican, did not relish the thought of a convention in a Republican state attacking a Republican president. If any of their anti-Roosevelt shots fell wide of the mark, they did not want to damage their own party.

The governor of Colorado, under direction from the state legislature, called the Denver Public Lands Conference for the purpose of responding to the Roosevelt/Pinchot successes in forest reservations.[6] Like the conference held in Salt Lake City 73 years later, the general objective was to loosen federal control and to turn the lands over to state control. Some commentators detected an antiadministration plan behind the entire meeting, but among the 861 delegates (500 from Colorado) there was fairly wide representation, including some from state forestry associations, industry groups, and even irrigation speculators. Pinchot, never one to turn down the chance for combat, agreed to attend and to address the meeting. His arch foe, Senator ''Slippery Tom'' Clark (Montana) was a chairman of the meetings along with a rabid antifederal Wyoming rancher. Other notables included congressional Representatives Mondell (Wyoming) and Taylor (Colorado) along with Senators Shaforth (Colorado) and Clark (Wyoming). In addition to Pinchot, Roosevelt sent Secretary of Interior James R. Garfield and Director of the Reclamation Service F. H. Newell.

Observers looking for a squabble were not disappointed. Even before the meetings began in summer 1907, there were bitter disputes about the credentials of state delegations, with conservationists charging the development interests with trying to pack the meeting. The developers generally won these preliminaries and excluded conservationists from speaking during most of the meeting. Much of the rhetoric was antiadministration, and there were frequent calls for the institution of state controls instead of federal regulation. The national forests were special targets and many delegates wanted the reserves disbanded. ''Are not men better than trees?'' demanded former Senator Henry Teller in a plea for accelerated development.[7]

Having invited Pinchot with the intention of embarrassing him, the organizers were chagrined when he appeared at the meeting prepared to speak. It was Pinchot at his best, seemingly outnumbered but a major force on the scene. He used humor to fight his hard-speaking opponents. When he started his presentation he grinned at the audience and began, ''If you fellows can stand me, I can stand you.'' When a heckler rose to challenge him, Pinchot sternly pointed and in his best Forest Service command voice joked, ''Give me that officer's name: he will be discharged at once!'' The convention collapsed in laughter.[8]

After dispelling the prevailing dark image about himself, Pinchot went on to make the case for the national forests by pointing out their beneficial effects in economic and resource terms. He contended that no agricultural lands were locked up by the forests, and that mining was permitted in the reserves. He pointed out that many western congressional representatives had approved the forest legislation, and he linked successful reclamation efforts, so dear to western hearts, with a successful forest policy. He concluded with the theme that the policies he and Roosevelt pursued were best for the long-term growth of the West. With his good humor, his bold and reasonable approach won much support, and from that time on the developers were in the business of trying to cut their losses rather than to use the meeting as a springboard for a broad-based national effort.

The final resolutions of the meeting that were supposed to ignite a national antifederal movement turned out to be relatively mild as the delegates failed to rally behind stern measures.[9] Some delegations, especially those from California, Oregon, Utah, and New Mexico, would probably have supported strongly defined alternatives to the national forests, but the developers had few well-defined programs to offer. Unwilling to return to the wasteful chaos that prevailed before the reserves, the undecided delegations were not won over by the sagebrush rebels. Resolutions that were approved merely stated opposition to new reserves in settled areas and regions that were not forested, while others called for state control of mining and irrigation. Western editorialists wondered how this mighty lion had turned out to be such a meek kitten. The *Arizona Gazette* bemoaned the lack of constructive alternatives and the *Salt Lake Tribune* darkly hinted at the work of federal subversive agents. The *Portland Oregonian* labeled the Denver Public Lands Convention appropriately: it was a fizzle.[10]

However, one far-reaching development did occur at the meeting. Pinchot met informally with Colorado ranchers and discussed whether or not the Forest Service had the right to restrict the numbers of sheep and cattle on the reserves. The legality of grazing fees had also been the subject of a long dispute, and the informal meeting moved to that topic. Fred Light, an early pioneer from the mountains near Aspen, was impressed by Pinchot and his assistant Albert Potter, but he could not agree that the new Forest Service had the right to charge fees for entry onto the Holy Cross National Forest where Light had grazed his stock for several years without charge. The men did agree that the question had significant constitutional overtones that should be resolved formally by the courts, so they planned a "friendly suit."[11] A short time later, at a prearranged location, Light released his stock on state land and allowed them to drift onto forest land. Forest Service officials dutifully served him with a notice that such trespass had been prohibited by a U.S. District Court Injunction, and (as planned) they went to the U.S. Supreme Court to uphold their rights to regulate the usage of forest lands.[12]

The case of *U.S. v. Light* became an important turning point in the development of resource administration by the federal government.[13] Light, his legal fees paid by the Colorado legislature in another informal arrangement, contended that the Forest Service had no right to administer the resources because the reserves had been formed without the consent of the states where they were located. He contended that his unrestricted grazing before the establishment of the forests constituted an implied license that should continue after the reserves were made. The case dragged on for many years, but finally in 1911 the Court ruled that the Forest Service could exercise control as it deemed fit, legitimizing Section 24 of the General Revision Act of 1891.[14] The Court did not accept the concept that long-term use of the land by local residents was a counter to the concept that the public lands were "held in trust for all the people of the whole country."[15] The only acceptable representative of that larger group was a federal agency.

The critical victory in the case of *U.S. v. Light* was not by luck or chance. Forest Service lawyers George W. Woodruff and Phillip P. Wells knew that Supreme Court cases usually relied on precedent where possible, so they doggedly established a series of favorable decisions in lower courts. They did not pursue cases between the Forest Service and trespassers when the presiding judge appeared hostile, but when judges appeared friendly to the federal position, they aggressively prosecuted.[16] By the time the test case reached the decision point, there was a large body of case law established that was favorable to the government position. Thus it came about that the most important development of the antifederal Denver Public Lands Conference was the initiation of a court case, and that case ultimately was decided in favor of the agency that the conference was designed to destroy.

After their victory at the Denver Public Lands Conference, the proforest groups gained a true sense of vitality and identity. The problem now was to find a readily identifiable label. The developing ethic was much more wide-ranging than "forestry," because it encompassed forests, water, soil, grazing, and minerals. With advice from his assistant Overton Price and John Wesley Powell's protégé WJ McGee,

Pinchot proposed to the president that the new philosophy should be called "conservation."[17] Roosevelt enthusiastically embraced the word and the concept, using both as political tools and as a general guiding principle. Pinchot was certainly not the first to identify the ideas inherent in "conservation," but he was the first national figure to use those ideas in changing American life.

In order to capitalize on the developing conservation movement as well as giving it added legitimacy, Roosevelt called a Conference of Governors on Natural Resources for May 1908.[18] Governors and their representatives from every state attended, along with congressional representatives, professionals, scientists, and industrial leaders.[19] Capitalists Andrew Carnegie and James Hill were there, as was R. A. Long, a major figure in the lumber industry. Noted conservationists also attended, but Pinchot wanted no competition for the limelight so there were some remarkable absences: John Muir, Edward Bowers, Charles Sargent, and Bernhard Fernow. The conservation movement was already showing some signs of maturity: internal conflicts and jealousies.

After three days of well-publicized speeches, the conservation ethic was firmly entrenched in the political arena. Its adherents transcended party, economic, and regional boundaries. In the long term, the meeting (later known as the First National Conservation Conference) spawned international meetings and established the United States as a proponent of conservation in international affairs. In the short term, it stimulated the formation of state conservation agencies, which put the state governments in the business of environmental management.[20] California was the only state with such an agency before 1908, but within a year of the conference, New Mexico, Utah, Nevada, Oregon, and Colorado established successful state conservation agencies. Success was assured because conservationists and industrial interests were fairly represented in the agencies. Failures occurred in Montana and Washington where the agencies were packed with industry representatives. In Washington, for example, Governor Albert E. Mead staffed his Conservation Commission with anticonservationists: State Land Commissioner E. W. Ross (an organizer of the Denver Public Lands Conference), former Senator John Wilson (implicated in land frauds), and J. J. Donovan (a lumber baron).[21]

At the level of the national government, the conservation movement was in full swing. Roosevelt appointed several commissions, including the Inland Waterways Commission and the National Conservation Commission. Pinchot was almost always appointed to the commissions that did not directly influence legislation but that provided the basic data required for sound decision making. The ponderous three-volume report of the National Conservation Commission represented the most accurate and comprehensive accounting of the nation's resources then available, all compiled without direct budgetary support.[22] However by 1908, time was running out: Roosevelt was in the last year of his presidency, and Pinchot had achieved the limits of progress in the fight to establish conservation. The final phase of the second sagebrush rebellion, a war of attrition, was about to begin.

Roosevelt personally approved of his successor, William Howard Taft. Pinchot and Roosevelt hoped he would continue the development of conservation policies, but Taft was subject to other powerful influences. Roosevelt was a Republican, but as soon as he left the White House the party put as much political distance between itself and the former president as possible. Roosevelt's conservation policies were at odds with the party's industrialist supporters, and with his departure they renewed their efforts to influence White House policies. Taft was easily moved from progressive to conservative positions, and Pinchot soon realized that little more was likely to be accomplished by conservationists inside the executive. His feelings of foreboding were quickly realized with the appointment of Richard A. Ballinger to the position of secretary of interior.

Ballinger had been a reform mayor of Seattle with well-developed connections to mining and industrial interests in the Pacific Northwest. He had been Roosevelt's Land Office commissioner for a little more than a year, cleaning up the remaining vestiges of nineteenth-century land fraud. He became so aggressive in dismissing patronage employees that complaints went all the way to the president. When the friend of one of the dismissed secretaries protested to Roosevelt, he shrugged and responded "I can't help it. Ballinger has discharged a classmate of mine!"[23] In the late days of his appointment, Ballinger became careless, and he transacted business concerning coal lands in Alaska with associates from Seattle. He maintained a running feud with Pinchot, whom he considered an interloper on Interior's business. Pinchot frequently won disputes as a result of his close ties with Roosevelt, so it is not surprising that when Ballinger later became secretary of interior under Taft, he and Pinchot argued publicly about resource management. Two areas of dispute surfaced: the question of water power sites and the question of the Alaskan coal lands.

In a throwback to the first sagebrush rebellion, the water-power sites were prime locations for hydroelectric dams that had been reserved from the public domain on western rivers by Roosevelt in 1906.[24] The argument for this approach was that it prevented the development of electrical power monopolies, a view ascribed to by Pinchot and F. H. Newell (director of the Reclamation Service). Ballinger was determined to return the sites to the public domain where they would be claimed and developed by private enterprise. Pinchot and Newell took their complaints about the new release policy to President Taft. Taft convinced Ballinger to delay action, and Pinchot was confident that the offensive secretary of interior would soon resign.

Pinchot was especially disenchanted with Ballinger because of the secretary's disapproval of national forest reserves, particularly in Alaska, but developments over coal lands were the major point of dispute between them. Early in his administration Ballinger was approached by an investment syndicate of Washington State businessmen that included former governor Miles C. Moore, financier J. P. Morgan, and mining magnate Daniel Guggenheim. They requested approval of coal tract claims in the Bering River area of Alaska, but deputies advised Ballinger that the claims were illegal and they would be consolidated into a mining monopoly if approved. To add to the aura of conspiracy, Ballinger himself had assisted in filing the claims after he left office as land commissioner.

Interior Department aides who feared a massive conspiracy alerted Pinchot to the situation. He sent several assistants to copy correspondence from Ballinger's files, and they found that the secretary had broken some of the arrangements between Pinchot's office and the Department of Interior. When he found that his antagonist had examined his files, Ballinger determined to make a fight of the issue of Pinchot's unwarranted influence in his department, and the two men asked President Taft to mediate.

The dispute soon became public and unlike the disagreement over water-power sites, it was not easily resolved. Taft adjudicated like a father of two rambunctious sons by saying he saw no illegalities on either side, but only a sincere difference in philosophy. Newspapers used the case for sensationalism. Western representatives hoped that Pinchot had finally met his match, and Democrats saw the case as valuable political ammunition. *Hampton's*, *LaFollet's*, and *Collier's*, popular weekly magazines, painted Ballinger as a corrupt official typical of the old Land Office days. The public pressure became so intense that Taft was forced to request a congressional investigation. Ballinger agreed, hoping for vindication, while Pinchot agreed, hoping for even more publicity for the conservation movement.

Taft instructed both men not to publicly discuss the case, but Pinchot refused to back down, and he wrote a letter of support for Ballinger's rebellious aids that was read on the floor of the Senate. Taft recognized a challenge and a gambit when he saw them. He thought of Pinchot as a "good deal of a radical and a good deal of a crank," and he felt he had been left with no choice.[25] On January 7, 1910, he asked for the chief forester's resignation. Pinchot read the letter of his dismissal while at a dinner party, and announced to the other guests simply "I've been bounced."[26]

In a real sense, both combatants had won. Ballinger was exonerated by the subsequent congressional investigation.[27] But, despite advice from his friend Ormsby McNary to "fight like hell" or see his "reputation shot,"[28] Ballinger and his lawyer failed to expose the true question—the issue of how resources on federal lands should be administered. He advocated the role of private industry with little restriction by the federal government, but instead of using the congressional hearings as a platform for his views, he merely answered questions in a defensive manner that made it appear to the public that he actually did have something to hide.

Pinchot was satisfied because he felt that the conservation movement was flagging and that it needed a rallying point that he was only too happy to provide. Rarely has there been such a self-satisfied martyr sacked from government employment. He solidified his position with eloquent policy statements at the congressional hearings, and he gave the conservation movement a definable focus as a democratic ethic of wise resource use for the people and not for the monopolists. It all made great press, and Pinchot emerged a popular hero. He set a trend in the debate over resource issues that has persisted from that day to this: the conservationists espoused a well-defined public policy of the greatest good for the most people as a theme that played well for public hearing, while the development interests pursued a classic profit motive in a secretive and defensive approach that generated little public sympathy.

After John Wesley Powell's resignation from the Geological Survey in the early 1890s, the agency was attacked by opponents of its irrigation survey work in a vindictive form of bureaucratic punishment. Fifteen years later when Pinchot resigned, a similar attack was made on the Forest Service. Pinchot's trusted assistant, Overton Price, was fired, and legal advisor Phillip P. Wells resigned in protest. Albert Potter, one-time Arizona rancher and now the Forest Service's chief grazing officer, guided the agency until its new chief, Henry S. Graves, arrived. Graves, dean of the Yale School of Forestry and close friend of Pinchot, assumed the directorship in a time of low morale and great external pressures, similar to the conditions in the Geological Survey a decade and a half before.[29]

Problems of the Forest Service in the immediate post-Pinchot era were twofold. First, budgetary problems threatened the ability of the agency to administer the newly won national forests. Appropriations from a suspicious Congress declined, and established budgets were cut. Democrats in Congress saw a campaign issue in Forest Service spending patterns that carried large amounts for travel, clerical help, and public education, which some considered little more than press relations and advertising. The agency survived the budgetary problems, but in a weakened condition, and for many years its expenditures exceeded receipts.[30]

The second problem facing Pinchot's successor was the ongoing sagebrush rebellion. The weakened condition of the agency invited attempts to return the forests to state control.[31] Congressional attempts at such a transfer became common over the next decade while Graves tried to establish his Forest Service as the only agency qualified to scientifically administer the forests safe from the intrigue of special interest groups. Under the experienced and patient Albert Potter, working arrangements with the Department of Interior were patched up, and Richard Ballinger's agency posed no threat to the Forest Service despite occasional bills introduced to Congress to shift the forests away from the Department of Agriculture.[32]

The natives were restless in the hinterlands. In June 1909, Governor John A. Shaforth and his advisor, J. Arthur Eddy, formed the National Public Domain League.[33] With strong vocal support from Senator Carter of Montana, the league represented itself as a grass-roots effort that was the only viable opposition by the people to "Mr. Pinchot's course of empire." Actually, the league was a publicity office staffed by a few individuals for purely anticonservation efforts. In a single month it mailed 75,000 pieces, mostly "bulletins" written by Eddy. Representatives of the league, including former Senator Henry Teller and Senator Patterson, attended public meetings and conventions in the West. At the Twentieth Trans-Mississippi Commercial Congress in August of 1910, they induced approval of resolutions endorsing the league, favoring the private development of hydroelectric power, and criticizing the Forest Service.[34] Late in the same month another league representative, D. C. Beaman, achieved similar publicity successes at the National Conservation Congress in Seattle. Beaman's credentials as a conservationist were questionable: he was an officer of the Colorado Fuel and Iron Company, which had fraudulently obtained more than 10,000 acres of coal lands from Colorado's public domain.[35]

In 1909, L. K. Armstrong, editor of Spokane's *Northwest Mining News*, founded another sagebrush rebellion group, complete with a constitution. With views similar to the Colorado group, his Western Conservation League's stated principles avowed state control of public lands and resources, and state cooperation with the federal government only when benefits reverted to the states.[36] Like its Colorado counterpart, the Western Conservation League was a publicity mill that generated bulletins and dispatched emissaries to carry their version of the "gospel." Although the state governor was unwilling to openly endorse the group, praises long and loud were forthcoming from Congressman William Humphrey, Senator Wesley L. Jones, and Governor Shaforth of Colorado. Intrepid rebel Senators Carter (Montana) and Mondell (Wyoming) also voiced approval. The true brand of conservation advocated by the league was revealed in its source of financing: Louis Hill, whose Northern Pacific Railroad had become one of the leaders in timber fraud in the Pacific Northwest.

The league and its supporters made vigorous anticonservation statements, but alternative suggestions were not forthcoming. The ranks of anticonservationists were dominated by those who stood to gain most from federal deregulation: state officials (expanded empires), journalists (whose papers relied on land advertisements for revenue), lawyers (who handled land and resource commerce), and investors (speculating for profit). J. J. Donovan, the Washington lumberman, declared he was ready to lead the "Rebel West" in its "warfare on conservation," but he made nothing more than rhetoric.[37] The Seattle Chamber of Commerce went into the business of advocacy publishing with a series of municipal monographs. Its first volume, *The Neglected West*, advocated state control with the honesty of private enterprise as a check on corruption. It was also the last.[38]

The National Public Domain League and the Western Conservation League led short, fruitless lives that generated little heat and even less light in the controversies beyond newspaper copy and self-aggrandizement for their few participants. Their failure can be traced to a number of factors. The organizations had neither highly visible and respected leaders nor broad bases of popular support. It was impossible for any organization, without either leadership or support, to succeed. Each organization was the brainchild of a few individuals with a limited number of self-serving activists for members. The organizations existed solely for the sake of publicity and failed to focus on solid issues or to seriously enter current debates. Beyond feeble cries of states' rights, they failed to offer meaningful alternatives to the blossoming federal role in public resource management, and, in the end, they had little impact on the course of events.

The final battle of the second sagebrush rebellion was at the National Conservation Congress held in St. Paul, Minnesota, in September, 1910.[39] Western anticonservationists planned to use the meeting as a platform for a counterattack, but they feared that the meeting would be packed against them, and they could not secure promises for floor time from the organizers. Western rebels met at Salt Lake City immediately prior to the general meeting to organize strategy. Utah Governor William Spry moderated the meeting with his profederal perspective and the premeeting produced no plans, no agreement, and only a little "cheap publicity," according to one observer.[40]

Pinchot had a strong hand in organizing the Minnesota meeting, but using his usual all-encompassing approach he provided some speaking time for his opposition. Governor Morris (Montana) defended the states' ability to manage resources, Governor Brooks (Wyoming) read the resolutions of the rebels' Salt Lake City meeting, and Frank Short (a water attorney representing California) attacked pervasive federal ownership. Governors Hays (Washington) and Shaforth (Colorado) added their antifederal perspectives.

Then Pinchot attacked. Every state sent loyal conservationists who were given preference over the opposition, and the broad base of support present for the conservation viewpoint overwhelmed the western rebels. With no support and the prospect of an assured loss in any vote, the rebels stalked out of the conference to continue their sniping from the fringes of the conservation movement. Although the issue of state control of the forests was brought up in each Congress for the next decade, conservationists were firmly in control, and they were not seriously challenged again. The second sagebrush rebellion was finished.

The strength of the conservationists' victory can be seen in President Taft's presentation to the Minneapolis meeting.[41] Although Roosevelt and Pinchot were unhappy with his commitment to conservation, Taft left no doubt about the significance of the movement. He stated that the danger of dissipation and waste of natural resources was not readily apparent to easterners, while westerners showed "sympathy with expansion and development so strong that the danger is scoffed at or ignored." He pointed out that the government owned only one-quarter of the nation's timber, and that of the privately owned remainder, only 3 percent was properly managed. To Taft, these figures were indicative of the likely fate of forests relinquished by the government to private owners, and he concluded that conservation was therefore a "great national ambition." Thus, in one short decade conservation had grown from a little-recognized, unnamed concept to the status of a national ethic.

With the second sagebrush rebellion at an end, the principal participants moved to other pursuits. The leading conservationists, Roosevelt and Pinchot, remained in the public eye for a time, but mostly with regard to other issues. Roosevelt, piqued that his hand-selected heir Taft did not follow progressive tendencies, formed his own party for a reelection attempt in 1912. His action split the Republican vote and ensured the victory of Democrat Woodrow Wilson. Always a man to stir passion, Roosevelt was shot during the campaign, but he recovered and remained a political activist until his death in 1919. Pinchot lost his bid to become a senator from Pennsylvania, but he won the governorship of the state twice. His progressive policies earned him great acclaim, and his success in improving the state's highway system was so great that in the 1930s paved roads there were known as "Pinchot roads."[42] His devotion to conservation never waned, and he produced a constant stream of articles on the subject of resource management. Throughout his distinguished career, he viewed his national forest policies as his greatest achievement, and until his death in 1945 he maintained close ties with the U.S. Forest Service. In his day, he was a villain to western developers, but to conservationists he was a hero in his own time and remains so even today.

The scientists who precipitated the second sagebrush rebellion made lasting impressions on their professions. Franklin Hough, whose dogged persistence brought the plight of the nation's forests to the public, died in 1885, long before the disputes he initiated were settled. He is revered as the father of American forestry.[43] After Bernhard E. Fernow was replaced by his protégé Pinchot in the role of national forester in 1898, he established the first school of forestry in the United States at Cornell University. When the governor of New York vetoed the program a few years later in a political controversy, Fernow moved to the University of Toronto to direct its forestry school. By the time of his death in 1923, he had published 250 articles and bulletins, and three books, and he provided a firm intellectual grounding for the technology of American forestry.[44] Although in later years many of his achievements were claimed by Pinchot or Pinchot's admirers,[45] it was the sophisticated, cultured, apolitical Fernow who set the course that the nation later followed under more aggressive and politically oriented leaders.[46] The botanist Charles S. Sargent and the geographer Henry Gannett influenced the courses of their respective sciences by leadership roles in scholarly societies, leaving historical traces of their work in public and scientific arenas.

The bureaucrats important to the second rebellion remained active in public life for several years after the controversy. F. H. Newell remained director of the Reclamation Service until 1914, and was responsible for directing the construction of all the early federal irrigation projects in the West. Outstanding as a speaker and well liked as an individual, he continued his career outside government service as a teacher at the University of Illinois. He later formed a water resources research service of consulting engineers and wrote several influential books. His counterpart in the Forest Service, Henry S. Graves, guided the service through a period of devastating budget cuts levied by a vengeful group of congressmen after the rebellion. He fought economic investigations of the Forest Service and strenuous efforts to return the forests to the scandal-ridden Department of Interior. If Pinchot deserved credit for building the national forest system and service, Graves deserved equal credit for saving them from an early demise.[47]

Albert Potter, the Arizona rancher drafted into the Forest Service by Pinchot, completed a distinguished career in the service. It was largely through his efforts that the service had the financial integrity to withstand searching congressional audits. He briefly served as acting director after Pinchot's departure, but because he was not a forester he soon gave way to Graves. Will C. Barnes, an active New Mexico and Arizona legislator, wrote to Potter that the rancher was the only western man who could understand the Washington job, and that no matter how good Potter was, eastern representatives would never allow a westerner to direct the service.[48] Potter's good will, however, had lasting benefits for the service and its rancher constituents. Richard A. Ballinger, legally absolved in the congressional hearings on his dispute with Pinchot, continued to serve as secretary of interior until 1911 when he returned to his native Washington. At least he had the satisfaction of having removed much of the corruption in the Department of Interior.

There were no truly outstanding rebel leaders in the forest lands controversy, no correlatives of Senators Teller and Stewart from the first rebellion. From among the more vocal rebels, Charles W. Fulton, the Oregon congressional member surrounded by scandal, disappeared from national politics after his first and only term in the Senate ended in 1909. Weldon B. Heyburn, Idaho's effective anticonservation spokesman, also left the Senate in 1909. He is most remembered for his inventions: designs for railroad cars that prevented telescoping of the car structure during accidents. Franklin W. Mondell, the Wyoming sheep rancher who parlayed anti-Washington rhetoric into a successful political career, remained in the House of Representatives until 1923, a reactionary conservative to the end. John F. Shaforth, the militant governor of Colorado, successfully built on his political base established by anticonservation efforts in the aborted National Public Domain League. He served first as governor and then as senator until 1919. His attacks on the Forest Service were relentless but to no avail.

The forest reserves over which the second rebellion was fought outlasted all the human participants in the controversy and became a part of the American landscape. The national forest system remained at about the same size as when Roosevelt left office (the system contained about 187,818,000 acres in 1978),[49] and it became a permanent feature of the economic, social, and political geography of the country (Figure 11). The importance of the system for wilderness preservation became apparent in 1964 when the first nine officially recognized wilderness areas in the world were in U. S. national forests. By the 1980s, the U.S. Forest Service had become the most advanced forestry management agency in the world, with a conservation ethic and a preservation ethic put into operation by a highly professional staff.[50] The Service graduated from an agency under heavy attack in the early 1900s to a bureaucratic power in midcentury able to withstand intense controversy. One of the agency's primary missions became protection and management of wilderness areas.

The resolution of the second sagebrush rebellion accomplished more than simply quieting the immediate controversies. It led to the settlement of issues that had brewed for half a century. Ownership of the public lands clearly lay with the federal government. Despite protests to the contrary, the Supreme Court declared that "all the public lands are held in trust for the people of the whole country,"[51] and the only viable governmental entity capable of representing the entire nation was federal. The Court also found that the federal government could reserve the public lands indefinitely,[52] and in *U.S. v. Light*, the Court affirmed the right of the federal government to charge user fees.[53] The principles of ownership, representation of the people of the entire nation, and user fees proved in later decades to be the legal basis of

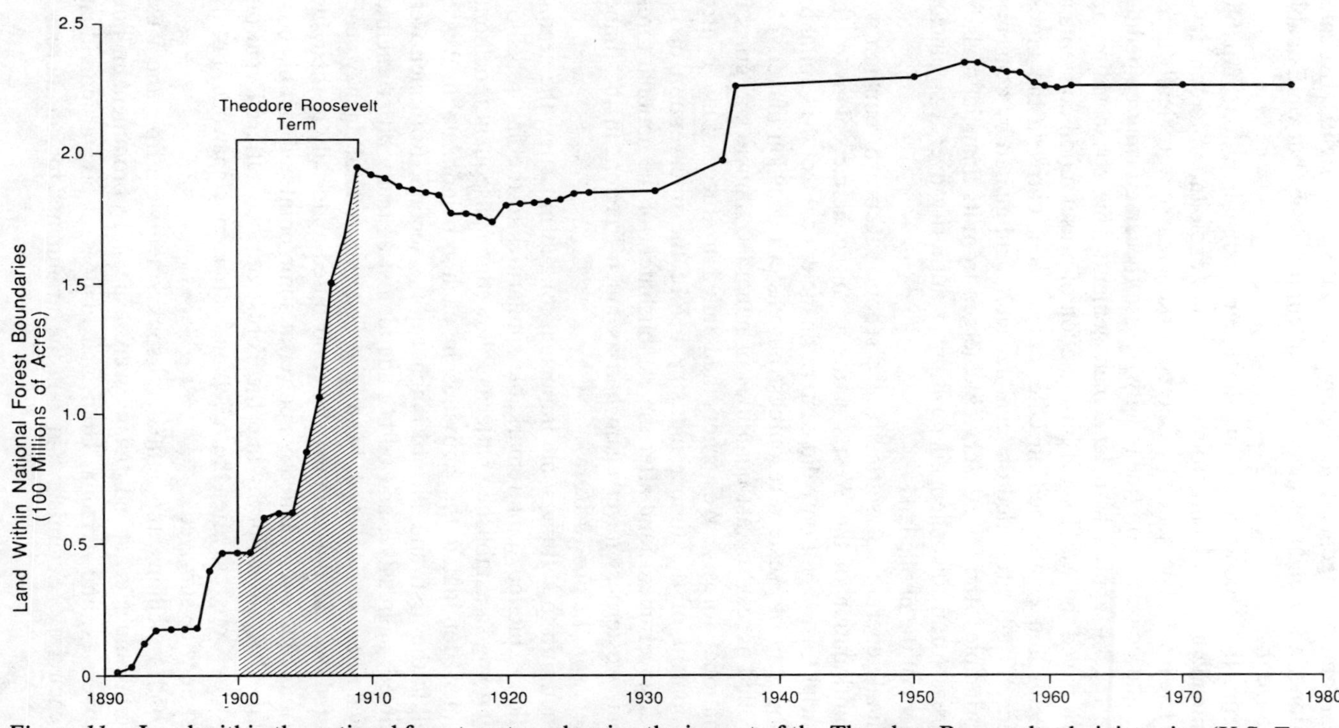

Figure 11. Land within the national forest system showing the impact of the Theodore Roosevelt administration (U.S. Forest Service data).

federal land management. Development and states' rights advocates questioned the principles in subsequent controversies but were unable to change them.

During the second rebellion, the federal administration of public lands made a radical shift from disposal and exploitation to conservation. The lessons preached by George Perkins Marsh were finally acted upon, and conservation (meaning wise use) became firmly embedded in the federal approach to land management. This conservation approach paved the way for the introduction of preservation concepts in the 1920s and established an elite management corps in the Forest Service. Firmly grounded on legal decisions and guided by conservation principles, the Forest Service emerged from its financial trials by Congress to an unchallenged position of authority over a significant segment of the public lands.

A final general issue settled by the second sagebrush rebellion was the continuation of the West's status as an economic colony of the industrial East.[54] Not only did eastern financiers control the capital for development of Western resources, but eastern decision makers also controlled access to vast quantities of minerals, timber, and grass in the national forests. Westerners had failed to successfully control development of the timber industry before the intervention of the federal government, and after the establishment of the national forest system, westerners never again had as much freedom in the timber market as in the pre-1890 era.

In addition to settling some important long-term issues, the second rebellion exhibited some remarkable similarities to the first rebellion, and in doing so established patterns followed in subsequent rebellions. One man dominated the activities in each rebellion: John Wesley Powell in the first, and Gifford Pinchot in the second. Both men were principled, aggressive, and both made almost as many bitter enemies as loyal friends. Each was head of a fledgling federal agency, each attracted dedicated assistants who contributed materially to their success, and each overreached his power at some point and lost his post. Both the Irrigation Survey and the Forest Service suffered vigorous attacks by Congress after the rebellion each agency spawned, but the Forest Service survived.

In both rebellions, the antifederal spokesmen were from the West. Colorado and Nevada politicians were especially prominent in both controversies, and Arizona, Utah, and Idaho sent pivotal individuals into each fray. Scientists laid out the arguments for an increased federal role in both cases, and they were notably absent from the ranks of the rebels.

Above all, in the second rebellion, stands the image of Gifford Pinchot. The devil incarnate to his opponents, Pinchot almost single-handedly forced the nation's government to embrace the newly named concept of conservation. He influenced an entire generation of young bureaucrats, won the loyal support of a few key lawmakers, activated a series of state-based conservation commissions, and operationalized a president's progressive environmental policy in an era when development-oriented ethics ruled the land. During his tenure in office, many events of modern American life became a reality, from the development of Novocaine and tape recorders to the first Wright brothers' flight and the theory of relativity.[55] Similarly, Pinchot brought the management of America's most valuable resource, its land, into a twentieth-century perspective that would have profound impacts on the nation and its citizens.

Part III

The Third Rebellion: Grazing Lands

Chapter 10

Sacred Cows and the Pastures of Heaven

The establishment and maintenance of the national forest system resolved the conflicts of the second sagebrush rebellion, but not even the visionary Gifford Pinchot could have foreseen the conundrum that next developed. The development of the national forest system would one day have overwhelming implications for a national wilderness system that would draw from the forests the first official wilderness areas. The 166 million acres of western public land outside the forests also contained other wilderness areas that one day would be subject to preservation. Mostly arid and semiarid grasslands of the remaining public domain were the focus of the third sagebrush rebellion that involved that sacred cow of western folklore, the range cattle and sheep industries. To understand wilderness preservation in the late twentieth century is to understand the grazing debates about the same lands several decades earlier.

The national forests became increasingly regulated for the grazing of cattle and sheep, but the remaining unreserved public land was unregulated, an invitation to managerial disaster. Disputes among various users of these unregulated lands involved cattle owners, sheep owners, and homesteaders. The 1909 Enlarged Homestead Act and the 1916 Grazing Homestead Act represented congressional attempts to deal with the increasingly clear issue of arid and semiarid limitations on the homestead concept. Both acts permitted homesteading of large parcels in the remaining public domain, and both led to the erection of hundreds of miles of fences on lands previously open for graziers. The frustrated president of the National Wool Growers Association in 1917 demanded the "final disposition of the public domain" to prevent further federal management,[1] but the plea went unheeded.

A major reason for the increased focus on the unreserved public domain was the developing control exerted by the federal government over grazing on the national forests. Prior to 1920, the Forest Service managed to forge effective working relationships with at least some of its lessees, even while gradually increasing fees. In 1914, the National Wool Growers Association for the first time passed a resolution praising the Forest Service,[2] and the service continued to hire and promote knowledgeable livestock managers. In 1916, Will C. Barnes, a widely respected and experienced Arizona and New Mexico rancher and legislator, was appointed chief of grazing, while Albert Potter, the venerable Arizona rancher, was promoted from chief of grazing to associate chief forester. Both men tried to increase grazing fees, but the pressures for production of meat and wool during World War I, unsympathetic secretaries of agriculture, vocal graziers, and postwar declining markets thwarted their efforts. Although graziers (usually sheepgrowers from Utah) made occasional attempts to shift the Forest Service from the Department of Agriculture to a more compliant Department of Interior,[3] the service remained secure.

During the 1920s, relationships between managers and users deteriorated. By 1924, senators and congressmen had introduced eight bills to regulate grazing, usually to the benefit of the users.[4] Generally the supporters of such efforts were the owners of large western cattle operations, while the opposition was supplied by the advocates of homesteading. Congressman Ferguson of New Mexico constantly complained of the efforts of the "big operator" and of western corporate fraud among cattlemen. Sheep owners were caught in a difficult situation: as nomadic operators who trailed their flocks, they opposed the cattle owners who worked from established privately owned islands in the ocean of public land. Sheep herders also suffered from the fencing activities of the homesteaders.[5] As mixed operations became more common and the federal government exerted more control, the sheep owners eventually threw their lot in with the cattle owners.

During the early 1920s, two issues prevented the solution of the dilemma of how grazing should be managed on the public domain exclusive of the national forests: who should do the regulating and what should be the economic basis of grazing fees? The old debate between the Departments of Agriculture and Interior resurfaced and centered on grazing lands instead of forest lands, as it had earlier in the century. As early as 1919, Agriculture Secretary Houston argued in a Salt Lake City speech that although Interior held title to the grasslands, administration should be by Agriculture.[6] Subsequently, the Coolidge administration made concerted efforts to reduce the controversy between the two agencies, and in 1924 Secretary of Agriculture Henry C. Wallace and Secretary of Interior Hubert Work agreed to a compromise. Agriculture would administer the grazing activities, but Interior would administer land title and disposal,[7] a clear indication that the ultimate fate of the land would be other than federal ownership.

Once the question of jurisdiction was answered, the two agencies pressed for legislation to control grazing activities. E. A. Sherman of the Forest Service and E. C. Finny, an assistant secretary of Interior, jointly drafted a bill for congressional approval of the working arrangements of the two agencies and for the establishment of grazing districts. Although the bill failed,[8] it represented another tentative step toward the eventual approval of the Taylor Grazing Act in 1934.

The second major issue related to the grasslands in the early 1920s was how to define grazing fees for users. In the Forest Service, Barnes and Potter campaigned for increased fees and argued that the fees should be based on the real value of the forage. While the service valued timber on the national forests at market price, it valued forage as a "reasonable use fee," resulting in undervaluation. Stock owners opposed any changes in this arrangement and effectively argued that the service had failed to establish any reasonable basis for changing the fee structure. In 1920, a Forest Service administrator, C. E. Rachford, convinced Chief Forester William Greeley that the value of rangeland needed to be tied in some way to "public welfare." Rachford proposed that the Forest Service should operate its grazing enterprise in the same way that it operated its lumber enterprise—as a business.

Commissioned by the chief forester to produce a comprehensive report on the matter, Rachford produced a monumental study of the economics of western grazing.[9] With the aid of Will Barnes, Rachford created a cogent description of the western range industry and made the case for the concept that forage was a commodity. He argued that federal agencies should charge users in a fee structure that generated enough revenue to cover administrative expenses plus a small profit, an approach ultimately tied to market values. He pointed out that the national forests were in public ownership and that they therefore were for the benefit of all. Subsidies to the graziers through artificially low fees represented unequal benefits, according to Rachford.

If carried to its logical conclusion, the Rachford report would have generated substantial fee increases. The reaction to the 1924 report was predictably mixed. The new secretary of agriculture, William M. Jardine, was skeptical, reflecting his ties to the ranching industry. Western ranchers were universally opposed to fee increases. Support for the report came from professional foresters in the American Forestry Association and the Society of American Foresters. Henry Graves, past chief forester and a professor at Yale, was especially vocal in extolling the virtues of increased fees. The final resolution was modest fee increases spread over several years, but falling market prices took most of the steam out of the entire effort.[10] However, Rachford had made his mark, and in 1928 he succeeded Barnes as chief of grazing in the Forest Service.

The Rachford report, the first salvo in the third sagebrush rebellion, touched off a debate about grazing fees on the public lands that produced, as potential resolutions, two major efforts to cede the public lands to the states and private owners. The debate extended roughly from the 1920s to the late 1940s, a period during which a new federal agency appeared (the Bureau of Land Management) and a major management plan developed for grazing lands (the Taylor Grazing Act). In 1925, resolutions were still far in the future when two senators answered the call to arms to protect the position of range users.

Senator Ralph H. Cameron, Republican from Arizona, was closely attuned to the opinions of western livestock owners. A frontier entrepreneur in northern Arizona, operator of a copper mine in what later became Grand Canyon National Park, and last territorial delegate to Congress from Arizona, the feisty Cameron considered himself the epitome of the independent westerner. In 1925, he introduced a resolution to investigate "all matters relating to national forests and the public domain and their administration" in hopes of forestalling the fee increases.[11] Senator R. N. Stanfield, Republican from Oregon, chaired the investigating committee. Stanfield had lost his own Forest Service grazing permit because of misrepresentation,[12] and had sparred with Forest Service policies for several years. After brief hearings in Washington, D.C., Stanfield took the hearing process west, primarily to generate a forum for complaints against the Forest Service.

Senators Cameron and Henry Ashurst, also of Arizona, led the field hearings on the Stanfield Bill beginning in June 1925. Both senators harbored an intense dislike for the Forest Service based on their own grazing experiences in Arizona,[13] and used the hearings to pry from witnesses every possible condemnation of the agency to bolster the case for cessation of lands to the states. Ashurst tried to convince

John Stephens, an Arizona rancher for forty years, that the federal government was the enemy of the grazier, but the rancher held fast and refused to condemn the managers.[14] The senators generally had better luck with disgruntled former permittees or former Forest Service employees. C. H. Finderer, for example, had progressed from an unhappy experience as a federal employee to a successful Prescott banker, and he testified that the service wrote, applied, and interpreted the regulations in truly dictatorial fashion.[15]

Seeing an opportunity to control their own destinies, cattle owners met in Salt Lake City in August 1925, and demanded that the federal government (1) recognize grazing rights of the usual user as his own, (2) recognize grazing rights as property rights, (3) hold graziers responsible only for willful damage to public lands, and (4) control grazing only through local boards staffed by the graziers themselves.[16] Based on the hearings he had conducted and on the demands by the graziers, Stanfield's bill incorporated the position of the cattle owners by including ten-year leases, preferences to present users, no large fee increases, and grazing districts under local control.[17]

This tour de force produced an equal and opposite reaction. Although the American Forestry Association and Society of American Foresters again resisted the western interests, the opposition to the Stanfield Bill was widespread. For the first time, eastern conservation organizations took up the fight and painted the controversy as a battle between owners of large ranching operations who were trying to raid the public trust, and small operations and homesteaders. The American Forestry Association received a $10,000 donation from private sources with the stipulation that the money be used to head off drives by looters of the public lands.[18] Opposition was strong in western urban areas concerned about watershed protection and the rise in corporate ranching. The *San Francisco Examiner* opined that the Stanfield Bill was "a noxious measure that should be killed."[19] Unable to overcome the opposition by an aroused conservation lobby, the Stanfield Bill died a slow death.

The Department of Agriculture recognized that the Rachford report was explosive material and that implementation of all its recommendations was unlikely. Desiring to avoid more encounters such as the Stanfield Bill and Hearings, the Department in 1926 assured Congress that an external review of the report would provide an evaluation of the proposed fee increases. The department selected Dan D. Casement, a Kansas livestock dealer and prior permittee, for the task. Casement reported that the Forest Service was involved in social engineering rather than in the development of market fees.[20] The Forest Service accepted the evaluation and increased fees only to a moderate degree, although in 1928 the fees were for the first time based on the principle of the animal unit month (that is, grazing by one cow on a given area for one month).

While several grazing bills appeared in Congress between 1925 and 1929, none survived because the legislators did not want to appear to cater to the cattle interests who were portrayed in public as monopolistic and anticonservationist. However, the 1928 election of Republican Herbert Hoover provided new reasons to attempt the same objective. Two precepts were the foundation of Hoover's policies in the West: water resource development and states' rights. He appointed Ray Lyman Wilbur, a medical doctor then president of Stanford University, as secretary of interior in March, 1928. In July, Wilbur formally proposed to the Western Governors' Conference in Boise, Idaho, that all federal lands, including the national forests, be turned over to the states.[21] In emphasizing the problems of overgrazing, which he attributed to mismanagement and neglect by the federal government, Wilbur linked the question of public land disposal to water resources. "We must replace homestead thinking with watershed thinking," he said.[22]

In August 1928, Hoover extolled the virtues of states' rights to the Conference of Public Lands Governors at Salt Lake City, and pointed out the connection with the public lands debate. Although Hoover's direct expression of the desire to cede federal lands to the states may appear to play directly to the audience of graziers, it is more likely that his own statement to the governors is a true explanation of his thinking.[23] He indicated that the western states "are today more competent to manage much of these affairs than is the Federal Government. Moreover, we must seek every opportunity to retard the expansion of the Federal bureaucracy and to place our communities in control of their own destinies."[24] Thus, within a few short months, the sagebrush rebels had gained powerful spokesmen who carried their banner for reasons totally unrelated to local ranching concerns.

Once begun, the Hoover-Wilbur machine rolled merrily along. Hoover proposed that a public lands commission be formed to investigate the proposed transfer of lands. Congressman Donald B. Colton of Utah, recognizing the connection with the sagebrush rebels, enthusiastically introduced the bill to establish the commission, arguing that overgrazing was causing silt problems for reservoirs in downstream areas.[25] Still in the honeymoon period with a new president, Congress dutifully passed the bill and established the third public lands commission in the nation's history as the Committee on Conservation and Administration of the Public Domain.

In appointing commission members, Hoover assured Congress that he sought broad regional and political representation (he promised the appointment of at least two women as well). The most important appointees to the commission were James R. Garfield, secretary of interior for Theodore Roosevelt (and well-known conservationist of the Pinchot mold) and William B. Greeley, (former chief forester of the Forest Service) secretary-manager of the West Coast Lumbermen's Association. The president charged the commission with addressing two primary issues: (1) overgrazing on the public domain that diminished land values and destroyed the water supply, and (2) determining the best method of applying water reclamation into the western public lands. The commission was also free to investigate other matters it thought relevant.[26] As the Stanfield activities had done,[27] the commission was more instrumental in gauging public opinion than creating it.

Almost two years later, in September 1931, the commission issued its final report that demonstrated the true nature of its activities. The commission spent little effort on water resource issues, but instead focused on the question of public land ownership and administration. The commission recommended that additional reserves be established for oil, coal, forests, parks, wildlife, and national defense, and that the remaining public domain (grazing lands) be ceded to the states. Boards of local citizens were to be established in each state to assess which lands were to be added to national forests or alternatively restored to the public domain. All the commission members signed the report with one notable exception: William Greeley, the ex-forester who had dealt with the grazing issue for so long recognized the report as a political instrument. He never made public his objections, but the glaring absence of his signature was a clear signal of his opinions.

Hurrying to cash in on the support of the commission and a sympathetic president, Senator Gerald Nye and Congressman John M. Evans (Colorado) proposed bills in early 1932 to carry out the transfer of lands to the western states.[28] President Hoover recommended that Congress do its duty and pass the bills.[29] As with the Stanfield effort, however, the opposition that had virtually been asleep during the early part of the process awakened with a vengence. Opposition to the plan developed from professional managers, the popular press, and even western representatives.

Professional foresters, who had attempted to manage the grazing on national forests, were vocal in their opposition to the transfer plans, but Hoover's commission never formally consulted the Forest Service in formulating its proposals. The most effective spokesman for the Forest Service position was one of its veterans, ex-chief forester Henry B. Graves, who had succeeded Pinchot. Graves, now at the Yale University School of Forestry, offered often, and in highly visible public fora, management alternatives to the disposal of the public grazing lands.[30] His advice to add grazing lands to the national forests went unheeded, but it gave opponents of the commission report an aura of scientific respectability.

The press sensed a hot topic in the public lands debate, and usually sided with the opponents of public land disposal. Western newspapers carried editorials claiming that the disposal plan would benefit only the large cattle operations and that the small operators, without access to livestock organizations and state legislatures, would suffer from unfair competition. Eastern liberal editorialists swung into action to defend the small operator and to forestall what appeared to be a potential raid on the national forests, labeling the disposal idea as the "Handout Magnificent" and claiming that "the plan, in essence, [was] one of monopoly and eviction, antisocial, undemocratic."[31] One widely read author had a simple suggestion for Congress: "shelve the report."[32]

While all these efforts were important, the essential opposition came from western representatives. Amazingly, the commission's report claimed that "most states expressed their desire to undertake the task" of administering the grazing lands.[33] Statements by governors and congressional representatives did not support this assertion. The state administrators correctly saw that they were likely to receive the grazing lands with their low potential for revenue generation, but not the more lucrative mineral, oil, and forest lands. The *Billings Montana Gazette* observed: "The West doesn't care much about getting the lid without the bucket."[34] Senator Borah of Idaho, veteran of a previous sagebrush rebellion, referred to the grazing lands "on which a jack rabbit could hardly live" as "skimmed milk."[35]

The disposal of the public domain by the Hoover administration was stalled by conservationist opposition and the unwillingness of the western states to undertake the unprofitable task of administration. The demise of the entire disposal effort was assured by the economic collapse of the Great Depression and the electoral success of the opposition Democrats. In 1932, the Republicans were swept from many national offices, and Franklin Roosevelt overwhelmed Hoover at the polls. The sagebrush rebels' lack of political support for Hoover is apparent from the fact that Roosevelt won every western state. Issues other than those related to the public lands had become more important to westerners. Surpluses plagued western farmers, but the Hoover administration took actions that stimulated more production; oil producers needed federal regulatory aid to stabilize prices that had fallen to ten cents per barrel but received none; social welfare needs in western states went unmet; and even funding for the highly touted water resource development projects was withheld. The result was overwhelming support for Roosevelt in western states.[36]

Keeping the western grazing lands in federal jurisdiction did not solve the ongoing problems of the western cattle industry that depended on those lands. During the 1920s and early 1930s, livestock growers faced the realization that they could not adequately control the public grazing lands themselves. The range economy was deteriorating, with cattle prices falling by 50 percent between 1929 and 1932, followed by another 40 percent decline by 1934.[37] Cattle ranchers were again at war with the sheep herders, and large operations were pitted against small ones. Illegal fencing of the public domain confused management and individual operators had no reason to control overgrazing on lands in which they had little personal stake. Erosion, partly resulting from overgrazing, threatened public water supplies and generated flood hazards. States and counties lost tax revenues. Chaos seemed to rule the western range, and when the move to transfer the lands to state and private ownership failed, even the most reluctant sagebrush rebels began to seriously consider agreement with some type of federal control to establish order.

Two factors, western grass roots activists and an eastern congressional effort, paved the way for the development of federal controls on the public grazing lands in the form of the Taylor Grazing Act of 1934. The first was the development of the locally controlled, federally owned grazing district. The Mizpah-Pumpkin Creek Project was an experiment that subsumed more than 100,000 acres of federal and private grazing land south of Miles City, Montana. The ownership pattern was fragmented, the forage productivity damaged from drought and overgrazing, and the future bleak until a local organization formed to assume control. The group included Nick Monte, a stockman who later became a regional administrator for the federal Grazing Service; Evan W. Hall, an agricultural development agent for the Milwaukee Railroad that served the area; and Montana congressional members Senator Thomas J. Walsh and Congressman Scott Leavitt.[38]

In 1928, Congress approved a special bill allowing the U.S. Department of Interior to enter into agreements with state and private groups to jointly manage grazing lands in the Montana district. Trades of federal and state lands were permitted to establish large contiguous parcels, and control of the combined areas was given to the local stockmen's association in a ten-year experiment. The federal lands were leased to graziers at $20 per section (a square mile) and taxes were to be paid for private lands that were included in the plan.[39] The significance of the Mizpah-Pumpkin Creek experiment was its obvious success within a short time, proving that cooperative efforts with only little federal control could bring order to a previously fractious problem. It gave congressional observers confidence to press ahead with grazing legislation.

The second factor that led to the Taylor Grazing Act was the establishment of a firm policy for forestry, including grazing, that guided the U.S. Forest Service. Senator Royal Copeland (New York) requested that the Forest Service provide an economic evaluation of all aspects of its management as a prelude to legislative action on the part of the Roosevelt administration. The resulting Copeland Report in 1933 positioned the Forest Service to advocate expanding federal controls on the use of the public domain, and recommended the acquisition of additional lands.[40] The report did not result in direct changes in the administration of western federal public lands, but it made subsequent changes easier by providing a framework for future actions.

Congressman Don B. Colton (Utah) and Burton L. French (Idaho) introduced a bill to establish federal grazing controls in 1932 before the end of the Hoover administration. With the support of Ray Lymon Wilbur, the secretary of interior, and Arthur M. Hyde, secretary of agriculture, the bill became the first piece of grazing legislation to pass either house of Congress.[41] It failed in the Senate. Western and eastern pressures for resolution of the grazing crisis, with a new Congress and presidential administration, began the development of federal controls on the unreserved western public lands.

In 1933, Representative Edward T. Taylor (Colorado) reintroduced the defeated Colton Bill. Taylor was a reformed sagebrush rebel who had served in Congress since 1909. He was a strong advocate of states' rights who originally had opposed reservations and leasing of the public domain as "un-American."[42] He later became concerned about the grazing issue, but in 1920, for example, he believed that public lands should be administered by the states.[43] The deteriorating environmental and management situation in the West and the defeat of attempts to transfer lands to the states during the Hoover administration convinced Taylor that federal controls on federal public lands were the only politically viable solution, and he set out to take a leadership position. He took complete command of the bill, dominated the hearings, and skillfully maneuvered the issue through a reluctant Congress.[44]

The Taylor Bill, H.R. 2835, represented a radical solution to the grazing issue by injecting federal management into the administration of lands that previously had been virtually ignored. The bill continued the jurisdiction of the lands in the Department of Interior, established grazing districts on the Mizpah-Pumpkin Creek model, provided for 10-year leases to graziers, and established a fee structure with the proceeds to be split among the Department of Interior, the individual states, and the federal treasury.[45] A most important provision was that all the remaining public lands would be reserved and classified by the executive branch of the federal government, effectively repealing the various homestead acts as they applied to the lower forty-eight states. The approval of Taylor's sweeping bill was not easily forthcoming.

Opposition to the Taylor Bill and its imposition of federal grazing controls on the unreserved public domain came from some western interests who still kept alive the flickering flame of the sagebrush rebellion. A most ardent opponent was Congressman Ayers (Montana) who introduced a substitute measure that would have transferred all the unreserved public domain to the states. In hearings, he belittled the Taylor argument that the western grazing lands required federal management: "In the last hearing you had here, a lot of men testified that overgrazing was the cause of erosion. Now I hope there is not any member of this committee that is going to be gullible enough to believe that."[46] He viewed the Taylor Bill as an imposition of eastern federal control on otherwise independent western lands, and commented that "the West does not need national parasites."[47] Other opponents were no less dramatic. Congressman Carter (Wyoming) compared the imposition of federal controls on the public domain to the situation in Russia, and other representatives from New Mexico, California, and Arizona argued for the rights of the homesteader or cowboy. The venerable Huey Long, perhaps tongue-in-cheek, pointed out that the lands in question were once the property of Louisiana, and that his constituents ought to have a special right in their disposition.[48]

Taylor enjoyed the support of powerful individuals in his squabbles with opponents of the bill. Some cattle owners testified on behalf of the bill, including F. R. Carpenter of Colorado, who argued that the bill was needed to protect small operators from being overrun by the large cattle corporations. Edward A. Sherman, chief forester of the Forest Service, contended that immediate controls were needed and, if not imposed, the public grazing lands would soon be fit only for the grazing of goats.[49] Franklin Roosevelt's executive branch was also strongly supportive. Secretary of Agriculture Henry Wallace had reached an agreement with his counterpart at Interior, Harold L. Ickes, and was supportive even though the management of grazing did not fall to his own Forest Service. Ickes was the most important supporter of the bill and was instrumental in its eventual approval.

During the 1933 debates on the Taylor Bill, Ickes successfully convinced Congress to approve the Emergency Conservation Work Act, which created the Civilian Conservation Corps. The Corps established work camps where jobless young people were employed in conservation-related projects on public lands. They constructed fences, trails, buildings, drainage works, roads, stock-watering systems, and planted thousands of trees. The Corps was valuable to congressional representatives eager to appear effective in garnering federal dollars for their districts and in improving the conservation of public lands. Ickes, desiring swift approval of the Taylor Bill, refused to assign any projects for the Corps on the public domain, ostensibly because the future of those lands remained in doubt. His actual reason for withholding the investment from some regions was revealed in his typically combative offer to Congress: pass the Taylor Bill and in return receive Corps projects in western districts.[50]

Despite such pressure, the bill did not pass, but its legislative momentum was sufficient for rapid reintroduction in the next Congress as House Resolution 6462. Ickes increased the pressure. He found through legal counsel that by means of an obscure law (the Pickett Act of 1910) he could withdraw all public lands, making them federal reservations without further legislation. Ickes's threat was effective because congressional representatives knew that the crusty interior secretary would carry it out, and because they knew he had the support of the president.[51] On April 11, 1934, the House passed the bill.

On May 11, while the Senate was considering the matter, raging dust storms in the West injected so much material into the atmosphere that dust falls occurred in New York and Washington, D.C.[52] Blown from eroded lands, the dust was characterized by one senator as "the most tragic, the most impressive lobbyist that has ever come to this Capitol."[53] A month later the Senate passed the bill and President Roosevelt signed it into law on June 28, 1934.[54]

At the time of its enactment, the Taylor Grazing Act applied grazing controls to about 80 million acres of the 165.7 million acres remaining in the public domain.[55] Local western demands for regulation produced congressional amendments that eventually extended the act to more than 146 million acres. More importantly for the story of land ownership, the act triggered two executive orders in late 1934 and early 1935 wherein President Roosevelt withdrew from the public domain all lands for classification purposes.[56] A few limited parcels were found to be suited for agriculture, but almost all the lands came to be included in the Taylor grazing districts or in other federal units. The "public domain" ceased to exist.

The significance of the Taylor grazing lands to wilderness preservation was that half a century later they would be administered by the Bureau of Land Management, and would be subject to review for inclusion in the National Wilderness System. Grazing lands were concentrated in the West, intensifying regional differences (Figure 12). Opposition to wilderness classification was strong for lands that had been the focus of grazing controversies. Ranchers would come to feel that they had suffered enough bureaucratic slings and arrows without having to fight one more battle against wilderness proponents.

In 1934, however, federal wilderness preservation was still far in the future, and passage of the Taylor Grazing Act depended on western representatives who still desired transfer of the lands to states. The support of these sagebrush rebels was critical to the success of the act, so Congressman Taylor included in his bill the phrase that the act was designed to regulate grazing of the western unreserved "public lands pending . . . final disposal." Many western representatives supported the bill because although they desired such a transfer, it did not appear possible at that time. Inclusion of the phrase threatened the support of Secretary of Agriculture Wallace, who disapproved of the clause to the point of urging the president not to sign the bill.[57] The phrase survived and became a clear statement in the final law that although the rebels had not won their objective of transfer to state and private ownership, they were still powerful enough to keep the hope alive.

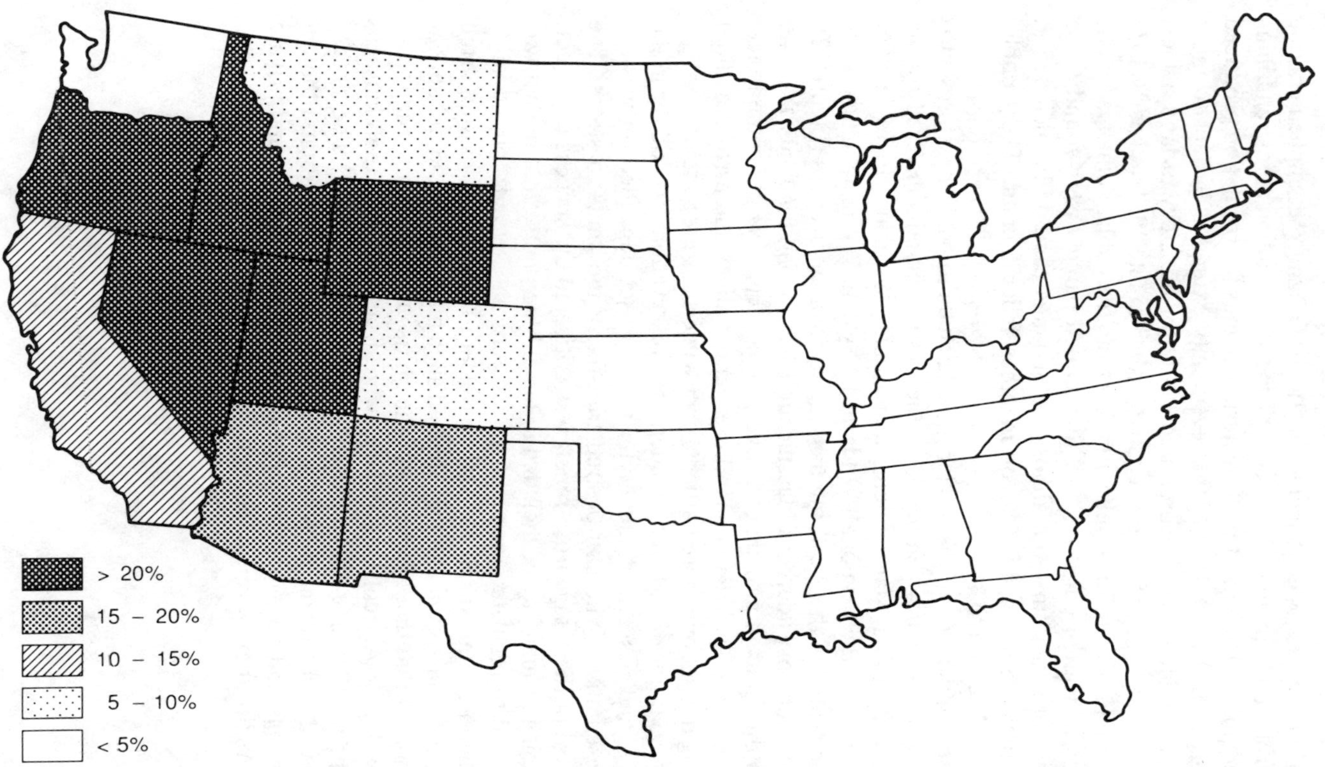

Figure 12. Percentages of total state area contained within grazing districts showing the concentration of grazing lands in interior western states (Bureau of Land Management data).

Chapter 11

The Wild West Show

While western cattle and sheep graziers struggled with the problems of unreserved public land during the period after World War I, changes were occurring in the administration of the national forests that ultimately produced the first formal wilderness areas. Chief Forester Henry Graves perceived that the usefulness of the national forests for the general population could be enhanced if the objectives of forest management could be broadened from the traditional base of timber and water conservation. In 1915, he signaled the first recognition of recreation by the granting of permits to construct second homes on national forests.[1] Graves also commissioned a landscape architect, Frank Waugh, to report on the scenic potential of the national forests. Predictably, Waugh indicated that the Forest Service should protect especially scenic locations from development and should place recreation on an equal management footing with timber and grazing.[2]

These trends away from exclusive attention to strictly economic resources by the Forest Service were triggered in part by the rapid development of the National Park Service. In 1916, the Park Service was formally established as a management agency in the Department of Interior in order to administer the growing series of park reservations. The aggressive administration of the new Park Service director, Stephen T. Mather, alarmed the Forest Service because new parks were created and old ones expanded mostly by the transfer of national forest land to the new agency.[3] Mather was a master salesman for his vision of a successful park system, and he pushed not only the American public to use his new parks, he pushed the managers of the facilities to think expansively. When dealing with a hotel operator at Yosemite in 1924, Mather urged rapid expansion by pointing out that although there were only 200 cars in the park, someday there would be a thousand. Just four years later, 130,000 entered the park.[4]

As the transfer of forest lands to park jurisdiction threatened to reduce the influence of the Forest Service, Graves moved to develop the recreation potential of the forest reserves.[5] Between 1916 and 1920, Congress appropriated $33 million for road development to provide access to the forests.[6] As a result, wilderness areas shrank under the onslaught of automobile-borne hunters, fishermen, and recreationists. A countercurrent of protectionism developed in the Forest Service with the intent of excluding development from some areas. In 1919, Arthur Carhart, a Forest Service landscape architect, recommended that Trappers Lake, Colorado, be excluded from plans for road and second home development in order to preserve the pristine scenery of the area.[7] The service accepted the recommendation, and thus established the first area of national forest set aside specifically to preclude development. At the same time Mather's Park Service was advocating the fullest possible development of the national parks for the automobile tourists, and preservation was largely absent from park administrative goals.

Carhart shared his preservationist opinions with a manager, Aldo Leopold, who was the assistant forest supervisor of the Gila National Forest in southwest New Mexico. Leopold was already concerned with intensive road-building efforts in the Gila forest, which seemed to be limiting the potential for a wilderness-hunting experience.[8] Because he was a manager in the Forest Service and a gifted writer who could articulate the preservation case effectively, Leopold's 1921 article "The Wilderness and Its Place in Forest Recreation Policy" in the *Journal of Forestry* was of singular importance.[9]

In his article, Leopold outlined a reasoned argument for wilderness preservation that went far beyond Carhart's effort to preserve "scenery." Leopold endorsed Pinchot's concept of management for the greatest good of the greatest number over the longest time, but he proposed that the concept should not be limited to industrial resource development. He wrote that democratic approaches protected the rights of minorities, in this case the minority of forest users who desired something other than strict adherence to production of timber, water, and grass. He offered the first widely recognized definition of wilderness as a "continuous stretch of country preserved in its natural state, open to lawful hunting and fishing, big enough to absorb a two-week pack trip, and kept devoid of roads, artificial trails, cottages, or other works of man."[10]

As a realist, he proposed that such areas would be small relative to the total forest system, that they would be areas not useful for other development, and that grazing would be permitted in the wilderness areas as an accurate historical feature. Specifically, he called for the setting aside of 500,000 roadless acres of the Gila National Forest. Leopold initiated the bureaucratic process of wilderness designation within the Forest Service, and was rewarded in 1924 when Chief Forester William B. Greeley designated an area in Gila National Forest to be preserved in its roadless condition. Carhart was also similarly successful in the Superior National Forest, Minnesota. Within a year four additional areas had been set aside.[11]

The Forest Service interest in such moves was partly to protect its own territory, as Stephen Mather continually urged Congress to transfer recreation lands into the Park Service. He proposed that the Forest Service be given no funds for recreation management, a position supported by the Sierra Club.[12] The Park Service also had the support of an increasingly recreation-oriented public, but the Forest Service had too many friends in Congress to be overcome, and wilderness management continued to develop in the forestry agency.

The first general inventory of roadless areas (that is, potential wilderness areas in the terminology of the post-1960s) was the product of the 1928 National Conference on Outdoor Recreation, headed by Assistant Chief Forester L. F. Kneipp. Using a lower size limit of 10 survey sections (230,400 acres), the conference identified 74 areas totalling 55 million acres, an area equal to about one-third of the entire national forest system.[13] The overly ambitious proposal received support from foresters and other managers who reacted with disdain to the commercial tourist development being fostered in the national parks.

The origination of the forest reserves without a management policy had caused problems for administrators in the 1890s; similar issues had arisen in setting aside the national parks and the left-over grazing lands. By the late 1920s, it was obvious that mere designation of roadless areas was insufficient without defined public policies for their management. The formal policy that the Forest Service developed through 1928 and 1929 became known by its bureacratic label, the "L-20 Regulations."[13] A direct outgrowth of Leopold's philosophy and Kneipp's administration, the L-20 Regulations represented a loosely defined set of requests for local foresters who had jurisdiction over designated roadless areas. The "regulations" were not strongly worded because many local foresters were opposed to the maintenance of roadless areas, believing that such an approach needlessly prevented economic development, a refrain heard often in subsequent years from other quarters. Because the service was a decentralized agency, the local objections carried considerable weight for national policy.[14] Some foresters wanted to compete with the national parks for public recreation, while others interpreted the Pinchot tradition in strictly econonmic terms related to commercial resources. Although the L-20 Regulations were weakly defined and haphazardly applied, they were the first formal wilderness policy.

During the 1930s, while Congress was slowly sorting out the problems of grazing on the public domain, the Forest Service continued to refine and enlarge its wilderness management. By 1934 (date of the Taylor Grazing Act) there were 63 primitive (roadless) areas encompassing 8.5 million acres in six national forest regions. By the standards of the 1980s, this first attempt at a wilderness system was slipshod: 23 of the areas permitted future logging, 8 had fire-road construction, and 22 had state or private inholdings.[15]

Preservation efforts moved forward, however, and a new generation of forest managers emerged with an appreciation of the value of primitive areas. Foremost among these young administrators was Robert Marshall, who combined pragmatism and idealism. With a Harvard master's degree in forestry and a Johns Hopkins Ph.D. in plant physiology coupled with his contribution to the Copeland Report (which defined forest policy for the Franklin Roosevelt administration), Marshall was in a unique position to combine intellectual and bureaucratic experience in speaking for wilderness preservation.

Marshall's general approach was similar to Leopold's in that both put forward their philosophical position in writing and then manipulated the Forest Service regulation structure to achieve their stated goals. Marshall's signal publication, "The Problem of Wilderness," appeared in the prestigious journal *The Scientific Monthly* while he was still a graduate student. He defined a wilderness area as "a region which contains no permanent inhabitants, possesses no possibility of conveyance by any mechanical means and is sufficiently spacious that a person crossing it must have the experience of sleeping out."[16] He also stipulated that such areas preserve the "primitive environment" without roads, power lines, and settlements.

Whereas Leopold's wilderness was a place where fishing and hunting were the major attractions, Marshall's wilderness areas were sources of personal rejuvenation through an intense psychological experience. For Marshall, wilderness areas represented retreat from civilization, thinking, adventure, and appreciation of natural beauty. He adopted Leopold's democratic arguments, saying that the minority of wilderness users had an inherent right to such places, even while yielding in other areas to the developments needed to satisfy the majority of automobile tourists.

Just as Leopold's work led to the development of the L-20 Regulations, Marshall's efforts, with the aid of his experienced Forest Service assistant John H. Sieker, led to a new and more stringent policy articulated in the Forest Service "U-Regulations" of 1939. The U-Regulations made the general policy of earlier operating rules binding on local administrators and provided for the classification of those lands designated primitive and roadless under the old L-20 rules. The new regulations specified three types of management zones: U-1 wilderness areas greater than 100,000 acres in extent designated by the secretary of agriculture, U-2 wild areas of 5,000-100,000 acres designated by the chief forester, and U-3 recreation areas that were roadless regions of varying size also managed for a variety of commercial resource uses.[17] By the 1940s, the U-Regulations had produced several wilderness areas, including the Bob Marshall Wilderness Area in western Montana.[18]

The availability of future wilderness areas was directly influenced by the disposition of the public domain under the Taylor Grazing Act and by the developing management strategies within the Forest Service. A third mechanism—wildlife refuges—was also important as relatively minor but significant additions to lands with potential for preservation. Sport hunting for all sorts of game and wild fowl was popular in the 1920s, and pressure on wildlife populations was intense. By the early 1930s, migratory bird populations, especially ducks, had declined precipitously. Habitat reduction by human activities, hunting, and drought resulted in a drop in the number of ducks in the Great Plains region from about 100 million in 1930 to only 20 million in 1934.[19] The National Audubon Society, and especially the Emergency Conservation Committee under the leadership of Rosalie Edge, repeatedly articulated the public case for lessening the pressure on wildlife from hunters, and for the establishment of specific areas for the protection of wildlife.

The 1932 presidential election was a turning point in federal wildlife management. A well-known Iowa political cartoonist and conservationist, Jay Norwood Darling, was an influential member of the Republican Resolutions Committee, and he worked to ensure that the party had a proconservation platform. Darling commanded a national audience through his cartoons, which were signed with the distinctive label "Ding." As an editorial veteran of the controversies between wildlife preservationists led by William Hornaday and the hunting lobby backed by gun and ammunition manufacturers, Darling served a clearly defined clientele when, in 1933, President Roosevelt appointed him to serve as a member of the newly formed President's Committee on Wildlife Restoration. Thomas Beck, president of Crowell-Collier Publishing Company, was the chair and Aldo Leopold rounded out the committee membership.

The president charged the commission with proposing a solution to the problem of the declining duck population. The three members (all enthusiastic duck hunters) proposed that as much as $75 million be made available to purchase 17 million acres for duck refuges.[20] When Roosevelt offered no immediate responses to the suggestion, Darling began to pen cartoons critical of the president's lack of wildlife policy. In 1934, Henry Wallace, the secretary of agriculture from Iowa who was acquainted with Darling, invited the editorialist to assume the directorship of the Biological Survey, the agency charged with management of wild game resources. With some reluctance, Darling accepted.

Crusader Rosalie Edge saw Darling's appointment as a perfect opportunity to push the federal government into the wildlife refuge business. She had previously castigated the Biological Survey for its predator control programs in the West, but she hoped that under Darling's informed leadership the predator control programs would be dropped and hunting seasons reduced. Darling found that public officials, unlike editorial cartoonists, needed to build consensus and to establish compromises that were the antithesis to Edge's position. The predator control programs remained in effect, changes in hunting seasons did not satisfy the bird enthusiasts, and Edge began to criticize her erstwhile champion as having sold out. A frustrated Darling reported that she knew "little about any angle of work except the techniques of poison dart making."[21]

Despite the problems with predators and hunting, Darling was successful in establishing refuges. In 1934, by sheer force of personality and adept bureaucratic maneuvering, he used Duck Stamp money and funds from the Federal Emergency Relief Agency as well as from the Public Works Administration to purchase marginal agricultural lands, mostly in the upper Midwest. His newly established Migratory Waterfowl Division created from the purchased lands 32 bird refuges totalling 840,000 acres, and managed to wring $8.5 million from the federal budget for wildlife management.[22] In 1935, Darling obtained another $6 million from Congress, and dunned Roosevelt for $4 million more. The president, reluctant to invest so much in wildlife preservation, refused with the observation that Darling was the first man in history to hold up the U.S. Treasury and get away with it.[23]

Roosevelt was clearly in the business of consensus and compromise, and his comparison of Darling's aggressive wildlife refuge program with water reclamation and flood control projects in the same areas suggested to him that jobs and economic development were more important than the refuges. Darling resigned his post as director of the Biological Survey in November 1935, realizing that though Roosevelt was friendly in principle to wildlife interests, the better-organized sporting arms manufacturers and developers had the president's ear.

Although he had been in federal service only 20 months, Darling's contribution to federal public land management and eventually to wilderness preservation was the appearance of wildlife refuges in a major way in the geography of the United States. The modest areas set aside in the 1930s would later be augmented by donation and (mostly) purchases, so that eventually the daughter agency of the Biological Survey, the Fish and Wildlife Service, administered more than 31 million acres.[24] These wildlife areas offered additional opportunity for wilderness preservation that otherwise would not have existed. More importantly, the Darling/Edge efforts in the mid-1930s brought the wildlife lobby directly into the debate about public lands. Because conservation groups were small and poorly organized, wildlife preservationists were left to provide the major organized opposition to those proposing disposal of the federal lands to states and individuals during the latter stages of the third sagebrush rebellion.

Management of the new wildlife refuges was the responsibility of the Biological Survey and its parent agency, the Department of Interior. Management of the Taylor grazing lands was the responsibility of the newly created Grazing Division (to become the Grazing Service in 1939), which was also in the Department of Interior. While the Departments of Interior and Agriculture had worked in an uneasy truce to support the Taylor Grazing Act, upon its passage competition renewed between the agencies for control of the land-management functions of the federal government.

The Forest Service supported the Taylor Grazing Act because the act was congruent with public policy established by the Copeland Report, the New Deal strategy statement by the service.[25] The foresters did not want competition for the management of grazing and believed that their own agency in the Department of Agriculture had a management record superior to any record of the Department of Interior. Secretary of Agriculture Wallace organized 35 authors to generate a report defining evidence of poor grazing management by his opponents in the Department of Interior and to make the Forest Service case.[26] Once created, however, the question was how to launch the report, contained in a volume known informally as "the green book."[27] Apparently with some gentle prompting, Senator George Norris (Nebraska) formally requested a report on grazing in 1936, and four days later the Forest Service dutifully presented Congress with the green book, now titled *The Western Range*.[28]

The Western Range was a full-scale attempt by the Forest Service to obtain exclusive control over all the western grazing lands. The report stated that the Grazing Division of the Department of Interior was a poor choice to manage the lands because of its questionable conservation record as opposed to the supposedly sterling record of the Forest Service. The authors argued that grazing was an agricultural activity that should be managed by the Department of Agriculture, and that Interior was wasting and destroying the resource. The report claimed that Interior management was so poor that a century would be required to restore the lands, and provided inflated figures of deteriorating conditions on the ranges. Not content with Interior-bashing, the report also criticized the management of state and private lands.

Western grazing interests reacted to the "green book" with alarm.[29] The Forest Service report threatened their hard-won local control now established in Interior's Grazing Division and did not want to see the well-developed controls of the forestry agency expanded. F. E. Mollin, secretary of the American National Live Stock Association, made the grazier's position plain. He argued that variation in rainfall was more important than land management on the western range,[30] and he wanted to avoid the expansion of authority of the Forest Service, which he viewed as "a bigoted, conceited bureaucratic setup."[31]

Not surprisingly, the most potent opposition to the "setup" was from Secretary of Interior Harold Ickes. Ickes was a consummate political infighter who wanted to build his agency into a comprehensive land management organization with the label "Department of Conservation," which he thought would enhance the agency's public image.[32] He sought to have governmental reorganization instituted at the cabinet level so that the new department would include the Forest Service, a move advocated in 1937 by a Congressional committee.[33] When Congress failed to act on the recommendation, Ickes went to work on his own. He established a Division of Forestry to deal with forested Interior lands from a perspective of sustained yield.[34] He convinced Roosevelt to transfer all the national monuments in the national forests to Interior jurisdiction, but only after many were enlarged at the expense of neighboring national forests. Finally, he established additional national park units, most of which were carved form National Forest holdings. Concerned about the position of the Forest Service, the retired but venerated Gifford Pinchot entered the fray, charging that Ickes's only interest was empire-building.[35]

The issue faded with the onset of World War II, but in retrospect it appears that in initiating the attack through *The Western Range,* the Forest Service received more than it bargained for. In most cases, Ickes successfully defended his agency through vigorous counterattacks. Although he managed to irritate almost everyone with his aggressive approaches, he was dedicated to the ideals of effective land management.[36] Many of the national park units he established and enlarged contained areas that one day would be prime candidates for inclusion in the national wilderness system. The failure of the Forest Service to gain control of the Taylor grazing lands meant that these leftovers from the public domain would continue to be the subject of ineffective management, and their wilderness potential would continue to decline.

The sound management of grazing promised by the passage of the Taylor Grazing Act was never realized, a failure that led to a second phase of the sagebrush rebellion focused on the western livestock industry. The experience of Ferry Carpenter, the first Grazing Service director was unfortunately typical. With the passage of the legislation, Carpenter met with warring sheep and cattle grazers in Grand Junction on the western slope of Colorado. Previously, the game was that the grass was available for those wily enough to outwit their competitors, but now it was under federal control, and each user wanted his fair share. Carpenter's account defines the problems faced by the new range managers. There was little help from Washington because "there wasn't anybody in the whole department that knew which end of the cow got up first."[37] The General Land Office could offer him no maps of the regions that he was expected to divide into grazing districts. And finally he was saddled with unfavorable local perceptions of national decision makers, a circumstance that Carpenter handled by siding with the locals.

Carpenter's experiences were typical rather than exceptional, and from its inception the Grazing Division had difficulty in dealing with those it regulated. The major sources of contention were the agency policies of reducing the numbers of livestock on public ranges, centralization of authority, and increasing user fees. Despite the lack of convincing evidence that reasonable stocking levels could be defined by scientific methods, direct observation of conditions indicated that reductions were the proper direction of change (Figure 13).

The centralization of authority in the Grazing Service was a major component of F. E. Carpenter's strategy as the agency's first director. In part, the centralization effort neutralized some of the control exerted by local management boards mandated by the Taylor Grazing Act, but it also allowed Carpenter to attempt to superimpose a strong control system in the new agency similar to the strategy successfully employed by Pinchot in the early Forest Service. The approach failed in the Grazing Service because of the political strength of the grazing lobby and because, when Carpenter resigned in 1939, his replacement, R. H. Rutledge, was committed to local authority. In that spirit, the Grazing Service moved its headquarters from Washington, D.C., to Salt Lake City in August 1941.[38] Cloaked in the guise of decentraliza-

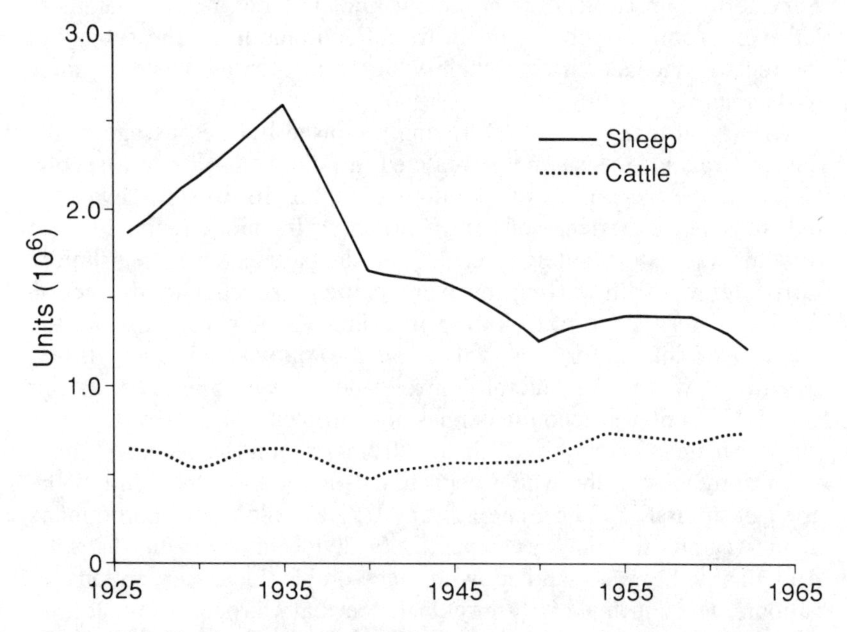

Figure 13. Numbers of cattle and sheep in the upper Colorado River Basin (exclusive of the Navajo Indian Reservation) showing the impact of the 1934 Taylor Grazing Act on sheep numbers but not on cattle numbers (U.S. Department of Agriculture data).

tion of government in the face of threatening world war, the move actually reflected the political need for a visible gesture to placate the increasingly restive sagebrush rebels.

The western rebellion against controls imposed by the Grazing Service drew most of its force from controversy surrounding proposed fee increases. No fees were assessed in the first year following the passage of the Taylor Grazing Act, but in 1936 a general meeting of advisory board members agreed to five cents per animal unit month (AUM) for cattle and one cent per AUM for sheep. Despite all the subsequent debates and posturing, the rates were not raised until 1947. The decision in the case *U.S. v. Light* sustained the Forest Service right to assess fees, but a new legal challenge, focusing on the Grazing Service, emerged along the same lines. In 1936, a Nevada rancher named Dewar challenged the right of the regional grazer, named Brooks, to set fees in the newly established grazing districts.

The arguments in *Dewar v. Brooks* were the same as those in the earlier case: that previously users had free access and uniform fees did not reflect natural variation in the resource.[39] The Nevada Supreme Court supported the ranchers, but on appeal the federal Supreme Court overturned the decision and established that the Grazing Service had the right to charge reasonable fees as outlined in the enabling legislation.[40] The Grazing Service (jointly with the Forest Service) immediately began studies aimed at better definition of a variable fee structure that resulted in a proposed sliding scale ranging from 18.5 cents per AUM in Colorado to 7.3 cents per AUM in New Mexico.[41] The ruling hit Nevada especially hard because the state had the largest area by percent of state total in Taylor grazing lands, had more animal unit months leased, and had seven of the ten largest permit holders.[42]

The firm establishment of fees brought the rebel contingents into the open in 1941. Senator Patrick McCarren, a Nevada states' rights Democrat who rode the crest of the New Deal politics into office in 1933, offered Senate Resolution 241 as a platform for a full-scale attack on the Grazing Service and related agencies.[43] In an attack similar to those of earlier years against the Geological Survey and the Forest Service, McCarren's resolution called for a sweeping investigation of U.S. land policy and the agencies charged with administration. Continuing resolutions supported McCarren's investigations until 1947. The committee authorized by Resolution 241 took testimony in Casper, Reno, and Las Vegas, where ranchers paraded their complaints about the existing and proposed fees. A cowed Department of Interior agreed to delay fee increases until the hearings were complete, a delay that lasted several years.

McCarren became the champion of western cattle interests through the Resolution 241 process that he had created. He hamstrung the agency operations, reduced appropriations, and attacked personnel. In 1944, Secretary of Interior Ickes (who typically had a continuing feud with McCarren) appointed Clarence L. Fosling, an experienced Forest Service manager as director of the Grazing Service. Ickes ignored the wishes of local advisory boards who preferred another candidate. When the new director proposed increasing grazing fees by 300 percent, McCarren exploded that the idea "struck consternation into the industry."[44] McCarren's hearings continued to damage the service.

An important countercurrent also prevailed in Congress during the early 1940s. Eastern representatives were dissatisfied with the Grazing Service because appropriations from the national treasury were not matched by grazing receipts. In 1941, for example, Congress appropriated $5.2 million to the agency, which collected less than $1 million in fees.[45] Eastern and midwestern representatives saw the Grazing Service as an unneeded subsidy for Western cattle empires, and they opposed continuing funds to the agency as vehemently as McCarren.[46] The agency was under attack from in front and behind by two factions, one that wanted higher fees and the other that wanted no fees.

By 1946, the two factions indirectly cooperated with each other in the destruction of the Grazing Service by reducing the appropriations to the level of the frozen fees.[47] This so-called "McCarren Reduction" also resulted in Fosling's transfer to the bureaucratic wastes of Arizona as a regional administrator. The Grazing Service was combined with the old General Land Office to produce a decidedly weak child, the Bureau of Land Management. The Grazing Service, an agency that experienced rapid, unplanned expansion and relatively poor management, was no more. Ironically for McCarren and his associates, BLM (known derisively as Bureau of Livestock and Mining) soon increased grazing fees.[48].

Buoyed by McCarren's successes and a new turn toward conservativism, cattle graziers and their congressional representatives again attempted to transfer federal lands to states and individuals. In 1946, G. Norman Winder (president of the National Wool Growers Association) and J. Elmer Brock (former president of the American Live Stock Association) established a small group of cattle and sheep owners to pool resources. The primary objective of their Joint National Livestock Committee was to organize a broad congressional effort to transfer public lands to private ownership.[49] They proposed to sell lands only to permittees at generous financial terms: 10 percent down for 30 years at 1.5 percent interest, with prices ranging from nine cents to $2.80 per acre. At August and December meetings in Salt Lake City and Denver, plans were made to include Congressman Frank A. Barrett of Wyoming as the leading congressional proponent of the transfer.[50] He agreed to present legislation if the entire industry approved of the move.

Before Barrett could act, however, Senator E. V. Robertson, also from Wyoming, stole the march. His 1945 senate resolution went well beyond the extravagant hopes of the graziers. The Robertson Bill would cede all unappropriated lands to the states, including the rights for oil, gas, and minerals.[51] State commissions were to examine all federal lands within their borders to determine which (in the state's opinion) could be best managed locally. The bill opened the door for dismemberment of national forests, parks, and wildlife refuges. Charles C. Moore of the Dude Ranchers Association testified that the bill was a case of "whole hog or none, with the none left off."[52] The bill inadvertently advertised the graziers' position and triggered an unexpectedly aggressive reaction among widely read editorialists.

The primary reaction was from Bernard DeVoto, who seized on the Robertson Bill as a true indication of the intent of western public land users to obtain control of the public property. DeVoto was the author of a monthly opinion column in *Harper's Magazine,* a publication that enjoyed an extensive readership and that carried a reputation for high quality. DeVoto's column, the "Easy Chair," was a platform for its controversial author to expound on what he considered infringements on the rights of citizens. Previously, he carried the editorial torch against censorship of books, restrictions on war-time reporting, anti-communists excesses, and what forty years later would be called consumer fraud. His wrath from 1947 until his death in 1955 was mainly directed toward western grazers, whom he considered little more than reincarnations of the nineteenth-century looters of the public domain.

DeVoto's first attack appeared in December 1946 in a preliminary article,[53] and then in January 1947, *Harper's* published his lengthy editorial against the Robertson Bill plus an extensive article entitled "The West Against Itself."[54] As a native Utahan and an accomplished historian, DeVoto's righteous indignation carried the added weight of scholarly arguments based on historical facts, insightful interpretations, and a presentation intended for a broad audience. He outlined how westerners themselves had exploited public resources in a destructive fashion and then asked the federal government for subsidies and support. He exposed the real intent behind the Robertson Bill as being an attempt by individuals to gain control of public lands rather than a states' rights effort.

DeVoto's influence was apparent in the immediate public response that flooded congressional mailbags on the subject. He had intentionally timed his publications to do the maximum harm to the legislative program of the sagebrush rebels,[55] and his description of the westerners' agenda as "get out and give us more money" struck a responsive chord in the general public as well as in congressional representatives.

While DeVoto was dealing with the third sagebrush rebellion at the most general levels of public policy in a media onslaught that surprised western interests, new forces entered the fight. Elmer Brock, the Colorado rancher who was a former president of the American Live Stock Association and prime mover behind the Joint National Livestock Committee, published in the *Denver Post* the details of the committee's plans for transfer of the public lands to private ownership.[56] His description of his opponents as bungling officials and poolhall conservationists, and his use of the anti-communist line ("all predacious and most of them tinged with pink or even deeper hue"), signaled an intensifying conflict.

Recognizing a valuable piece of ammunition when he saw it, Arthur Carhart copied the Brock's *Post* editorial and airmailed it to his friend Kenneth Reid, director of the Izaak Walton League. Reid, then attending the North American Wildlife Conference, distributed copies of the article to conference attendees with the added note "If you don't believe your federal estate is in jeopardy, READ THIS!"[57] In his closing address to the conference, Aldo Leopold claimed that the debate over public land disposal, now suddenly in the public spotlight, was the biggest conservation battle since the Ballinger days, and that if conservationists gave in on the grazing lands, the national forests and national parks would follow. For the first time, an organization of nonprofessional conservationists had entered the debate.

By March 1947, DeVoto had so publicized the Robertson Bill and the organized efforts behind it that Secretary of Agriculture Clinton Anderson believed that DeVoto had singlehandedly stopped the rebellion.[58] This opinion failed to recognize the efforts of Reid and Carhart and turned out to be premature. The rebellion was hardly stopped.

The grazing lobby regrouped and on April 17, 1947, Frank Barrett secured House approval of his House Resolution 93 authorizing his

Subcommittee on Public Lands to hold public hearings on grazing practices of the Forest Service. Barrett, in association with the National Joint Livestock Committee, rigged the committee's hearings to favor graziers who were given unlimited time before the committee to attack the Forest Service, and to advocate disposal of the public lands. As chair of the hearings, Barrett relegated opposing witnesses to strict time limits and to the end of each hearing session. In September, 1947, he took his show on the road, with hearings held before enthusiastic audiences at Rawlings, Grand Junction, and Salt Lake City. Amid foot-stomping, catcalls, and boos, the new local opposition witnesses fared poorly. The *Denver Post* labeled the proceedings as "The Wild West Show," a handle gleefully spread by Kenneth Reid's Izaak Walton League.

As the show continued to other cities, however, the tables slowly began to turn as Carhart and Reid went to work. Carhart called Benton Strong of the National Farmers Union, asking him to get the small operators out to testify on behalf of the Forest Service. Strong's response was helpful: "How many do you want at the courthouse, Art? Will two hundred irate farmers do?"[59] Two hundred did not show up, but enough appeared at the hearings to make the point that the transfer was the favored idea of the large operators. At the hearings in Ely, Nevada, the hostile witnesses attacking the disposal idea included conservation groups, mining corporations, and union representatives. Reid had so successfully organized the opposition to Barrett that the congressman canceled a final hearing to be held in Phoenix.

Carhart, Reid, and their associates within the Izaak Walton League kept a steady stream of information flowing from the Barrett subcommittee field hearings to Bernard DeVoto, who, in turn, continued to publish on the subject. DeVoto enlisted the support of highly visible fellow editors, including newspaper columnist Elmer Davis, *Collier's* Lester Velie, and syndicated columnists Marquis Childs and Joseph Alsop. To counter the efforts of the Joint National Livestock Committee, the conservationists formed the Natural Resources Council of America as an effective umbrella organization.[60] As had happened in the national forest controversies, the sagebrush rebels adopted an extreme position that ultimately triggered an effective response.

Faced with organized opposition in the field, lack of unified support among western graziers (with the small operators pitted against the large ones),[61] and a full-scale media campaign at the national level, Barrett's offensive collapsed. He did not propose legislation, but instead sent a letter to Secretary Anderson with six demands for relatively minor changes in Forest Service procedures. Anderson refused all of them. The third sagebrush rebellion ended with a clear victory for the newly powerful conservation lobby. In addition, by the late 1940s the federal government had firm policies and unchallenged jurisdiction over potential wilderness areas in national forests, national parks, wildlife refuges, and finally, the Bureau of Land Management areas. For the first time in the history of the West, federal ownership of future wilderness areas was complete.

Chapter 12

Cowboys, Columnists, and Conservationists

One route to understanding the course of events in the third sagebrush rebellion concerning the grazing lands is to understand the backgrounds and motivations of the people involved. The first two rebellions involved a few major players, but in the third rebellion there were no dominant individuals. Instead, groups of lesser-known activists drove the processes. To a remarkable degree, the proponents of federal lands transfer to state and private ownership were cast from the same mold: they were western ranchers who became lawyers and politicians. Their positions on public policy reflected their personal and professional interests. The conservationists and preservationists in public positions were all eastern urbanites. The originators of wilderness concepts and their regulations had experienced the wilderness of the West, but they were not native westerners. The newspaper and magazine writers who played a major role during the latter stages of the rebellion were eastern-based and are best described as professional critics of the West.

The three primary congressional representatives who were active opponents of federal policy in the 1920s, Senators Cameron, Ashurst, and Stanfield, illustrate the common backgrounds and motives of the sagebrush rebels. Ralph H. Cameron had immigrated to Arizona in 1882, working as a sawyer and grocer before entering the cattle business near Flagstaff. He raised cattle on the northern Arizona range and then drove them to Kansas feedlots for sale. He was Arizona's last territorial delegate, and served in the U.S. Senate from 1921 to 1926. His antagonism toward federal regulation was deeply rooted in his experiences as a ranch operator and particularly as a miner. Along with his brother, he discovered the Last Chance Mine located below Grand View Point in what is now Grand Canyon National Park. The establishment of the national park in 1908 led to a decades-long dispute between Cameron and federal officials over ownership and operation of the mine. Ironically, modern visitors to the canyon use the Bright Angel Trail more than any other, but it originally was the Cameron Trail and provided access to the diggings.[1] Cameron's attack on the Forest Service during the Stanfield hearings was a logical extension of antifederal opinions.[2]

Cameron's Arizona colleague in attacking the Forest Service during the Stanfield hearings, Senator Henry Fountain Ashurst, also had a ranching background. A son of Nevada pioneers, he too lived in the Flagstaff area, and at age 15 was already an experienced cowboy. He became a lawyer in the ranching community of Williams, Arizona, and at age 21 embarked on a political career by being elected one of the state's first senators. He retained his northern Arizona perspective on federal land ownership, and took the Stanfield hearings as an opportunity to reduce federal influence on the management of western lands.[3]

Robert Nelson Stanfield was the most successful rancher of his day: in 1920, he was said to be the proprietor of the largest sheep operation in the world.[4] His family were pioneers in northern California and Oregon in the middle 1800s, and at age 19 he assumed the role of manager of the family ranch when his father died. A shrewd businessman, he quickly learned to take advantage of the scheme of owning limited amounts of private land and using the surrounding government land to support his herds. By 1925, the date of the initial efforts of the third rebellion, he had more than 400,000 sheep grazing on three million acres.[5] His attacks on the Forest Service and attempts to divest the federal government of land management responsibilities can be traced to a history of friction with the Forest Service managers who administered much of the land used by his herds. Prior to the hearings, the Service had revoked some of his grazing permits in a misrepresentation dispute.[6]

The resolution of the early phase of the third rebellion was due mostly to the efforts of Edward T. Taylor of Colorado. Taylor was a lawyer and an educator. He was also an administrator in public schools on the mountains and western slope of Colorado, and so became familiar with the problems of ranching communities.[7] Faithfully representing his constituents after his initial election to the House of

Representatives in 1909, he opposed withdrawals from the public domain and any charges on use of federal lands. Originally, he felt that the public lands should be under state control, but after several years of observation and debate he concluded that the federal government would have to assume the management responsibility.[8] Taylor was the most prominent western representative to change his position regarding land ownership and not suffer politically: he was reelected to his congressional seat and when he died in 1941 had served 16 terms. At 83, he was the oldest member in the House.[9]

Patrick A. McCarren and Frank A. Barrett were the primary forces behind the second and final phase of the grazing lands rebellion. McCarren was the most effective rebel, given his successful attack on the Grazing Service, which eventually destroyed the agency and forced the creation of the weaker Bureau of Land Management. Like his predecessors in the rebellion, his family were pioneer ranchers. His father was an Irish Catholic immigrant who settled near Reno, Nevada, and developed a sheep ranch. Young McCarren attended the University of Nevada, but did not complete his formal education because he left school to return to the family ranch when an injury disabled his father.[10] His parents infused him with a fighting spirit, and at age 26 he won a seat in the Nevada legislature. By age 29, he had become a practicing lawyer and his career took him progressively further away from his rural roots. He soon became an associate justice of the Nevada Supreme Court.

In 1920, his prowess as a Nevada lawyer was given a boost when he successfully represented actress Mary Pickford in a difficult and widely publicized divorce proceeding. His greatest ambition was to serve as a United States senator, and after two failures he was elected in 1932 in the Democratic landslide. Although he was a Democrat, he was at odds with Roosevelt and the New Deal from its inception, and his conservative politics thrust him into a coalition opposed to Roosevelt's policies. By the 1940s, the time of his successful attacks on the Grazing Service, McCarren had become one of the most powerful men in the Senate. His constituents approved of his antifederal policies and elected him three times despite the lack of administration support. His effectiveness in hamstringing federal management of the grazing lands was a product of his skill as a lawyer, his encyclopedic knowledge of the statutes, and his positions as chair of the Judiciary Committee and the Appropriations Subcommittee.[11]

McCarren's position in the third rebellion represented a facet of his militant antifederalism. He was greatly disturbed by trends away from traditional rural values and toward an increasing emphasis on liberalism, urban programs, and federal bureaucracy. He described his politics with reference to his days as a football player at the University of Nevada, explaining that he began at left tackle but then moved to right tackle, "and I have never been left of center since."[12] His personality lent effectiveness to his efforts. A short, stocky man with sharp blue eyes, a large pink face, and flowing white hair, he was warm-hearted and loyal to friends. To political opponents, who grudgingly respected him, he was a relentless attacker who nonetheless maintained a cordiality and an easy smile even when he asked cutting questions.[13]

While McCarren commanded the respect if not the affection of the conservationists of the 1940s, they viewed Frank Aloysius Barrett as a dangerous buffoon. The architect of the "Wild West Show" was a stocky man with short-clipped brown hair and a face that became beet red when he was angry. Barrett was rough and aggressive, and he showed none of the courtesies of McCarren's public demeanor.[14] Although he was the son of two Omaha school teachers, Barrett became a rancher like the other sagebrush rebels in the grazing controversy. He established a law practice in the oil boom town of Lusk, Wyoming, and in 1924 he acquired a sheep and cattle ranch. He became active in the most powerful political groups in the state: the stock growers' and sheep growers' associations as well as in the Rocky Mountain Oil and Gas Association.[15]

Like McCarren, Barrett fairly represented the interests of his constituents in his antifederal attitudes. He is the only person in Wyoming history to have been governor, congressman, and senator.[16] He was part of a strongly developed conservative wing in Congress so that his attempts to transfer ownership to states and individuals were part of a well-established philosophy. Even after his 1947 failure in the rebellion, he continued to adhere to that policy throughout his political career, especially when he served as general counsel for the Department of Agriculture in the Eisenhower administration.

The primary forces for federal land disposal in the executive branch during the third rebellion, Herbert Hoover and Ray Lyman Wilbur, were also westerners. Hoover was born in West Branch, Iowa, to a Quaker family, but after the early deaths of his parents he spent his teen years with ranching relatives in Oregon. He was a member of the first graduating class from Stanford University, and his degree in geology launched an extensive career as a mining engineer.[17] By 1910, he had become a world-renowned expert on mine development and management. While working in China as chief engineer of that country's bureau of mines in 1900, he also developed an interest in providing food and other relief to civilian populations endangered by the Boxer Rebellion. Eventually his abilities as a relief organizer became better recognized than his engineering expertise, and during World War I he directed food relief to Europe through a private foundation. He seemed a logical choice for President Harding's secretary of commerce, and by 1928 he was so well known in the United States as a bureaucratic manager that he had no serious opposition as the Republican nominee for president.[18] His round, full face and friendly but no-nonsense demeanor were popular with Americans in the 1920s.

Hoover's perspective on the management of western public lands developed from his experiences in the West as a youth and later as a policymaker for resources. His work in the ranching industry and mining enterprises gave him a distrust of intervening federal regulations, and his lifelong policy was that less government was better government.[19] His lengthy involvement in attempts to negotiate an agreement among southwestern states for the distribution of the waters of the Colorado River led to the 1922 Colorado River Compact among six (later seven) states to develop the river for irrigation and power production.[20] The experience educated Hoover to the problems of overgrazing and the resulting pollution of water resources by sediment. His plans to transfer the public domain to states and private individuals were founded on his belief that the federal government had not managed the lands well, that private individuals had a greater stake in good land management, and that the route of privatization was the only salvation for successful water resource development in the West. All of his public statements in support of the position of the sagebrush rebellion dissolve into an expression of the need to protect water resources.

Hoover's secretary of interior also strongly supported the divestiture of the public domain. Ray Lyman Wilbur was born and raised in Riverside, California, became a medical doctor, and met Hoover at Stanford. While Hoover was active in international relief circles, Wilbur started the Stanford Medical School, became its first dean, and eventually was the president of the university.[21] The president trusted Wilbur as a personal friend, and because he represented the West and was adept at complex administration, appointed Wilbur secretary of interior. More importantly, both men shared the same general perspective on the federal government: that it was at a crossroads leading either to individualism or collectivism.[22] The public lands issue clearly fit into their calculus as a debate over which direction the federal managers would take, and neither Hoover nor Wilbur had been in office more than a few months before he made efforts to reduce the "collective" approach to land management.

The advent of the Great Depression sidetracked the Hoover administration's efforts to dispose of the public assets, and the new Roosevelt administration quickly took an opposite approach to the nation's economic problems. Public land policy and conservation in the Roosevelt administration were extensions of the efforts to reestablish a stable economy.[23] The public land policies were also diametrically opposed to those of the Hoover administration and resulted in an even firmer grip on land management by the federal government. Roosevelt and his secretary of interior, Harold Ickes, had backgrounds that help explain their divergence from their predecessors. Roosevelt was an eastern urbanite who grew up in a family accustomed to New York high society. His father was vice president of the Delaware and Hudson Railroad, and his youthful mother was strong-willed.[24] The childhood of Franklin Roosevelt had some similarities to the childhood of his cousin, Theodore Roosevelt, in that both had consuming interests in collections and in birds. Franklin admired his cousin and hoped someday to follow him to the White House.

Franklin Roosevelt's public land policies developed from his Harvard and Columbia educations and his experiences in New York politics, where he was the champion of the "little guy."[25] He fought Tammany machine politics for several years, and learned to deal with the Washington scene as Woodrow Wilson's secretary of the navy. He was a vice-presidential candidate for the Democrats in 1920, but the Harding landslide did not seriously impair his political career. In 1921, he suffered an attack of poliomyelitis, which robbed him of the use of his legs, and he remained out of politics during much of the turbulent 1920s. When he became governor of New York as a moderate progressive in 1928, he unknowingly positioned himself for the 1932 Democratic presidential nomination.

During his administration, he favored conservation-oriented policies, but his first priority always was to stabilize the economy and provide for economic development. His unwillingness to aggressively pursue the wildlife refuge program of Ding Darling's Biological Survey because of conflicting interests in public power and irrigation development was symptomatic of this internal policy conflict that was rarely settled in favor of conservation or preservation. Nonetheless, Roosevelt's support for the Taylor Grazing Act was important to passage of the critical legislation, and once empowered to reserve the remaining public domain for classification, Roosevelt became the president who reserved the last of the public lands. Eventually, some of those parcels would be set aside for wilderness preservation through the auspices of the Bureau of Land Management.

Roosevelt's major associate in public land matters was Harold LeClair Ickes, who was also a conservation-oriented urbanite. Raised in Chicago by an aunt after the early death of his mother, Ickes gained political experience as a newspaper reporter, lawyer, and Democratic campaign manager in the rough and tumble political world of the city. He was not so much a candidate but rather a kingmaker behind the scenes, a manipulator, a manager.[26] His strong progressive views on Indians and conservation brought him to the attention of Roosevelt in 1933 at an economic conference, and the new president quickly appointed him to the Interior post. The square-jawed, determined face of the new secretary soon became a fixture on the Washington scene. From a distance he was easy to respect, but close up he seemed to irritate almost everyone.[27] Even he referred to himself as a "curmudgeon."[28] Although he submitted his resignation several times, it was not accepted until 1947 when he had served in the position longer than any other secretary.

Ickes's Chicago experience made him a fearful political infighter whose rough treatment of Congress in the passage of the Taylor Grazing Act was typical rather than exceptional. He failed in his attempts to create a Department of Conservation, which would have included all the resource-based agencies including the Forest Service.[29] His relations with Gifford Pinchot over the matter became heated, and at one stage Ickes asked an assistant to dig up some unsavory information about his opponent. The aide reported that "I truly do not see anything that could be used to discredit the Governor—at least not without embarrassing you at the same time."[30] Ickes was successful in developing a powerful Interior department and he significantly raised the standards of quality in the agency, thus benefiting the management of much of the nation's public lands.[31]

While the third sagebrush rebellion was in progress, Aldo Leopold and Robert Marshall were exerting a profound influence over the development of the new concepts of wilderness preservation and management. Neither man was a westerner by birth, but each adopted the western environment as a special personal interest. Rand Aldo Leopold grew up in Burlington, Iowa, where his father was a factory manager. As a boy, Leopold developed an interest in nature through his explorations of the Mississippi River bottomlands near his home, and his observations of migratory birds led to a lifelong interest in wildlife.[32] After receiving his masters' degree from Yale University, Leopold worked with the Forest Service on the Apache and Gila National Forests in Arizona and New Mexico. He came to appreciate the wilderness experience there and in the Colorado Delta in Mexico, and he developed a sense of concern when road-building threatened previously undisturbed areas in the American Southwest. His simple question, "of what avail are forty freedoms without a blank spot on the map?" is an apt summary of the reason for his efforts to establish wilderness management in the national forests.[33]

With his interest in wilderness areas, Leopold developed expertise in managing the wildlife that inhabited them. In his early years, Leopold favored the elimination of predators from the national forests in the hope that deer would become more plentiful, but when deer populations grew explosively after predator elimination programs in northern Arizona, Leopold shifted his perspective to one of seeking a balance among natural controls on wildlife populations.[34] In 1933 he published the first textbook on wildlife management,[35] and assumed the position of professor of Wildlife Management, which had been created specifically for him at the University of Wisconsin. Most modern concepts of wildlife management derive from his work,[36] and his *Sand County Almanac* represents a conservationist land ethic that remains a vital statement forty years after its publication.[37]

While Leopold approached wilderness concepts from the perspective of a manager and policymaker, Robert Marshall's perspective was more a philosophical and theoretical one. Marshall came from a wealthy New York family whose interest in the West began from reading about the Lewis and Clark expeditions.[38] As a teenager he climbed all the major peaks in the Adirondack Mountains, and after earning degrees at Harvard and Johns Hopkins Universities, he joined the Forest Service in search of adventure. Instead, he found a tiresome existence broken by long leaves of absence during which he explored western and Alaskan wilderness areas on foot.[39] In 1934, he became Director of Forestry for the Office of Indian Affairs, where he tried to establish wilderness areas on Indian reservations. Later, he moved back to the Forest Service as head of the Recreation and Lands Division where he was better able to influence policy, including the U-regulations of the late 1930s.

Marshall viewed the wilderness experience in a vein similar to John Muir's, with an emphasis on the transcendental experience offered in remote environments. A tireless worker, he was also a tireless walker: after his move back to the Forest Service he visited as many western primitive areas as possible, making 200 day-hikes of 30 miles or more. In one case, he made 70 miles in one continuous walk.[40] His friendly, open face was a familiar backcountry sight in western forests, but his death from a heart attack in 1939 at age 38 probably delayed the development of the wilderness movement several years.

Leopold and Marshall were instrumental in initiating an important organizational force in the field of preservation that long outlived them. In 1934, when Marshall was on a day hike in the Great Smoky, he developed the idea of a Wilderness Society. His companions were Benton McKaye (a forester who originated the Appalachian Trail), Harvey Broom (a Tennessee lawyer and expert on the Great Smoky), and Harold C. Anderson (of the Potomac Appalachian Trail Club).[41] Joined later by Leopold and Ernest C. Oberholtzer (conservationist-preservationist from the upper Midwest), this group established the Wilderness Society with the aim of perpetuating the ideals of wilderness preservation and use. Although the new society did not participate in turning back the third sagebrush rebellion, it later grew to powerful proportions and materially influenced the enactment of the Wilderness Act of 1964. In the 1970s and 1980s, it became a major force in defeating the fourth rebellion.

Like Roosevelt and Marshall, preservation activist Rosalie Edge was born and raised in a well-to-do New York City setting. The product of a genteel finishing school, she spent several years in England before returning to the United States in 1913 as a dedicated feminist. In the suffragist movement she learned the political skills of organization, activism, stump-speaking, and the inside operations of public policy.[42] She became interested in birds during visits to a family summer home on Long Island Sound, and by 1930 she had combined her political activism and conservation interests. At first she attempted to participate in the Audubon Society, but the internal strife in the virtually all-male organization stymied her efforts. Outraged at the collusion between the supposed conservationists leading the organization, and trappers in an Audubon bird refuge in Louisiana, she fought the leadership for several years in a well-organized and highly publicized campaign.[43]

By 1933 it was clear that the Audubon Society did not allow Edge to exercise her considerable political talents to their fullest, so she organized the Emergency Conservation Committee. The committee had no dues, no continuing funds, and no paid officers, but with a membership of about 900 sometime supporters and Edge at the helm, it became a significant force in conservation policy in the 1930s. Her attempts to influence Ding Darling in the question of setting aside wildlife refuges represented a consistent theme of her conservation work—the setting aside of portions of the landscape for the exclusive purpose of preservation. In 1932, for example, she urged the Audubon Society to purchase more than 1,600 acres of Hawk Mountain, Pennsylvania, to prevent the sport shooting of migrating hawks in the region. When the society failed to act, she purchased the property with her own funds and staffed it with a warden to protect the birds.[44]

A short, trim woman with a relaxed demeanor in private, Edge was a vicious debater who used every skill at steamrolling opponents who generally regarded her with a mixture of confusion, fear, and respect. She was the first prominent woman in conservation-preservation activism, "the most belligerent of all the conservationists,"[45] and according to DeVoto, "a one-woman army."[46] In many ways her methods of operation were similar to those of Harold Ickes, and though neither paid attention to the sensibilities of their opposition, both usually obtained the desired objectives. Edge occupies a unique place in the story of wilderness and the sagebrush rebellions, because she pursued the preservation of specific areas for the protection of wildlife at a critical time when the idea was not popular.

During the latter phase of the third rebellion, another activist also emerged. Kenneth Reid, director of the Izaak Walton League, was singularly effective in the use of the same publicity-oriented tactics as those used by Rosalie Edge. Reid recognized that the best method of dealing with Frank Barrett's "Wild West Show" was to expose the excesses of the sagebrush rebellion to public scrutiny. Reid and his organizaion performed important connector functions between local observers in the West and the nationally syndicated columnists in the East. Arthur Carhart provided information directly to Reid about western opinions and the moves made by the livestock and woolgrowers' associations. Local members, especially in Barrett's home state of Wyoming, provided valuable advance intelligence about the sagebrush rebellion. Charles E. Piersall and Harry L. Miner of the Casper chapter of the Izaak Walton League, for example, published a pamphlet highly critical of the rebellion. Piersall paid dearly for his activities because local ranchers boycotted his oil distribution firm and eventually drove him out of business.[47]

Reid and his organization funneled the information they obtained in the West directly to Bernard DeVoto. It was Reid who coined the term "land grab" to describe the sagebrush rebellion,[48] but it was DeVoto who gave the label a national audience. DeVoto's role as a publicist in the defeat of the third rebellion was instrumental because his column in *Harper's Magazine* was widely read and highly respected. He was born in Ogden, Utah, and as a "gentile" in the Utah Mormon community he adopted a critical, odd-man-out perspective from childhood.[49] After his education at Harvard, a nervous breakdown, a stint as an instructor at Northwestern University, and occasional teaching duties back at Harvard, he settled down in Cambridge, Massachusetts, as a professional writer. Writing serial stories for popular magazines under the pen name of John August to earn a living, his real love was to write as a historian of western American culture (one of his works was a Pulitzer Prize winner) and as a "controversialist."[50] He adopted many causes in his criticism, including literary topics, censorship, restrictions on war information, excesses of the federal investigative agencies, and McCarthy's witch hunts for supposed communists.

Late in life, DeVoto took up the counterattack against the sagebrush rebels and gained a national reputation as a conservationist. He was particularly well suited to deal with the western protestors because of his knowledge of western history and his sharply critical writing skills. He accurately assessed and expressed the split personality of the westerners who wanted less federal control but more federal money. His general opinion was that the only problem with the West was that it was "lousy with Westerners."[51] His approach to stopping the sagebrush rebellion was simple: to attack. Within a few short years he wrote 40 pieces on the subject, convinced other nationally syndicated columnists such as Joseph Alsop and Elmer Davis to publicly join his side of the issue, and managed to have sympathetic articles published in popular magazines, including *Atlantic Monthly, The Saturday Evening Post,* and *Collier's*. With the combined efforts of the Izaak Walton League and Bernard DeVoto, Barrett's Wild West Show was doomed.

Chapter 13

The West in Transition

During the period of the third sagebrush rebellion, from the 1920s through the 1940s, the West was a region in transition. Every aspect of the West's culture and environment experienced fundamental changes that influenced the relationship of the region to the federal government. Most importantly for the story of the rebellions, the transfer of lands from federal to private ownership through purchase and a variety of modified homestead acts ended, and a new system of land valuation based on rights of use rather than rights of ownership appeared.

The number of acres entering private ownership from the public domain steadily increased from 1898 to its last major peak in 1910 when more than 26 million acres were converted (Figure 14).[1] Thereafter, the acreage of entries declined except for minor reversals related to the Stock Raising Homestead Act of 1916 and to the elevated commodity prices shortly before 1920.[2] Franklin Roosevelt's withdrawal of remaining public domain lands for classification under the Taylor Grazing Act virtually terminated the transfer process, so that from about 1935 onward the map of private versus public ownership remained relatively unchanged. Graziers who saw increasing restrictions appear for the use of first the national forests, and then for the Taylor grazing lands, no longer had the option of purchase.

In fact, this limitation was not a severe one, as the primary water sources and developable parcels had been taken out of the public domain by the 1930s. There was so little flexibility in the land tenure system that graziers attempted to exert some control through the state land selection process.[3] When western territories became states, they were given the option by the federal government of selecting varying amounts of land from the federal public domain for support of educa-

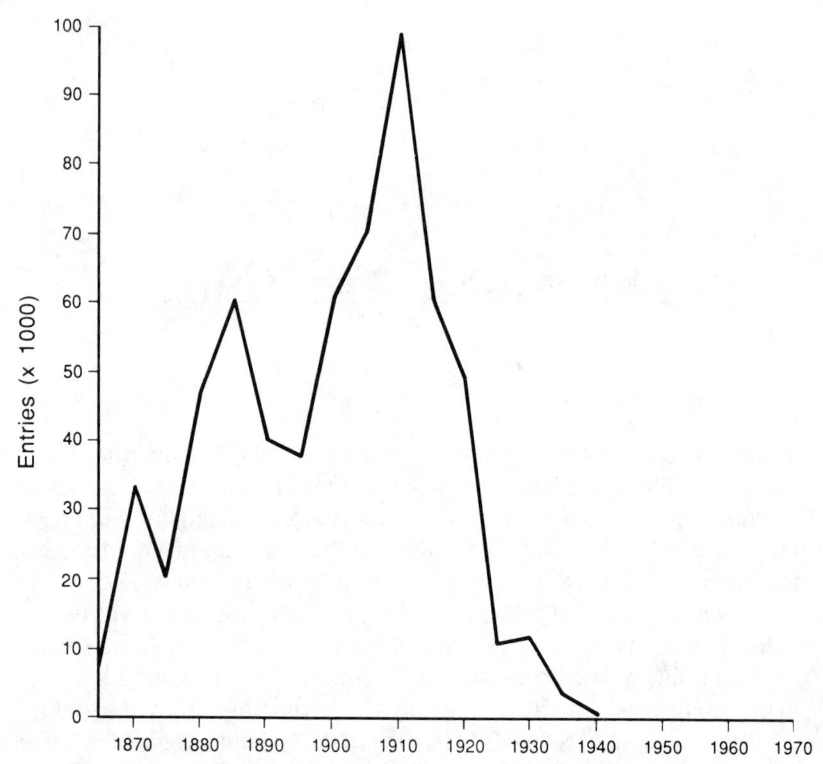

Figure 14. Entries under the Homestead Act and various revisions showing cyclic activity and a final land rush between 1910 and 1920 (U.S. Department of Interior data).

tion. Newly formed states established selection boards that chose varying numbers of "school sections" in addition to those automatically granted by the federal government.

The Arizona experience with state selection boards was typical. Originally, the board determined to select those parcels likely to be made agriculturally productive through reclamation.[4] Grazing lands had a low priority, and during the first two years of statehood, only about 2 percent of the selected lands were used for grazing. In 1914, the newly formed Arizona State Land Department took over the selection process, and stock raisers controlled the choices for several decades. Selection of state lands to benefit graziers through control of water sources or as a means of parcel consolidation was also common in New Mexico, Idaho, Wyoming, and Nevada. In Wyoming, a Forest

Service official reported that reservations by state authorities were at "the request of persons who desire to lease or purchase the tracts which they ask the State to select."[5]

If stock owners could not secure direct ownership of grazing lands or if the investment proved too expensive, an alternative system was made available by the development of a unique banking system in the West. Most ranches, including the largest ones, had only a small percentage of their total areas in private ownership, with the major grazing areas being on public lands. The Taylor Grazing Act formalized the use of the public lands for grazing by instituting a permit system. When the owner of a ranch applied for a loan from the local bank, there was little privately owned property that could be used for collateral. Western banks depended upon their surrounding economy for their livelihood, and developed an innovative system to increase the security for loans by accepting as additional collateral the grazing rights to public lands.[6]

Thus, grazing permits became similar to property in the western economic community. Although the Taylor Grazing Act specifically stated that issuance of a permit did not create any right, title, interest, or estate,[7] in reality permits performed all of these functions. Stock growers consistently lobbied for long-term permits by citing the need for stability in the industry and the likelihood that a long-term leaseholder would be a better land manager, but an important underlying reason was that long-term leases were worth more as collateral. When federal managers tried to assess fair market value for lease fees, this role of the permit system in regional economics became a factor that could not be adequately taken into account.[8]

From the 1920s through the 1940s, the western cattle industry continued its consolidation into fewer and larger operations, but new forces were at work in the West that diversified the economies and public policies of the western states. Urbanization and massive industrial development began the slow process of diluting the influence of the grazing interests. The western states had always been largely urban from the perspective of population distribution, but urban concentration accelerated in the 1930s and 1940s as a product of the Great Depression and defense industries related to World War II.

The advent of World War II brought to the West numerous military installations that had obvious implications for local economies throughout the region. More importantly, industrial expansion in the interior West occurred as part of a deliberate policy to disperse American factories to offer less concentrated targets for enemy attack.

Vehicle and aircraft assembly along the West Coast were supported by hundreds of machine and foundry industries in the interior, all drawing workers and their families into the growing cities.[9] Denver even took on the role of a port city by serving as a construction point for navy gunboats that were built at the foot of the Rocky Mountains and then shipped to the Pacific by rail.[10] At the end of the war, many of these industries turned their efforts to consumer products, resulting in a continued industrial growth. Within a few short years the cities of the western agricultural states took on the appearance of industrial centers, and the portion of their economies resulting from manufacturing increased several fold; e.g., in 1939 New Mexico's value added by manufacture was about $9 million but by 1947 the figure had climbed to $55 million; in Colorado the increase was from $90 million to $280 million.[11]

The West's natural resources played a part in the industrialization of the region. Some development was related to the abundant electrical power generated by high dams built in the 1930s. Production of strategic metals, particularly aluminum and magnesium, required huge amounts of electricity so that the erection of processing plants in the states of Washington and Oregon, as well as in the cities of Las Vegas and Phoenix, added to the industrialization of the region.[12] The president ordered a $200 million expansion of the tiny iron and steel plant in Provo, Utah. The resulting industry was the largest of its kind west of the Mississippi River and took advantage of the secure inland location, iron deposits in southwestern Utah, and nearby high-quality coal deposits.[13] Similar circumstances prevailed at Pueblo, Colorado. The growth of these industries brought about profound changes to their host states because after decades of unchallenged control, the stock growers now had to contend with industrialists, union members, and an increasingly mobile urban constituency. By the late 1940s (the period of the Wild West Show), it was impossible for western political representatives to depict their states as monolithic in support of antifederal positions.

Economic and political forces from outside the West continued to exert the almost colonial control of earlier decades when mining, ranching, and railroad empires derived their livelihood in the West but their management from the East. The aluminum industry was totally controlled either by the Reynolds corporation (an eastern concern) or by the Kaiser corporation (based on the West Coast). Henry Kaiser took special pride in his development of the Utah iron and steel industry,[14] but he correctly saw it as a piece of an immense industrial kingdom whose interests were national rather than regional in scope.

The influence of the Second World War on the development of the West may be overstated because the region was already undergoing significant adjustments.[15] The war may merely have accelerated established trends. Politically, the West of the 1920s was conservative Republican, and it was reasonable to expect that the region would support Herbert Hoover. However, Hoover's policies alienated his natural constituency and led to widespread western support for Franklin Roosevelt. When the New Deal appeared, it was embraced by the West partly out of preference, but mostly out of necessity.

The natural environment of the western grazing community was also changing. Historical records of the Palmer Drought Severity Index, based on temperature and precipitation measurements, show that during the period of the third sagebrush rebellion, the geography of drought and moisture conditions changed radically in the region.[16] During the early 1920s, the interior West enjoyed relatively moist conditions and ranges were productive for forage.[17] In the mid- and late-1920s, drought conditions became common in the intermountain area, causing decreased forage production and increased political pressures to resolve the confusion over land management of the public domain.

During the early- and mid-1930s, public land states experienced near normal rainfall conditions while the western Great Plains suffered the worst drought on record. Palmer Drought indices from the period show that during the development of the Dust Bowl on the Great Plains, states with large areas of public grazing lands (Nevada, Utah, Arizona, and western New Mexico) had only brief periods of dry conditions. it is ironic that dust storms generated from the Dust Bowl area, which was mostly private land, were used as propaganda to aid the passage of the Taylor Grazing Act addressed at public lands not experiencing the same conditions.

Despite the lack of severe drought conditions in the interior West, erosion of slopes and stream channels continued from the early decades of the century. Gullies and arroyos continued to expand during the 1930s, destroying hillslope soils and eroding into flood plains that had previously provided prime grazing land.[18] In New Mexico, for example, the Puerco River (Rio Puerco del Oeste) flowed through an arroyo tens of feet deep, and during flood periods it vigorously eroded its banks, destroying the best grazing land.[19] Similar conditions extended from Arizona to Montana and from Colorado to California.[20]

The sediment eroded from the small and medium-sized streams in the interior West moved downstream into the large rivers where it was deposited along channels and in reservoirs. In this process, which was especially prominent in the 1930s, valuable agricultural lands were buried and channels were unstable. Along the San Juan River in New Mexico and Utah, the destructive burial and instability had begun in the 1920s and resulted in abandonment of farmlands and the relocation of settlements.[21] Reservoirs, designed to store water and regulate its flow, were made much less efficient by the inflow of sediment that occupied reservoir space intended for water. The newly constructed Boulder Dam appeared to have a shortened useful life because of sediment inputs from the Colorado River Basin,[22] while Elephant Butte Dam on the Rio Grande appeared threatened by sediment eroded from public grazing lands in New Mexico and Colorado.[23]

The early 1940s saw moist conditions dominate the entire western and southwestern United States, and grazing lands became more productive. The moisture surpluses were the product of a shift in the synoptic circulation patterns that dominate weather conditions in the western United States. In the 1940s, air flows in the lower atmosphere of the region were more often zonal (that is, west to east), resulting in increased transport of moisture from water sources in the Pacific Ocean to the continental areas.[24] Precipitation-producing low-pressure systems were more common in the region after 1943 than in the decades prior to 1943.[25] The Palmer Drought Severity Index reflects these atmospheric adjustments, because the index shows a major increase in moisture throughout the sagebrush region after 1943.[26]

The change in atmospheric circulation and surface moisture resulted in profound adjustments in surface conditions. The rapid erosion that was dominant before 1943, when occasional rainfalls eroded unprotected soils, gradually slowed. The increasingly dense vegetation cover apparently resulted in a reversal of the erosion episode that had produced the gullies and arroyos, and after 1943, deposition of sediment began occurring in the channels. New flood plains began to form and continued to store sediment that once was flushed through the river systems.[27] These deposits continue to grow along such streams as the Little Colorado River,[28] the Paria River,[29] and many representative streams in the Colorado Plateau.[30]

The reduced rates of erosion on public lands of the interior West after the early 1940s resulted in reduced inflows of sediment to the region's reservoirs. The early estimates of inflow rates for Boulder Dam, for example, quickly became obsolete as post-1943 data appeared and reservoir managers took a more optimistic view of the life expectancy of their structures.[31] This radical adjustment in atmospheric and surface processes in part explains why the early phase of the third rebellion (in the 1920s) emphasized the role of land management in water resource protection, while the later phase (in the late 1940s) took little note of this connection.

The connection between land management and water resource development was especially important for Herbert Hoover, and his argument that public lands should be ceded to states in order to improve their management was a direct outgrowth of his experience with water resources. Hoover headed the commission that negotiated a 1922 agreement whereby states of the Colorado River basin (except Arizona) established a legal division of the basin's water.[32] This Colorado River Compact was the result of several years' work, and during that time Hoover became intimately familiar with the importance of reservoir construction and maintenance to the economic development of the Southwest. The signing of the compact paved the way for the ultimate construction of Boulder Dam, though numerous congressional funding debates delayed its start until the 1930s.[33] Once the funding was won and the dam was built, the threat to its usefulness through sedimentation became a cause for widespread concern.

By the time the dam was complete, Franklin Roosevelt's administration was in power, but interest in the connection between land management and water resources became even more firmly established. The control of overgrazing as a way of preventing rapid erosion and the resulting sedimentation problems became the centerpiece for soil conservation research and investment in the Southwest. The controls were a continuing source of unhappiness among western ranchers. Although the relative importance of climatic controls versus grazing controls had not been firmly established, the scientific establishments of Soil Conservation Service and Forest Service pursued numerous investigations into the role of grazing in influencing runoff and erosion.[34] Their work was used to justify restrictions on grazing of public lands.

All of these investigations were on small scales, often with experimental watersheds that were, at most, a few square miles in extent. The investigations clearly defined the role of grazing in reducing vegetation cover and producing increased amounts of runoff and erosion.[35] The Soil Conservation Service undertook a wide-ranging campaign to educate ranchers and farmers about the grazing-erosion-water resource connection with a veritable blizzard of publications, each promising an improved environment with reduced grazing.[36]

Subsequent research confirmed that reduced grazing resulted in improved vegetation cover with reduced runoff and erosion.[37] The problem was that human management and overgrazing occurred in relatively limited areas. The vast drainage area of more than 200,000 square miles above Boulder Dam, for example, was so diverse and had so many areas not subject to human management that, as an entire system, it did not respond to grazing management. In the 1980s, long-established records showed that the water and sediment yield of the basin responded mostly to climatic changes rather than to changes in grazing policy (Figure 15).[38]

Part of the federal preoccupation with controlling western environmental processes through land management was a derivative of eastern experiences. In the Piedmont of the southeastern United States, for example, land management had reversed catastrophic erosion and gully development,[39] so it seemed likely that a similar strategy would work in the West. The drainage basins in the East were smaller, however, and the humid climate, nearly total coverage of basins with manageable lands, and dominance of crop agriculture were conditions not duplicated in the West. In the end, the lessons from one region were not directly transferable to the other.

Sectional differences between East and West that extended beyond the respective environments continued to influence the issues of public land management during the third sagebrush rebellion as they had in previous rebellions. While the West experienced profound changes in culture, economics, population, and politics, the Eastern view of the region remained static, almost mythical. The development of the motion picture industry and its seemingly endless production of Western movies translated the romantic literature of Zane Grey and others into popular culture.[40] The Eastern view was still dominated by the image of the independent cowboy, the ranching family surrounded with its hard-won herds of cattle, and a rural lifestyle. The reality of western towns, that by the 1920s had streetcar systems, automobile garages, department stores, and movie houses (that slso showed Western films), was invisible in the popular mind.[41] Western politicians used the myth of the "Old West" when it served their purpose, especially when arguing against increased federal controls of the supposedly independent cattle operators. The argument seemed not to appear when these proud independent westerners asked for federal dollars to build dams and roads or to fund Civilian Conservation Corps activities.

Movies were not the only means by which western images reached the East. Western painters continued to depict the striking landscapes

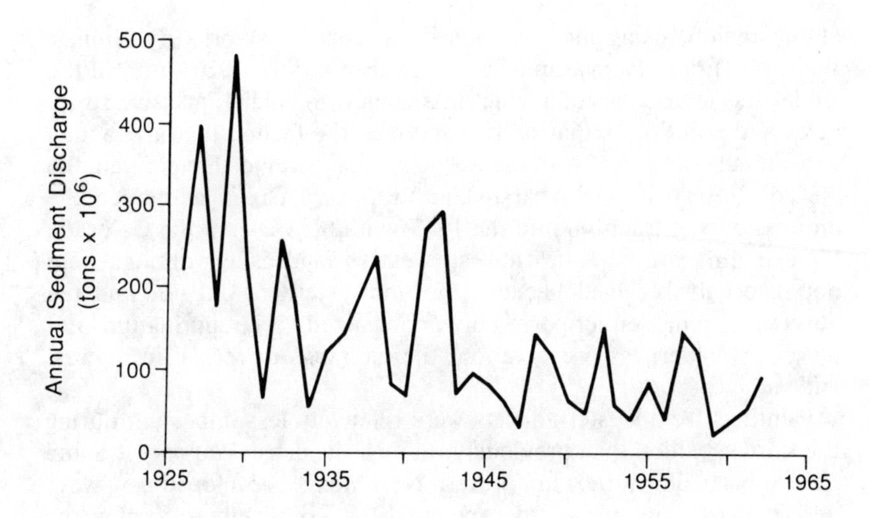

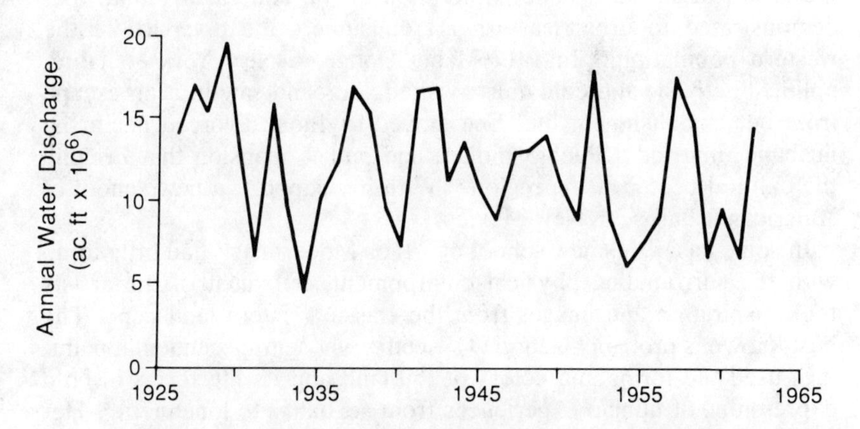

Figure 15. Water and sediment discharge data for the Colorado River in the Grand Canyon, Arizona, showing hydrologic processes in the Upper Colorado River Basin (1926–1962). Despite wide annual variations, mean annual water discharge did not significantly change, while mean annual sediment discharges declined, indicating decreased erosion or storage of sediment in flood plains (U.S. Geological Survey data).

of the region, using more impressionistic styles. Works by Gunnar Widforss, Edgar Payne, and Carl Oscar Borg in the 1920s infused the wilderness landscapes of northern Arizona with blocky, massive structures and soft colors that carried more of the feeling than the actual appearance of the environment. James G. Swinnerton painted the Nevada desert in stark, harsh lines. Maynard Dixon continued the impressionistic tradition into the 1940s with his views of the Colorado Plateau. In the East and Midwest, western landscape paintings were not favorably received because the nonwesterners did not find the desert appealing, either personally or artistically,[42] but during the later drive for wilderness preservation, the paintings increased in interest and value.

Paintings from western artists were relatively less important during the third rebellion than previously, though the artists' colony at Santa Fe and particularly Taos in northern New Mexico developed new ways of viewing the western landscape. In 1915, six members formed the Taos Society of Artists, and the number soon exceeded 100. Artists of the group began an artistic exploration of the American Indian that demonstrated to urban eastern art consumers the diversity of the western population.[43] In 1917 Mabel Dodge, a New York socialite, political activist, and columnist, visited Taos and saw in it an escape from her establishment life. She moved to Taos, divorced her artist husband, married a Pueblo Indian, and built a mansion that became the cultural center of the region.[44] With her support, a new school of American art arose in New Mexico.

In some cases this new school of "Taos Modernists" had little to do with the surrounding physical environment, but most of the artists took inspiration and images from the classic western landscape. The best known is probably Georgia O'Keeffe, whose transcendental paintings used the forms and colors of the land and its life forms in bold expressions of human experiences from sexuality to loneliness.[45] Her paintings became widely known and admired in the nation's centers of culture, and she thus brought to a broad and sophisticated audience a view of the West that was radically different from the nearly photographic renditions of earlier periods.

One explanation of the demise of the photographic style of landscape paintings of the West was the development of high-quality photography. O'Keeffe's friend, Ansel Adams, was the leading exponent of landscape photography as art, and his work also had a profound impact on the perceptions of nonwesterners about the region.[46] Adams's sharply defined black-and-white images of the natural environment of the West emphasized the form and texture of the landscape, bringing to the eye what John Muir's writings had brought to the imagination. Most of Adams's work was purely western, reflecting in part the influence of the New Mexican artists who worked in different mediums. Although Adams was just beginning his career and solidifying his position in thc art world during the third sagebrush rebellion, he was destined to become a central figure in the wilderness preservation movement and the fourth rebellion.

Rather than learn about the West vicariously through art, many Americans experienced it directly in that peculiarly Western institution, the dude ranch. As early as the 1870s, Colorado ranches were accepting paying guests, but the concept reached its greatest development in the 1920 to 1950 period when some ranchers recognized that they could turn greater profits by herding people instead of cows.[47] Because the visitors expected to find the "Old West" as opposed to the new West, which was similar in many ways to the cities from which they came, many entrepreneurs were more than willing to supply false-front buildings, staged gun fights, and the obligatory chuckwagon cookouts. When the visitor returned home to the East, it was often with memories of a mythical West that no longer existed except in a contrived way.

The dude ranch (the preferred term was the less perjorative "guest ranch") industry was a significant economic fact of Western life during the third sagebrush rebellion. By 1937, there were 102 operations in Wyoming alone that catered to the temporary westerner.[48] The dude ranches influenced the course of the rebellion and the grazing controversy in two ways. First, many of the patrons were conservationists who wanted to see firsthand the western environment,[49] and who took away with them a renewed fervor for preservation as opposed to exploitation. They were more effective debaters for having had experience in the country that was the focus of the grazing debate. Second, the owners of the dude ranches formed powerful commercial associations representing western interests much different from the traditional ranchers. When Congressman Barrett suggested disposal of the federal grazing lands, some of the most important opponents were the dude ranchers who saw the proposed move as the end of their livelihood. These dude ranchers were active western allies for the conservation groups, and they contributed to the division of Western opinion about federal control.

During the period of the third rebellion, the West matured and became more similar to the rest of the nation while maintaining only some distinctive characteristics. The growth of western cities, their increasingly industrial nature, and the integration of western images into American culture brought the West closer to decision makers in the East. Internally, western economic interests and politics became more diverse, so that the region did not speak with one voice at the national level. When stock owners, powerful as they were, sought control of the federal grazing lands, they required a unified homefront. It was not to be, and the unsuccessful end of their efforts ensured that more than 140 million acres of western land remained under federal jurisdiction. Just 17 years after Congressman Barrett's aborted Wild West Show, the signing of the Wilderness Act set the stage for the inclusion of many of those acres in a national wilderness system, and the transition from disposal to preservation would be complete.

Part IV

The Fourth Rebellion: Wilderness Lands

Chapter 14

The Law of the Wilderness

The Wilderness Act of 1964 and its subsequent administration set aside large areas of mostly western lands for preservation, and triggered a fourth sagebrush rebellion. The lands included in the National Wilderness Preservation System were available because of their involvement in the national forest or grazing lands controversies, while the pattern of political action and reaction was one established by the irrigation lands debate in the 1800s. The wilderness debate brought significant changes in the management and perception of the western federal lands.

As the controversy over grazing on public lands wound down in the late 1940s, the drive for statutory wilderness areas carved from federal public lands began to develop in earnest. After World War II, Americans had more time and money to spend on recreation than ever before, and outdoor recreation became a national pastime not limited to particular regions or classes. A part of this interest in the natural environment as a playground focused on the remaining roadless areas of public lands. The effort to afford these lands legal (rather than merely administrative) protection from economic development required 30 years from the time of its initial proposal, 9 years of legislative maneuvering, 18 separate congressional hearings, and the introduction of 65 different wilderness bills.[1]

Robert Marshall was the first to formally propose legislative action for wilderness areas. His work, along with that of Leopold, Carhart, and others, ensured that federal agencies would administer some lands as wilderness, but in 1933 Marshall proposed a bolder, more lasting solution by suggesting the establishment by law of a network of primeval reservations.[2] Marshall proposed a series of classifications, including "superlative" areas like the backcountry of Yellowstone and Yosemite National Parks as well as "primitive" and "wilderness" that were roadless forest areas. In formal memoranda to Secretary of Interior Harold Ickes, Marshall also recommended a wilderness planning board to oversee the administration of the reservations.[3]

Five years later, Marshall's proposals received support form H. H. Chapman, an influential professional forester and professor at Yale University. Chapman recognized the administrative hazard of dealing with a wilderness board in addition to land management agencies, so he urged that wilderness management become part of the responsibilities of the host agency.[4] In 1940, Harvey Broome of the fledgling Wilderness Society investigated the legislative means for using Marshall's and Chapman's suggestions in protecting the recognized wilderness areas on the national forests and Indian reservations. In the same year Irving Brant, a maverick journalist, solicited and received Ickes's support for a wilderness bill that had been introduced in Congress.[5] Shortly thereafter, the world war consumed the attention of the nation, and no action was forthcoming on the wilderness issue.

After the war, economic and recreational pressures on remaining wilderness areas increased, and some conservationists worried that the federal administrative agencies would not protect roadless areas from exploitation. The national parks provided a case in point: between 1930 and 1945, areas administered by the Park Service grew in number from 55 to 180, in area from 10 million acres to 24 million acres, but in visitors from 3 million to more than 30 million. The Park Service was oriented toward providing facilities for these visitors, and the appointment of landscape architect Conrad Wirth as Park Service director in 1951 ensured a development attitude not conducive to wilderness preservation. Bureau of Reclamation and Army Corps of Engineers water projects threatened several areas, while grazing and lumbering activities increased.[6] Uneducated overuse of some wilderness areas in the High Sierra, California, began to result in serious damage to ecosystems.[7] Howard C. Zahniser, executive director of the Wilderness Society, and David Brower, executive director of the Sierra Club, became convinced that although the Forest Service recognized 8 million acres as wilderness, it would be unable to protect them from the pressures of development.[8]

In 1949, the drive for wilderness legislation became more of a public issue. Kenneth Reid, venerable survivor of the grazing wars in the days of Congressman Barrett's Wild West Show, proposed to the Natural Resources Council a program to establish a legislated wilderness system. The Council, preoccupied by other concerns, including the environmental implications of the rapid expansion of water development projects in the Colorado River system, did not endorse the idea.[9] In the same year, Howard Zahniser of the Wilderness Society had better luck.

Zahniser's approach was twofold: first, convince the conservation lobby to accept a preservationist perspective, and second, begin the legislative process by a congressional fact-finding exercise. The Sierra Club provided the platform for addressing the conservation lobby in the form of its first Sierra Club Wilderness Conference in 1949, a meeting that drew together about 100 federal and state administrators, club members, and professional wilderness outfitters. In his address, Zahniser outlined the major threats to wilderness preservation: commercial development, road building, and unwise overuse, and he sketched the outline for a national wilderness system.[10] At the same time, Zahniser convinced Congressman Raymond H. Burke (Ohio) to request a report on the status of American wilderness areas from the Legislative Reference Service of the Library of Congress (the principal data-gathering arm of the legislative branch). In the early stages of developing a wilderness bill, the resulting report became a valuable lobbying tool, because it noted the disjointed programs for wilderness preservation in several governmental agencies. The report included a survey of federal, state, and citizen's organizations that indicated widespread concern and support for a wilderness bill.[11] It is perhaps not surprising that the director of the Reference Service at the time was Ernest Griffith, the treasurer of the Wilderness Society.[12]

The wilderness conferences became biennial affairs, and in the 1951 version Zahniser presented a fairly complete legislative proposal.[13] While the attendees at the California meeting were enthusiastic, the idea was received less warmly elsewhere. When Zahniser broached the subject to his luncheon partners at the Cosmos Club, a liberal Washington group, Ovid Butler of the American Forestry Association responded "What's the matter, aren't the wilderness areas there anymore? Why save them twice?"[14] As it turned out, Zahniser would have to save the wilderness areas many more times than twice. He recruited help from wherever he could find it, tapping John Baker of the Audubon Society for help in gathering support in the conservation community, and John Oakes of the *New York Times* editorial staff for educating legislators and the public.

Before Zahniser was finished, he had won the important influence of Karl Menninger, Eleanor Roosevelt, and Adlai Stevenson. He built at the grass roots as well, so that between 1956 and 1964, Wilderness Society membership increased from 7,600 to 27,000.[15] On May 24, 1955, when he presented a specific and detailed legislative plan including a draft bill in an address to the National Citizens Planning Conference on Parks and Open Space, the idea of a national wilderness system seemed to be an historic inevitability. His friend, Hubert Humphrey, inserted Zahniser's presentation into the Congressional Record.[16]

Zahniser's case was strengthened by the work of James P. Gilligan. The case for wilderness preservation was difficult to make because of a lack of specific information about where the areas were, how large they were, and how they were presently being administered. In 1954, Gilligan completed his Ph.D. dissertation in the School of Natural Resources at the University of Michigan on the subject of wilderness areas.[17] On October 26, 1954, he presented his results in a formal paper before the Milwaukee meeting of the American Society of Foresters.

Gilligan's data gave a profile of the administrative and physical aspects of American wilderness areas in the 1950s. The Forest Service and National Park Service favored administration of their lands for the masses to the near exclusion of the interests of smaller user groups such as wilderness proponents. Decision making emphasized dollars rather than social values. Gilligan's research indicated that in the 77 wilderness, wild, and primitive areas in the West there were 200 miles of roads, 145,000 acres of private land, 400 to 500 mining claims, 60 mines, 24 air strips, 140,000 sheep, 25,000 cattle, and 90 small dams. Gilligan concluded that wilderness preservation was failing in its administrative methods, and that congressional action was required to save the remaining areas.[18] The combination of Gilligan's work, the Library of Congress report, and Zahniser's draft bill was sufficient to initiate the formal pursuit of a law for wilderness areas.

Senator Hubert Humphrey (Minnesota), in response to persuasion by William Magie of the Minnesota Friends of the Wilderness, introduced the first modern wilderness bill on June 7, 1956.[19] His nine cosponsors included only two westerners, Senators Richard Neuberger and Wayne Morse, both from Oregon. The proposed bill had been drafted by a consortium of conservation organizations headed by Zahniser, Brower, and Ira N. Gabrielson (chair of the Citizen's Committee on Natural Resources). Participating groups included the Wilderness Society, Sierra Club, National Parks Association, Izaak Walton League, Council of Conservationists, Wildlife Management

Institute, Citizen's Committee on Natural Resources, and Federation of Western Outdoor Clubs.[20]

The proposed bill created a national wilderness system from areas in national forests, national parks and monuments, wildlife refuges, and Indian reservations. It was a monumental wish list. The legislation would ban farming, logging, grazing, mining, prospecting, roads, and motorized vehicles from the wilderness areas, which would be administered by the parent land management agencies. The bill identified for immediate inclusion in the system the 13 million acres of Forest Service wilderness, wild, and primitive areas (from the U-regulations of the late 1930s). An additional 20 million roadless acres in the national forests and 20 million acres in wildlife refuges would be reviewed for possible inclusion. Indian lands would be included with consent of the various tribes. Additions to the wilderness system would be made by presidential executive order if not vetoed by both houses of Congress. Finally, the bill would establish a Wilderness Council to serve as a presidential advisory board made up of agency representatives and public participants.[21] Congressman John Saylor (Pennsylvania) introduced an identical companion bill in the House of Representatives.[22]

The issues raised in the first wilderness bill brought about a clash between supporting and opposing individuals that required nearly nine years for resolution. The basic issues of contention were the same through a succession of six wilderness bills between 1956 and 1964. The proponent's positions were easily stated: wilderness areas should be preserved because of their aesthetic qualities and their social value. The aesthetic qualities included the preservation of natural beauty, the ability to experience a completely undeveloped environment, and the opportunity to encounter an environment similar to the one that dominated the early history of the nation. The social value derived from benefits associated with recreation, wildlife, and scientific research.[23] Led by Zahniser, the preservation groups usually articulated their positions in a unified effort, and they effectively organized support for the wilderness bills. Opponents argued that there was no need for a wilderness bill to achieve these ends, that the forest and park services could administratively provide the required opportunities, and that in any case there was an abundant supply of undeveloped areas.

The opponents' case was more fragmented and was put forward by a variety of individuals and organizations whose financial interests would be restricted by the passage of a wilderness bill. The strongest opposition came from water managers who feared that wilderness designation of watershed areas would prevent the development of water storage and delivery systems to rapidly expanding western cities. These managers were the direct administrative descendants of the urban representatives who supported national forest reservations to protect mountain watersheds in the early 1900s, but now they were concerned with possible restrictions on construction and manipulation of those areas. Senator Kuchel (California) was especially concerned for the development of Colorado River water for the southern area of his state. William L. Berry, chief of the California State Department of Water Resources, tried to accommodate the wilderness proponents by proposing that "reservoirs can add to the beauty and recreation potential of these areas,"[24] a position obviously at odds with the intent of the bill. During early hearings, Colorado representatives were prominent, with the Colorado Water Conservancy Board, Colorado State Watershed Conservation Association, and Upper Colorado River Commission speaking against the bill. In later debates, Colorado congressional representatives would be the most difficult legislative obstacles for the Wilderness Act.

Water specialists in Arizona and New Mexico also opposed the wilderness bills. In Arizona, the construction of the Central Arizona Project Canal to tap Colorado River water was only one method of improving the state's water resources. During the 1950s and 1960s, there were concerted efforts to manage mountain watersheds to increase runoff for use in lowland agriculture and urban areas.[25] An important aspect of this management was the conversion of large areas from one vegetation type to another (chaparral to grassland, for example) to prevent losses from evaporation and to increase the amount of precipitation delivered to channels.[26] Richmond Johnson, executive secretary of the Central Arizona Project Association, testified against the wilderness bill because it would preclude such management.[27]

The New Mexico State Engineer, Steven Reynolds, consistently opposed the wilderness bills out of concern for the impact that the bills might have on the state's water resource plans.[28] The anticipated Hooker Dam on the Gila River would impound waters that would extend into the Forest Service's Gila Wilderness, an arrangement prohibited by the proposed wilderness bills. He also emphasized the management of watersheds for improved water yield and predicted that New Mexico would eventually develop much of its wilderness and primitive lands for water or tourism.[29] Reynolds's opinions were especially important because of his close connections with New Mexico

Senator Clinton P. Anderson, chair of the Senate Interior Committee, during much of the important debate about the wilderness bills.

The timber interests also lobbied against the bill because they wanted to leave as much of the national forests available for harvest as possible. The issue was critical in northern California, Oregon, and Washington, where communities were economically dependent on saw mills and timber supplies from nearby national forests. Industry representatives testified to the local impacts of restrictions on timber cutting. Stuart Moir of the Western Forestry and Conservation Association was perhaps the most enthusiastic: he castigated proponents of the bill, who he described as "professional forest lovers" who wanted to deny the homeless of lumber for their houses at reasonable prices.[30]

Mining interests had a special difficulty with the wilderness bill because many of the areas slated for preservation (and elimination of mining activities) had not been adequately explored for their mineral potential. The original proposal called for no mining and no prospecting in wilderness areas, a point that became a bargaining chip that was traded and compromised on during the political trade-offs in the drive for a final bill. The American Mining Congress ensured that witnesses were available for every hearing for every version of the bill in order to point out what they thought was folly in eliminating large areas from potential but unproved production.

Livestock graziers, flush with their success in the demolition of the Grazing Service and pleased with their control of the newly formed Bureau of Land Management, saw the wilderness bill as an attack on their domain from a new direction. Although relatively little grazing occurred in the Forest Service wilderness areas, some primitive areas were important summer ranges. Cattle and sheep owners also were concerned that if the wilderness system were to be established, it would be expanded out of their control to extend its ban on grazing to other, as yet unidentified, public lands. Many graziers saw the bill as one more example of eastern politicians attempting to dictate to western land users, charging that the bill was the work of "Eastern dominated sportsmen, wildlife and historical organizations."[31]

As the debate and compromise over the issues in the bill continued, western livestock interests became more militant. They hired a Denver public relations firm to carry their message to Washington, though in some cases the message was one of misinformation. For example, Farm Bureau and livestock associations in Colorado and New Mexico claimed that "the bill stops all grazing on wilderness areas and will be used to chase stockmen off the national forests altogether. Local people would be deprived of all livelihood. Three eastern states have tried this bill and it didn't work.''[32] None of these points was correct.

In previous controversies concerning management of public lands, conservationists and federal management agencies were natural allies opposing groups of land users interested in unrestricted commodity development. Early in the wilderness bill debates, however, the preservationists found their position opposed by the National Park Service and the Forest Service. The Park Service saw itself serving the auto-touring public, and viewed the development of roads and visitor facilities in the parks as a logical extension of its mission.[33] Park Service Director Conrad Wirth testified that roads did not end wilderness, but rather, made it accessible for users,[34] a position similar to the one espoused by the water resource specialist who saw reservoirs as improving the wilderness areas. The service did not want to give up to an external board the management options for undeveloped lands within the parks.

The Forest Service also did not want to see restrictions on its management of the primitive areas. While the service was willing to oversee the already-identified wilderness and wild lands, it wanted the freedom to continue its slow review of primitive areas. If some lands were to be dedicated to what the service professionals saw as a single-use wilderness, the multiple-use ethic for all national forest areas appeared to be in danger. The Forest Service was concerned too with the possibility that eventually the wilderness areas might come under the jurisdiction of the National Park Service, something that the foresters were especially sensitive about because most national parks had been carved from national forests.[35]

Although all of these issues were points of lengthy debates and eventual resolution, the underlying difference of opinion about the wilderness bills was in the perceptions of the role of public lands in the American democracy. Wilderness preservationists saw the roadless areas in light of their social value. Wilderness areas were worth preserving, even if for the use of only a few. Newton Drury, a former Park Service director, testified regarding such areas, that ''surely we are not so poor that we need to destroy them, or so rich that we can afford to lose them.''[36] Opponents of the wilderness bill argued that the public lands should contribute to the economic development of the entire nation, and that each area should be put to its highest economic (rather than social) use. Thus, the common thread (though almost always hidden) to the wilderness debate was a difference in moral judgements, and therefore the debate was not susceptible to complete resolution.

In 1956, any resolution seemed remote. The bill had been drafted by preservationists with attempts at compromise introduced late in the congressional session, and it expired shortly thereafter without hearings. In 1957, Senator Humphrey again introduced the bill with minor modifications, and two days of hearings were held in Washington, D.C. The hearings, where the water lobby was prominent, brought out the opponents of the bill and exposed many of the points where compromise would be required.[37] Zahniser and his associates returned to the legislative drafting board and recast the bill, giving as little to the opposition as possible. Senator Humphrey introduced the fruits of their labor as a third bill on June 18, 1958.

The 1958 version relaxed some provisions of earlier efforts in an attempt to win a broader base of support. The bill permitted motorized boats and grazing in national forest wilderness areas where such uses had already been established, reaffirmed the superiority of state water laws in wilderness areas, permitted insect and disease control, and downgraded the wilderness advisory council's influence.[38] In a strategic error, the revision of the bill included the provision that the Department of Interior need only consult with Indian tribes concerning their wilderness lands, instead of the language of the original bill that required Indian consent for inclusion of the lands in the wilderness system.

The Senate Interior Committee held field hearings for the 1958 bill in states of its key members: Richard Neuberger's Oregon, Arthur Watkins's Utah, Thomas Kuchel's California, and Clinton Anderson's New Mexico. The hearings gave local commodity groups opportunities to air their displeasure, but also allowed the senators a chance to assess the political waters back home. Despite the negative rhetoric from commodity groups, the Departments of Interior and Agriculture, parent agencies of the Park Service and Forest Service, came to cautious support of the bill, satisfied with the concessions made by the preservationists. The bill died at the end of the legislative session without being reported out of committee.

Senator Humphrey introduced still another revised wilderness bill on February 19, 1959, while Congressman Saylor offered a duplicate version in the House. Zahniser and his associates recognized the success of compromises made to win support of the federal management agencies, and in order to solidify that support they made further adjustments. Activities permitted in forested wilderness areas were extended to include fire control and wildlife management, and instead of immediately including primitive areas in the wilderness system, the Forest Service was granted 20 years to classify them and make recommendations. The Departments of Interior and Agriculture were also granted permanent seats on the wilderness council. Concessions were not forthcoming for the western commodity groups who clearly would not support any wilderness bill.[39] "We have a fight on our hands," declared Ira Gabrielson.[40] Field hearings in Seattle and Phoenix showed that the westerners agreed.

Uncomfortable with complaints about the bill from his northern Minnesota constituents, Humphrey sought a broader base of support for the bill. He asked Clinton Anderson to serve as a cosponsor, but Anderson declined, perhaps partly in a self-protective measure. The New Mexico legislature had memorialized Congress to protest the bill, claiming that it would extend wilderness areas without limit, eventually end all grazing on public lands, and bring an end to predator control.[41] Based on advice from his legislative assistant, Anderson made it known that in order for him to support the bill, water development rights would have to be protected, the wilderness council would have to be eliminated, and expansion of the wilderness system would have to be at the behest of Congress rather than by presidential order.[42] Facing reelection from New Mexico in 1960, it is possible that Anderson sidestepped the issue temporarily.

By 1960, the battles over the wilderness bills had become the biggest public land debate since the grazing controversy. Passage of some sort of wilderness act began to appear inevitable, but two other related pieces of legislation affected its fate. The Outdoor Recreation Resources Review Commission Act slowed progress on the wilderness bill, while the Multiple Use-Sustained Yield Act cleared an important obstacle. In 1958, the Outdoor Recreation Resources Review Commission (ORRRC) was authorized to conduct a sweeping investigation of the explosive growth in American outdoor activities, to assess the available facilities, and to recommend avenues for improving opportunities for public recreation.

The National Park Service was concerned about the Commission because the service did not want to lose its perceived position as the leading federal agency for outdoor recreation,[43] while other agencies and interest groups used the anticipated reports of the Commission as an excuse to delay the action on the wilderness act into the early 1960s. Because wilderness was strongly associated with outdoor recreation, opponents argued that action on wilderness preservation would be premature without the benefit of the Commission reports, especially one planned for an assessment of the availability of wilderness recreation. Clinton Anderson was a major proponent of the ORRRC because he saw it as an opportunity to assess accurately the recreation value added to water development projects, a position that allowed him to offset objections to his later support for the wilderness bill.

Although the ORRRC became a vehicle for delaying the wilderness bills, the Multiple Use-Sustained Yield Act of 1960 was instrumental in generating an increasingly broad base of support for the wilderness position, especially in the Forest Service. The foresters sought to establish their management principles in law, protect their agency's position in the management of recreation, forestall transfers of land to the National Park Service, and counter the single-use perspective that dominated the philosophy of the wilderness bill.[44] The agency drafted the multiple-use bill that was introduced by Senator Humphrey to a supportive Congress.

According to the draft bill, the mission of the Forest Service was to manage "outdoor recreation, range, timber, watershed, and wildlife and fish." Interest groups assumed that the listing was in order of preference, and an alphabet game ensued.[45] The service wanted the list in alphabetical order, so the game was to generate terms with first letters that put one's own preferences in order. Foresters wanted recreation first, so that "outdoor" as a modifier would ensure its preeminent position. Wild game was largely the responsibility of the respective states, and so the usual term "fish and game" was changed to the awkward "wildlife and fish" and relegated to the end of the list. Livestock groups preferred "grazing" or "forage" to highlight their position, and lumber companies undoubtedly saw the term "forestry" as vastly superior to the lowly "timber" with its third position.

The inconsequential tempest over terminology did not significantly divide commodity groups, but a split developed among preservationists. David Brower of the Sierra Club thought the multiple-use bill would destroy the wilderness initiative because of its requirement for multiple use on forest lands, including wilderness areas. Howard Zahniser of the Wilderness Society saw the bill as critical to the continued support of the Forest Service for the wilderness bill, but he recognized that the infamous list of management responsibilities did not include "wilderness." Eventually the bill included the statement that "The establishment and maintenance of wilderness are consistent with the purpose and provisions of this act.''[46] The bill became law in 1960.

The year 1960 was critical for the wilderness bills for several reasons in addition to the passage of the Multiple Use-Sustained Yield Act. Two major senate supporters of the wilderness bills were lost: Richard Neuberger of Oregon died, and James Murray (Montana), the chair of the Senate Interior Committee, retired. Clinton Anderson was reelected and eventually would become chair of the committee. Hubert Humphrey phased himself out of the management of the wilderness initiative to concentrate on securing the Democratic presidential nomination. The general election brought John F. Kennedy to the presidency. He became the first president to endorse the wilderness bill,[47] and his administration was symnpathetic to the preservation perspective. Finally, proponents redrafted the wilderness bill again, and before he left the Senate, Murray introduced a modified version that offered more compromises by eliminating wilderness areas on Indian reservations, eliminating the National Wilderness Council, and after an initial period of 15 years, permitting additions to the system only by a positive act of Congress.[48]

Murray's bill died at the end of the congressional session in 1960, but it represented essential refinements. On January 5, 1961, Clinton Anderson's first action as new chair of the Senate Interior Committee was to introduce the revised bill again. Satisfied with its form and secure in the knowledge that the next senatorial election was five years distant, Anderson became a vigorous sponsor and skillful manager for the wilderness bill.[49] He arranged Washington hearings to coincide with the annual meeting of the National Wildlife Federation to ensure an abundance of supportive witnesses, and aided in the publication of informative (though positively slanted) articles in popular magazines. In managing the hearings, he emphasized that he would accept testimony on specific points such as how much land to preserve and how it would be managed, rather than general perspectives and ''philosophy.''[50] His position eliminated fundamental debates by presupposing that there would be some sort of wilderness system. Despite these efforts, three major technical objections remained, all posed by western interests.

First, Senator Frank Church (Idaho) argued that the process of setting aside new areas for the system by the president without a positive action by Congress was a loss of constitutional power for the legislature. Eventually, a compromise was struck whereby the Forest Service would review and recommend action for primitive areas, but the positive approval of both houses of Congress would be required for additions to the system. Part of this debate was purely geographic. Wallace Bennett (Utah) wanted to ensure that both houses would be involved in the reservation process to blunt the voting power of more populous eastern states. His fear was that "we in the West could be effectively preserved for the East."[51]

Second, water development interests were still concerned about reserving potential dam and reservoir sites. The water lobby continued to argue that reservoirs were compatible with wilderness preservation, but Anderson would not agree and aggressively engaged Federal Power Commission witnesses who tried to prove the point. The eventual compromise was that water development projects could occur in wilderness areas if the president certified that such work was in the national interest.

Third, original formulations of the bill prohibited mining in wilderness areas, but Alan Bible (Nevada) and Henry Dwarshak (Idaho) proposed an amendment that would have extended general mining laws to all Forest Service lands (including wilderness areas). Eventually Frank Church broke the impass with an amendment that mining surveys by means "not incompatible with the preservation of the wilderness environment" would be permitted. Church's actions were difficult for him because he took positions opposite those of the mining and timber industries in his home state of Idaho.[52] In September of 1961, the Senate approved the wilderness bill by a vote of 78 to 8, testimony to the power of the preservation lobby and the legislative skill of Anderson and Church.

In the Senate, wilderness proponents enjoyed the support of a powerful friend in Anderson as chair of the Senate Interior Committee, but in the House of Representatives an opponent occupied the position of authority. Wayne Aspinall (Colorado) vowed to take no action on the issue until the Senate completed its work, even though companion bills had consistently been introduced in the House by John Saylor. With the Senate bill in hand, he indicated a further delay in order to await the publication of the Outdoor Recreation Resources Review Commission reports.

The final general report of the ORRRC appeared on January 31, 1962, offering few surprises and little substance, though its call for a separate federal agency to manage recreation resulted in the creation of the Bureau of Outdoor Recreation in the Department of Interior.[53] One of the 26 other volumes in the ORRRC report series focused on wilderness, but it was three months late. When it finally appeared, the wilderness report lent little support to Aspinall's position, a predictable outcome given that the report was prepared under the direction of James Gilligan whose research provided the impetus for the initial wilderness bill.[54] Gilligan's Wildland Research Center at the University of California, Berkeley, was hardly a boost for proponents, however, because the work had been quickly done, provided little data on wildlife or aesthetic values, and used an outdated definition of wilderness for its surveys. Clinton Anderson blasted the report as being the product of a group who had little information and no understanding of the necessary planning process.[55]

With a Senate bill and the ORRRC reports available, Aspinall was true to his word and held hearings on the companion House bill. The hearings, held before the Subcommittee on Public Lands and chaired by Gracie Pfost (Idaho) resulted in statements that duplicated Senate testimony, except that Secretary of Interior Stuart Udall and Secretary of Agriculture Orville Freeman reflected their politically liberal administrations by asking for an even stronger bill.[56] Instead of reporting the original bill to the House for consideration, the committee offered an Aspinall substitute that completely changed the original. In the substitute, primitive areas were not considered, mining would be permitted in wilderness areas for 10 years, and review with potential reclassification of wilderness areas would occur every 25 years; further amendments extended mining laws to wilderness areas until 1987 and permitted the development of a ski area in the San Gorgonio Wilderness area of southern California.[57] The entire legislative effort in the House ended in a parliamentary dispute among Aspinall, Saylor, and House Speaker John W. McCormack (Massachusetts).[58]

Undeterred, Anderson reintroduced the bill that had previously passed the Senate in the following session. In hearings, the American Mining Congress representative complained that "66 million acres having high potential for mineral deposits" would be locked up. Recognizing the statement was completely false, Anderson wrote to the miners charging dishonesty on the part of the witness, saying that he "often wondered if witnesses really regard members of the committees so stupid that they are influenced by statements like that one."[59] A significant adjustment from previous bills was the inclusion that review processes for wilderness designations would include public input. Hearings, reports, and votes went quickly, and with support from all sections of the country the bill again passed the Senate in April, 1963, on a vote of 73 to 12. Prominent opponents included the Colorado delegation.[60]

In the House, Saylor again introduced a companion bill, and at first there was little prospect of passage. Wayne S. Baring, chair of the Public Lands Subcommittee, promised the American Mining Congress that no bill would be approved without major concessions from the proponents.[61] Aspinall concluded that a major issue with the bill was that it conflicted with the legal management guidelines for the Forest Service, and therefore a major review commission for land laws was required.[62] In a reflection of the delaying strategies surrounding the ORRRC reports, the American Cattlemen's Association gathered some support for its contention that the wilderness bill should not go forward until the report of such a review commission.[63]

Mid-1963 was a time when forces were at work to ensure movement on the bill despite these objections. The Kennedy administration was applying pressure to key congressmen to secure passage of the bill in order to bolster the administration's record in the environmental area. Aspinall was in somewhat of a compromise mood because he was losing support on the issue: in his most recent election his portion of the popular vote slipped to 58 percent from 71 percent just two years earlier.[64] The lumber industry worried that further delay might result in wholesale reservations of timber lands for wilderness by the executive branch. Finally, Anderson was ill, and unsure whether he would be able to run for reelection.

The final key to passage was a simple political trade between Aspinall and Anderson. Aspinall permitted the wilderness bill to move forward, and Anderson promised to support Aspinall's drive for a Public Land Law Review Commission.[65] The review commission was important to Aspinall because it offered a possible avenue to weakening any wilderness bill that might succeed, and it allowed him to serve conservative constituents who wanted federal land management agencies restricted. He also saw the chairmanship of the commission as his logical postretirement position. Floor debate in the House focused on the ski area proposed for the San Gorgonio Wilderness. Saylor successfully eliminated the development with an amendment.[66] The House bill passed by a vote of 373 to 1 on July 30, 1964, with Joe R. Pool, a Texas Democrat, casting the only nay vote.[67]

Differences between House and Senate versions were resolved by a Conference Committee chaired by Anderson. The final adjustments in the bill, mostly induced by the more conservative House version, included the requirement for affirmative action by both houses for additions to the system.[68] Primitive areas were to be managed without development until they were reviewed within 10 years, and the wilderness system was begun with only Forest Service "wilderness," "wild," or "canoe" areas amounting to 9.1 million acres. The lower size limit for wilderness areas of 5,000 acres was replaced by the term "sufficient size." The issue of mining in wilderness areas was settled by a final compromise between Anderson, who agreed to allow prospecting to continue, and Aspinall, who agreed to terminate prospecting in 1984.

When President Lyndon Johnson signed the Wilderness Act on September 3, 1964, the stage was set for the fourth sagebrush rebellion. With large areas set aside for preservation from commercial development, many western residents began to detect a heavy hand extending from Washington. Immediately after the passage of the law, there was little sign of a revolt, but the wilderness system was not static. Preservationists immediately began pressing for additions to the system, which ignited local controversies, while additional national legislation added to the restrictions on use of the public lands. Each controversy and each new restriction brought closer the inevitable reaction from the West.

Chapter 15

Winning the Wilderness Game

Opponents of the Wilderness Act, including the National Park Service and the lobbying groups supporting the interests of water development, logging, mining, and grazing soon became entangled in vigorous application of the precepts of the act. Each of these groups saw significant erosion of its position as a direct result of the Wilderness Act, and the period between 1964 and 1976 became a highway to inevitable conflict resulting in the fourth rebellion.

The National Park Service was slow to respond to the spirit and letter of the new law. Park Service Director George B. Hartzog recognized the traditional constituency of the service as the motoring public, and he foresaw no real changes resulting from the Wilderness Act.[1] Wilderness designations in parks would restrict management options, prevent facilities development, limit potential visitorship, and perhaps curtail budget increases. The service created its own exclosure zones where the Wilderness Act would not apply, including lake surfaces for motorboats in Crater Lake and Yellowstone National Parks, and 22 nine-acre exclosures in Kings Canyon and Seqüoia for rain gauges.[2] Because operating rules for the service would not have prohibited the gauges in wilderness areas, the exclosures appeared merely to be a means of preserving development options.

In 1966, the new realities of environmental management became abundantly clear in a controversy surrounding the wilderness portion of Great Smoky Mountains National Park in Tennessee and North Carolina. The huge park was bisected by a major highway, leaving two large potential wilderness areas on either side. The Park Service proposed the development of an extensive road network that would have left only six small wilderness areas.[3] Preservation organizations flexed their newly developed muscles to oppose the plan, and combined efforts with the Wilderness Club, Sierra Club, National Audubon Society, National Parks Association, Izaak Walton League, and the Appalachian Trail Conference. Effectively under attack, the Park Service retreated, and eventually developed a new plan that eliminated the offensive highway in favor of a park perimeter road.

While the Park Service was slowly learning the political power of the new prowilderness lobby, the water development interests were learning the same lesson. Water interests were successful in protecting their right to construct dams and reservoirs in wilderness areas, at least in theory. The Wilderness Act specified that these developments could take place upon the determination by the president that they would be in the national interest. Although such a decision in the political climate of the late twentieth century seemed unlikely, preservationists sought protection for rivers in addition to that offered by the Wilderness Act. Preservation influence in Congress was demonstrated with the passage of the Wild and Scenic Rivers Act of 1968.[4] The act prohibited the impoundment of rivers designated by Congress, identified eight streams for inclusion in the system immediately, and set aside 27 additional streams (18 administered by the Department of Interior and 9 by the Department of Agriculture) for further study.[5] The act also provided for future expansion of the National Wild and Scenic Rivers System by a process similar to the one used for wilderness.

The passage of the rivers act showed that Congress was willing to restrict water development in order to preserve wilderness characteristics. There were consistent annual additions to the original 773-mile system, which by 1976 had grown to 1,610 miles (Figure 16).[6] Clearly, the preservation lobby had the upper hand regarding wilderness rivers.

Preservationists also enjoyed considerable success in forcing the Forest Service to adhere to strict interpretations of the Wilderness Act. Immediately after its passage, it appeared that the act would have little effect on management of the national forests, as the Forest Service sought to preserve the status quo much in the same way as the Park Service.[7] The foresters held extreme positions on what constituted wilderness, and would not accept even slight deviations from completely pristine conditions in wilderness areas. The service resisted pressure from geologists in the Geological Survey who wanted to use motorized access for mineral exploration in wilderness areas.[8] When two hikers died in a wilderness area, the regional forester refused to allow their bodies to be removed by helicopter, arguing that

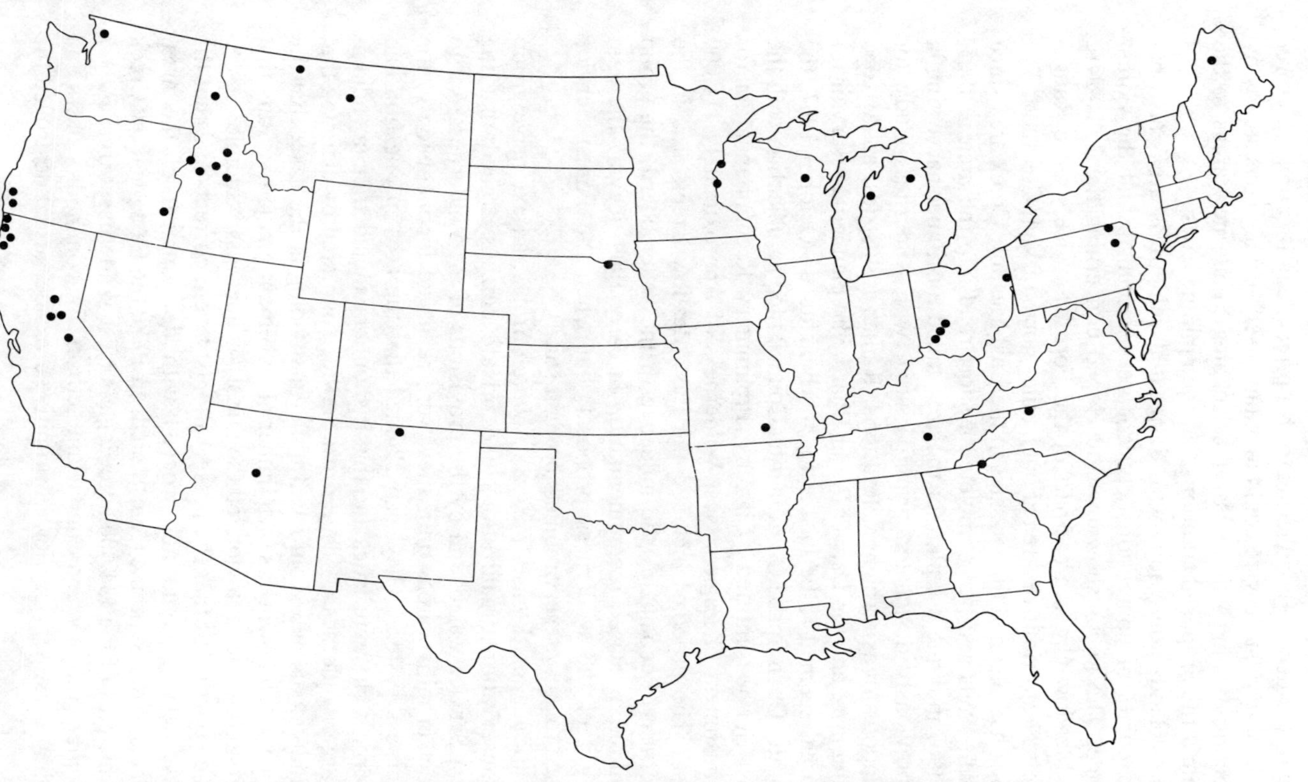

Figure 16. The distribution of units in the Wild and Scenic Rivers System in 1984 at the end of the fourth sagebrush rebellion (National Geographic Society data).

wilderness designation did not allow motorized transport for either the living or the dead.[9] By adhering to this principle in its purest interpretation, the Forest Service could maintain a self-serving consistency by protecting existing wilderness and declining to add new areas to the system if the proposed additions had even minute flaws.

Forest lands contiguous with designated wilderness areas also became points of contention between preservationists and the Forest Service. The major showdown developed concerning the 2,400-acre East Meadow Creek area, part of the Forest Service's Gore Range-Eagle's Nest Primitive Area about 8 miles north of Vail, Colorado. The service had planned timber sales in the East Meadow Creek area since the late 1950s and had completed a short road into the area in 1965. The 1967 multiple-use plan for the area called for timber harvesting in the previously designated primitive area, with wilderness study for contiguous areas at higher elevations. Kaibab Industries, an Arizona logging firm, and the Denver Water Board who planned a reservoir for the area, seemed to be in line for the benefits. In October 1967, the Colorado Open Space Coordinating Council filed an objection with the Forest Service entitled "Citizens' Preliminary Recommendation for Establishment of Eagle's Nest Wilderness of Approximately 111,000 acres." The proposal included the East Meadow Creek area in the wilderness. Recognizing the influence of the water board, the group submitted a revised recommendation in early 1969, leaving the dam site out of the wilderness proposal, but when the Forest Service refused to act, preservationists brought suit.[10]

In the landmark case, *Parker v. United States,* an extensive group of conservation organizations and private citizens argued that the Forest Service was bound by the provisions of the Wilderness Act to protect primitive areas until they were reviewed for inclusion in the wilderness system. They also claimed that the area in question had wilderness characteristics and that it should be included in any adjoining larger wilderness.[11] The Forest Service countered by claiming that the individuals and groups had no legal standing in the case, that the service was following its own timber management regulations, and that the road into the area violated the purist's wilderness principle.

The court ruled that the preservationists were correct, and that the Forest Service could not proceed with development of the area. Appeals to the Supreme Court failed. In 1976, the East Meadow Creek became part of the 134,000-acre Eagle's Nest Wilderness Area. *Parker v. United States* was an important setback for commodity users and for the Forest Service view of primitive areas because it destroyed the purity principle by stating that the road was inconsequential to the overall wild character of the area. The case reaffirmed that primitive areas would have to be reviewed before they could be used for other purposes.[12] It was the first of several legal and administrative victories for preservationists opposed to western development interests.

Logging in or near wilderness portions of the Boundary Waters Canoe Area of northern Minnesota was the second major point of contention between the Forest Service and preservationists. Boundary Waters was a special case with a wilderness zone and a "portal" zone, and its managment was the subject of special sections of the Wilderness Act that permitted logging of some virgin timber at the margins of the wilderness zone. In *Minnesota Public Interest Research Group v. Butz,* preservationists argued that the virgin timber was protected, but after hearings and decisions in several courts, the Forest Service was allowed to continue writing contracts for harvesting.[13] In a demonstration of the increasing interaction between legislature and the judiciary in wilderness matters, however, Congress passed the Boundary Waters Canoe Area Act of 1978. The new act repealed portions of the Wilderness Act relating directly to the canoe area, and overturned the court decisions by explicitly stating that neither harvesting of virgin timber nor prospecting on federal lands was to occur in the canoe area.[14] it was another major victory over logging and mining interests for the preservationists.

By the early 1970s, the Wilderness Act had produced chaos in the Forest Service. Preservationists were successfully attacking the service on legal grounds, Congress was putting more areas into wilderness than the service thought prudent, primitive areas were frozen until reviews could be completed, areas next to potential wilderness areas were likewise frozen, and private citizens were collecting more information on potential wilderness areas than could the beleaguered Forest Service staff. The purity principle was a defensive mechanism,[15] but it was not enough. To forestall criticism by Congress and the executive branch, the service established its own ten-month Roadless Area Review and Evaluation (RARE) program in June 1971.

The RARE program suffered intense pressure from the timber lobby, was in existence for too short a time to adequately consider many areas, and made some indefensible decisions.[16] For example, of the 950,000 roadless acres in Bridger National Forest, Wyoming, the service recommended only 13,500 (1.5 percent) for further wilderness study. After some delay, the RARE results, announced in January 1973, recommended 11 million of the 56 million roadless acres in the forest system for further wilderness consideration.[17] The remaining 45 million acres would be immediately released for multiple-use management. An environmental impact statement accompanied the report. After 8,000 letters of complaint and opposition were received by Congress, the service revised its recommendations upward to 12.3 million acres.

The Sierra Club brought suit, charging that the Forest Service should be required to protect all 56 million acres until a decision on their wilderness potential had been made by the president and Congress.[18] Federal Judge Samuel Conti granted the Sierra Club an injunction to stop Forest Service planning processes, and in an out-of-court settlement the Forest Service agreed to generate an environmental impact statement for each individual area before it was released to multiple-use management.[19] The significance of the Conti decision in *Sierra Club v. Butz* was that no roadless areas were released for non-wilderness purposes, thus extending the *Parker v. United States* decision far beyond East Meadow Creek. The effect was to completely negate the entire RARE program. By the mid-1970s, commodity interests and the Forest Service were hamstrung with all the roadless areas in limbo.

Miners, as well as the timber interests, encountered special problems growing out of the Wilderness Act. Politics in Washington State and a legal case in Minnesota defined the true dimensions on mining in wilderness areas. In the early 1900s, prospecting on Miner's Ridge in the Glacier Peak area of the Washington Cascade Mountains revealed significant copper deposits. By the 1950s, there were 17 patented claims in the area, though its remote location prevented development. In 1960, the secretary of agriculture designated the entire area as wilderness, and it was included in the National Wilderness System as an original member by the 1964 Wilderness Act. In 1966, reassured by the language of the Wilderness Act allowing development of valid claims in wilderness areas, Kennecott Copper announced plans to develop an open pit mine on its claims serviced by a 15-mile road through the wilderness area.[20]

A widespread public effort to prevent mining of the copper (with an unmined, in-place value of $351 million) included objections from Agriculture Secretary Freeman, Interior Secretary Udall, the entire Washington State congressional delegation, and aggressive preservation groups.[21] Some opponents, recognizing the provision of the Wilderness Act that permitted mining on valid claims, advocated outright purchase of the claims by the federal government,[22] but the $50 million price tag was prohibitive. Despite a massive public education effort by Kennecott, the mining industry was unable to convince key decision makers to support its position, and the Forest Service imposed restrictions on the road-building and mining operations that forced the development costs beyond the value of the mine. It had become apparent that though the industry had won protective language in the Wilderness Act, political reality made it a hollow victory.

The case of *Izaak Walton League v. St. Clair* translated those political realities into legal restrictions. St. Clair, a mining company, sought to develop its valid claims to a copper deposit in the central portion of the Boundary Waters Canoe Area. The plaintiffs (the league and the state of Minnesota) argued that the defendants (the Forest Service and the mining company) were putting the values of mining ahead of the values of wilderness preservation.[23] The case illuminated a basic contradiction in the Wilderness Act that specified the legality of mineral development, but also specified that such development must be consistent with the wilderness environment, an obvious impossibility. The court ruled that wilderness objectives outweighed the mining objectives,[24] a decision that virtually sealed wilderness areas from mining activities even on valid claims.

Graziers also had increasing difficulty with the federal land managers. The National Environmental Policy Act of 1969, signed by President Nixon in January 1970, had far-reaching implications for ranchers that Congress had not foreseen. The act required the preparation of an environmental impact statement for all major actions by the federal government likely to affect the environment; alternative actions to the one proposed, including no action at all, were required for the statements.[25] It was this act that triggered the settlement of the Conti decision in the *Sierra Club v. Butz* case, wherein the Forest Service agreed to generate such impact statements for each of the cases in the RARE program. The National Environmental Policy Act affected graziers in a similar fashion.

In the early 1970s in eleven western states, the Bureau of Land Management administered 171 million acres, with grazing on 150 million acres, that accounted for 14 percent of all livestock grazing in the nation. About 24,000 grazing leaseholders operated in 52 separate grazing districts that were further subdivided into more than 8,000 allotments.[26] The Bureau proposed to generate a single, all-encompassing impact statement for the entire program. Fearful that regional variation in environmental conditions and that potential wilderness areas would be unaccounted for in so general a document, preservationists pressed suit to force the bureau to generate more detailed environmental impact statements on smaller portions of the entire system. In the resulting case, *Natural Resources Defense Council v. Morton,* the bureau argued that its grazing management program was not a major federal action affecting the environment, and that the Taylor Grazing Act provided for sufficient environmental protection.[27] In a written decision that thinly disguised his scorn for the federal position, the judge in the case ruled that the bureau must prepare impact statements on smaller geographic areas of its system, with the total coming to 144 statements instead of one.

The implications of the decision in *Natural Resources Defense Council v. Morton* were staggering for the grazing lobby.[28] For more than two decades, graziers had enjoyed a sweetheart arrangement with the bureau without undue publicity or significant challenge from other groups. Now consideration of uses other than grazing, such as recreation and perhaps wilderness preservation, were required by law. Still worse was the requirement that impact statements be revealed for public comment, offering points of entry into the process for the now well-funded and well-organized environmental groups.

To make matters even more difficult from the standpoint of the rancher, the bureau that had once been a lackey took on new authority and a mandate for an exhaustive study that would result in designations of new wilderness areas. The Federal Land Policy and Management Act of 1976 was intended to streamline the federal land system, which operated under more than 3,000 outdated regulations. The law provided the Bureau of Land Management with an organic act, similar in philosophy to the organic acts for the Forest and Park services.[29] The act embodied a liberalized perspective on the bureau with far-reaching implications. The act dropped the implication that federal ownership of the public domain was temporary; it mandated the bureau to adopt a long-run, sustained-yield, multiple-use philosophy; and it required advisory boards for the bureau to include representatives of all users instead of only graziers.[30]

The Public Land Law Review Commission had recommended that Congress extend the provisions of the Wilderness Act to Bureau of Land Management areas, and authors of the Federal Land Policy and Management Act obliged. With the support of the Wilderness Society, Sierra Club, Wildlife Federation, the secretary of interior, and the director of the bureau, sponsors of the bill emphasized the need for a wilderness study process for bureau lands similar to the Forest Service process.[31] The act (referred to as "Flip-ma" in reference to its acronym, FLPMA) required the bureau to inventory roadless areas greater than 5,000 acres in extent and make recommendations for additions to the National Wilderness System within 15 years. The general processes and definitions outlined in the Wilderness Act for national forest lands were to apply to the BLM lands. Suddenly, the graziers were thrust directly into the wilderness issue because the lands that were most important to them were now directly involved.

A final blow to the graziers that contributed to the fourth sagebrush rebellion was a drastic increase in grazing fees on Forest Service and Bureau of Land Management ranges (Figure 17). Between 1936 and 1970, the fee per animal unit month on BLM land increased from $0.05 to $0.32, but between 1970 and 1976 it increased to $1.50[32] The gap between Forest Service fees and BLM fees had once been $0.55, but by 1976 grazing fees in the two agencies were nearly equal. The Federal Land Policy and Management Act provided a new formula for determining the fees, but it offered only further (if not so rapid) increases.

Wilderness preservation also played a key role in the ignition of the fourth rebellion. The hurried evaluation of roadless areas by the Forest Service in the RARE program brought complaints from the preservation lobby and negative comments from congressional supporters of the service. When President Carter took office in 1977, he called for an expansion of the wilderness system. He endorsed all the proposals by the previous administration and requested additional proposals (Figure 18).[33] He installed administrators in the Departments of Interior and Agriculture who were supportive of wilderness preservation and who were not directly connected with the western commodity developers.

Fearful of losing valuable budgetary support, the Forest Service conducted an entirely new RARE program, known as RARE II. Under the watchful eyes of a new assistant secretary, M. Rupert Cutler, the service reported that weaknesses had been found in the original RARE program: an incomplete inventory, lack of public participation, and methodological errors.[34] The final RARE II report, published in early 1979, found that the total roadless area was 67 million acres (11 million more than the previous report), and recommended 15.4 million acres for wilderness study.[35] The release of lands not immediately included in the wilderness study process was highlighted as an issue that needed resolution. Commodity developers saw the wilderness issue as completely out of control.[36] The congruence of forces related to grazing fees and allotments, the Federal Land Policy and Management Act, and a ballooning wilderness system produced the ultimate reaction in the West: one more rebellion.

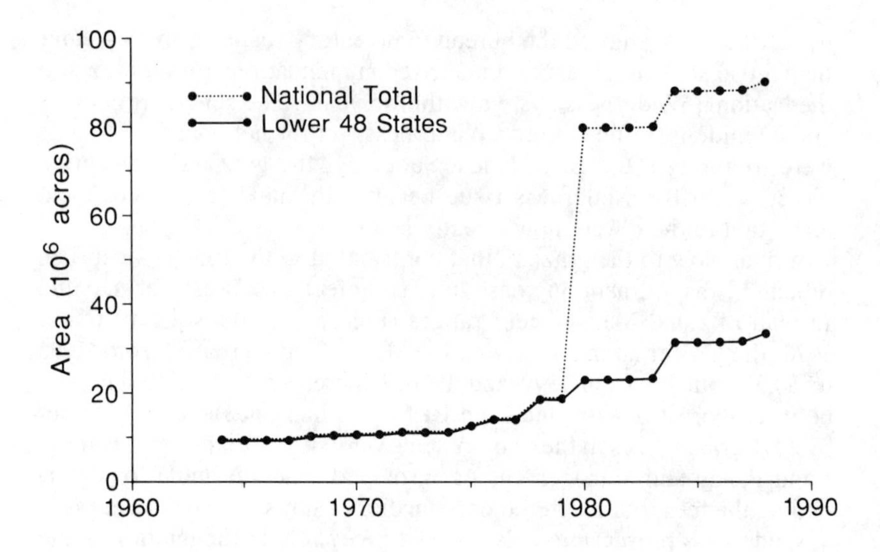

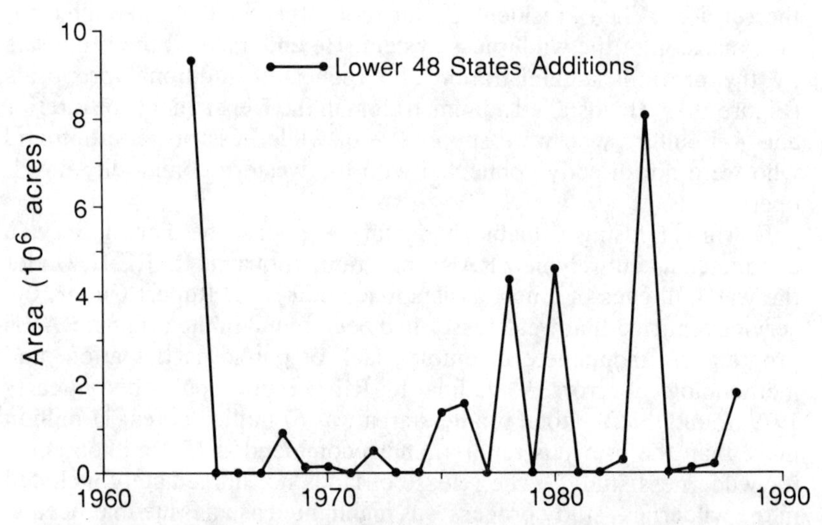

Figure 17. Grazing fees per animal unit month on public lands during the grazing controversy and the early part of the fourth sagebrush rebellion. Note the rapid increases in fees during the mid-1970s that raised concerns by graziers about wilderness reservations (Forest Service and Bureau of Land Management data).

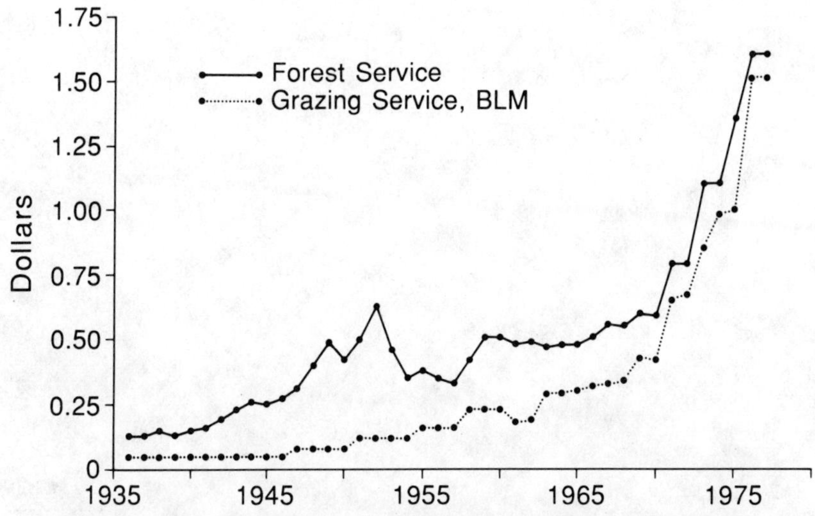

Figure 18. Total land area (above) and annual additions (below) in the National Wilderness Preservation System, showing the relative area in the lower 48 states along with the national total that includes Alaska. The additions to the system dramatically increased in size in the mid-1970s at a time when graziers experienced rapidly increasing fees as shown in Figure 17 (Wilderness Society data).

Chapter 16

Sagebrush Rebellion, Inc.

As during previous incidents, Nevada was in a position of leadership in the fourth sagebrush rebellion. Throughout the early 1970s, the Nevada Legislature's Select Committee on Public Lands explored the possibilities of transferring federal lands to state ownership. This activity was the result of complaints by ranchers with grazing permits for Bureau of Land Management allotments who claimed that their financial security was threatened by the changing administrative climate. Through the early and middle 1970s, federal actions consistently reduced management options for the graziers, and documents created later by rebellion leaders specified the particular issues that led to the push for the transfer of lands.

In addition to documenting the increasing size of the BLM bureaucracy, rebellion leaders also showed that the Federal Land Policy and Management Act (FLPMA), problems with the Forest Service roadless area and wilderness review processes, and the impending wilderness review in the bureau, were the most immediate concerns.[1] The wilderness-related processes appeared to the graziers to be direct threats to their access and use of lands over which they had heretofore exercised unchallenged control. They now found the new challenge, cloaked as wilderness preservation, to be unacceptable.

In summer of 1978, an informal meeting between Barry Goldwater, the conservative Arizona senator, and John L. Harmer led to the development of a new organization to channel the energies of the fourth rebellion.[2] Harmer, who had been lieutenant governor of California during the Reagan years, was an accomplished attorney with extraordinary organizational skills. The new organization was dubbed the League for the Advancement of States Equal Rights, or LASER, and under Harmer's aggressive direction it rapidly gained strength and influence. A second organization was formed in Idaho under the direction of Vernon Ravenscroft and took the name Sagebrush Rebellion, Inc.

A third rebellion organization, the Public Lands Council, had been in existence for several years by the late 1970s. The council, consisting of graziers who were BLM permit holders, used the organization as a means to express their increasing dissatisfaction with the policies of the federal agency. Unlike LASER, which was a publicity machine, the Public Lands Council focused on influencing the Nevada legislature. Council President Dean Rhoads was a member of the legislature, and in 1979 he successfully brought to fruition several years of legislative committee work by generating the first modern sagebrush rebellion bill.

Rhoads's Assembly Bill 413 recited the history of public lands in Nevada and targeted Bureau of Land Management lands for transfer to state administration. The bill created a state lands board to oversee the areas that would be received from the federal government, and asserted state control over the surface access and mineral rights of the lands. The bill proposed a multiple-use perspective that ensured the preservation of wilderness areas (though no such areas had been declared on the lands in question), and protected established lease agreements between the federal government, and individuals or companies. Provisions were made for the sale of land to individuals. Finally, the bill empowered the state attorney general to pursue legal action related to the proposed transfer.[3]

Rhoads and his Public Lands Council associates had prepared the way for the bill in the legislature. Rhoads, who bore a startling resemblance to the "Marlborough man," gave a booming floor speech in support of the effort, and his fellow legislators could hardly resist tweaking the nose of far-off federal administrators. The Nevada Assembly approved the bill with only a single dissenting vote, and the Nevada Senate approved by a margin of 20-3.[4] John Rice, a newspaper reporter assigned to cover the usually placid legislature, captured the sound and fury for his readers, and it was he who coined the term "sagebrush rebellion."[5]

The underlying premise of the action by Nevada's Assembly was not to take immediate control of the lands in question, for no money was appropriated for their administration. The true intention was revealed by the appropriation of $250,000 for the initiation of legal proceedings to challenge the federal jurisdiction in court.[6]

Much to the delight of their frustrated rural constituents, local, state, and national political leaders from the West saw a popular cause and joined the rebellion. Senator Orrin Hatch of Utah found the issue particularly appealing, and he took the lead in introducing the primary "sagebrush rebellion bill" to Congress in summer 1979. His cosponsor list showed the dimensions and distribution of the rebellion, which covered the interior West with support from Senators Goldwater and DeConcini (Arizona), Cannon and Laxalt (Nevada), Garn (Utah), and Simpson and Wallop (Wyoming). Senators Helms (North Carolina), Stevens (Alaska), and Jepsen (Iowa) represented archconservatives who supported the bill on more general principles.[7]

Hatch's bill bore a remarkable similarity to the Robinson bill introduced in 1946 at the height of the third rebellion. One commentator irreverently observed that it appeared as though one of Hatch's staff members had merely dusted off the text of the 23-year-old proposal and updated it with a few "environmental" words.[8] Some parts of Hatch's bill were nearly identical to wording in the Nevada Assembly bill (especially parts relating to state management and possible sale). The proposal was a blueprint for the transfer of BLM lands to the respective states, and included the formation of a Federal Land Transfer Board. Congressman James Santini of Nevada introduced a companion measure in the House of Representatives and garnered thirty cosponsors.[9]

Throughout 1979, the rush by western congressional representatives to play to the constituents continued unabated. Western senators on the appropriations committee used their positions to play to the voters in field hearings concerning the "new" sagebrush rebellion. Harrison Schmitt (New Mexico) hosted hearings in Albuquerque, while his Senate colleague Peter Domenici held forth in Farmington and Las Cruces. Elsewhere, Jake Garn plowed fertile electoral ground in Utah, Paul Laxalt did the same in Nevada, as did James McClure in Idaho. Ranching, mining, and local government groups were invited to testify; environmental groups were not.[10]

Santini also used the issue to conduct hearings in western states through his position on the House Subcommittee on Mines and Mining. There was a predictably strong emphasis on the testimony from rebellion spokesmen. A typical example was Arizona State Representative Joe Lane, a rancher from Wilcox. Speaking on behalf of the National Cattlemen's Association, National Wool Growers Association, and Dean Rhoads's Public Lands Council, Lane pinpointed a common theme of proponents of the land transfer. In putting forth the case for state control he emphasized the need for multiple-use management: "What is multiple-use about wilderness? You can't manage wilderness, that is contrary to the concept of wilderness. You can't extract minerals, cut timber or graze livestock economically in wilderness areas."[11] Lane apparently was unaware of or chose to ignore the Multiple Use-Sustained Yield Act of 1960 that specifically stated wilderness preservation was compatible with multiple use. The same point was the focus of extensive congressional discussions prior to passage of the Wilderness Act.

The central issues in the fourth sagebrush rebellion regarding the proposed transfer of the public domain from federal to state jurisdiction were mostly modern reincarnations of positions from previous rebellions. One new feature was a heavy reliance on the legality of federal management brought to the forefront of the argument because the newly enacted Federal Land Policy and Management Act stipulated that the national government would administer the lands in perpetuity, specifically that "the public lands be retained in Federal ownership."[12] This statement was not the same as the one contained in the Taylor Grazing Act, which foresaw management by the federal government pending the ultimate disposal of the lands.

Sagebrush rebels argued that the federal government must "return" the lands to the various states, because when states were admitted to the union all of the area within their borders should have been state-owned, as in the thirteen original colonies. Rebels pointed to a long history of court decisions that specified equal treatment for all states,[13] and claimed that western states were not treated equally because they were left with large areas of federal jurisdiction. This legal argument had as a flaw the fact that the federal government had obtained the western lands through purchase or by international conflict. Western states carved from these federal lands had not been previously owned by the states.[14] Additionally, when states were admitted to the Union, their constitutions or the acts making them states contained the specific provision that "all right and title to the unappropriated public lands lying within said territory, shall be and remain at the sole and entire disposition of the United States."[15]

Because of this debate about the relative roles of the state and federal governments, the rebellion was often cast as a "states' rights" issue. Although rebels often cast their position in the light of this lofty concept, their public comments and published literature showed that their true objection to the status quo was the management of the federal lands by administrators far removed from the local scene.[16]

Western ranchers in particular had lost considerable influence in the Bureau of Land Management,[17] but often the gain was made by western urbanites who represented recreation or preservation interests.

Opponents of the rebellion emphasized the need to protect public access to the public domain, focusing on the question of the ability of states to manage the lands. The track record of western states in administering their own public lands (as opposed to the federal public lands) was not impressive despite comments of the rebels to the contrary. Experience with the state lands in the West had been variable: Nevada retained only 1 percent of its area in state owenership, having sold almost all of its indemnity lands (those granted to the state when it was admitted to the Union), while 13.2 percent of Arizona was administered by the state.[18] Administration of the state lands in every case received less financial and staff support than federal lands on a per-acre basis. In Colorado, for example, only four field appraisers oversaw the administration of more than three million acres, so that each staff member was responsible for more than 1,150 square miles. The evidence concerning state management became a millstone for the rebels.

In a larger perspective, the opponents of the rebellion made much of the fact that the federal public lands belonged to everyone in the nation rather than simply to those who happened to live next door to the Bureau of Land Management or the Forest Service. Conservation and preservation organizations drew on national clienteles who had little understanding of the view that fourth-generation ranchers had a vested right to use public land. As with all the rebellions, this national versus local perspective was a strong undercurrent, even if not always stated.

Federal administrators and preservationists did not exactly flee before the advancing rebel hordes. The East seemed to ignore the issue, and preservation groups tried a low-key response in hopes of not fanning the publicity fires. For a time, there was a rebellion, but only one side showed up for the fight. The slumber outside the West was disrupted in 1980 by three events related to the rebellion: passage of rebellion bills in several states, a widely publicized rebellion conference, and the election of conservative Ronald Reagan to the presidency.

Although the associated processes required more than a year in some states, 1980 was the highwater period for state legislation for the rebellion. Wyoming,[19] Utah,[20] and New Mexico[21] passed bills similar to the Nevada model. Wyoming upped the ante by including a claim to the National Forests as well as to BLM lands. Arizona placed itself firmly in the mainstream of the rebellion by passing a bill overriding a governor's veto,[22] and sustaining the entire effort in a popular referendum. The Washington state legislature passed a rebellion bill,[23] but it was voided by a 60 percent "no" vote in a referendum.[24] In Montana, Idaho, and Oregon, sagebrush rebellion bills were defeated, while vetoes by Governor Jerry Brown in California, and Richard Lamm in Colorado, stopped bills in their states. This flurry of legislative activity brought the rebellion to the attention of the nation and indicated that support for the rebellion was widespread.

The legislative votes on the sagebrush rebellion bills indicated the nature of the political support for the movement. Support was strongest in states with strong grazing lobbies and much BLM land; support was weakest in those states with mostly Forest Service land. In the eight major rebellion states, 86 percent of the Republican legislators supported the rebellion while only 49 percent of the Democrats did so. The rebellion bills were supported by 92 percent of the rural Republicans, but by only 30 percent of the urban Democrats. One reason for the strong rebellion showing was the disproportionate representation of ranchers in legislatures: in several states ranchers held one-third of the legislative positions.[25]

With the legislative initiative rolling merrily along, visibility of the rebellion was further heightened by the second conference of the League for the Advancement of States Equal Rights. The three-day meeting in Salt Lake City attracted 600 participants for John Harmer's well-organized and updated version of Barrett's Wild West Show.[26] Presentations by speakers representing the commodity users, panel meetings hosted by leading rebels (Dean Rhoads and Calvin Black among them), statements by congressmen and senators, and a formal hearing by the House Subcommittee on Mines and Mining (chaired by James Santini) provided a news-media feast. Oratory was hot and heavy, and Senator Orrin Hatch even placed the rebellion in the mainstream of Mormon history as a logical outgrowth of God's command to subdue the earth.[27] At the very least, the conference subdued the American press, and the rebellion enjoyed its finest moments.

The LASER Salt Lake City meeting in late 1980 marked a change in strategy for the rebellion by realigning its course from a legal one to a political one. It had become apparent that a court-based challenge of federal ownership would not be successful. In previous rebellions the legal route had been abandoned because western spokesmen admitted that their claims to the public domain rested not in law, but in a sense of equity. During the Hoover effort to dispose of the lands, Governor

Dern of Utah testified before Congress that the western claims were not "legally true."[28] What was untrue in the 1930s was untrue in the 1980s. The Supreme Court, in the case *Kleppe v. New Mexico,* had recently reaffirmed that there was virtually no limit to federal authority on federal public lands,[29] and the rebels saw little hope of success in the courts. Even in Nevada, Governor Robert List allowed the legal initiative to wither.[30]

A major reason for the optimism regarding the political route was the election of President Reagan shortly before the LASER conference. Reagan wired his support to the meeting, sending his "best wishes to all my fellow Sagebrush Rebels. I renew my pledge to work toward a Sagebrush solution. My administration will work to ensure that the states have an equitable share of public lands and resources."[31] Rebellion leaders were hopeful that the new administration would stop the flood of new wilderness designations and provide support for the effort to transfer lands to the states.

The conservative new administration significantly slowed the rate of additions to the National Wilderness System. In the first Reagan term, 1980-1984, the system grew by only about 8.5 million acres,[32] and the administration left to individual state congressional delegations the leadership role in the process. At the LASER conference, the fervent hope was often expressed that the new administration would appoint a sympathetic secretary of interior to replace preservation-oriented Cecil Andres. Wishes seemed to be fulfilled when, in early 1981, the president appointed James Watt to the post.

Throughout 1981, while sagebrush rebellion legislation stalled in Congress, Watt carried the rebellion into the Department of Interior and sparred with aroused environmental activists. He immediately put a moratorium on purchases of new lands for parks, arguing that the federal government was having trouble administering the lands that it already had.[33] Les Line, editor of the Audubon Society's magazine, charged that Watt had developed a "hit list" of wildlife refuges that were to be turned over to states under pressure from grazing and mining interests. Line indicated that the list included two of the largest wilderness refuges in the lower forty-eight states, Kofa and Cabeza Prieta in Arizona, with others in the major sagebrush rebellion states.[34] Watt continuously kept under consideration proposals to explore for oil and gas in wilderness areas, especially in and near forest wilderness areas in northern Montana. Responding to constituents' concerns, the House of Representatives voted 350 to 58 to stop the mineral explorations in wilderness areas, and the Department of Interior backed down slightly. Watt was an accomplished baiter of environmental spokesmen and often precipitated heated debates. He also stimulated substantial increases in the memberships of the Sierra Club and Wilderness Society as citizens who never considered the issues seriously began to worry about the ability of the federal government to protect public lands from its own officials.

In 1982, Secretary Watt's department seriously investigated a number of avenues for disposing of federal lands. An alarmed eastern commentator stated that a fire-sale mentality prevailed in Washington.[35] Senator Charles Percy (Illinois) and Congressman Larry Winn (Kansas) introduced bills to inventory and sell public land as a debt reduction measure with the strong support of President Reagan and Budget Director David Stockman. Sagebrush Senator Paul Laxalt opined that at least 100 million acres of BLM's 360 million would sell, though the proposal did not go forward.[36] The Forest Service began surveying 140 million of its 191 million acres for possible sale, saying that up to 18 million acres might go on the block.[37] In the ultimate challenge, Watt announced his desire to remove 800,000 acres from wilderness protection for oil and gas development, though his ability to do so rested with the consent of Congress.[38] By 1982, it appeared that the western rebels had won, and that there were few visible limits to their success.

Chapter 17

Druids and Rebels

As with earlier rebellions, the fourth rebellion triggered by wilderness preservation was driven by the activities of a few individuals who personified the opinions of many. To explore the leading individuals on both sides of the issue is to explore some of the roots of the issues themselves. In the origination and passage of the Wilderness Act of 1964, the activities of three diverse individuals, Howard Zahniser, David Brower, and Clinton Anderson, were critical. The leading rebels were more numerous: James Watt, John Harmer, Wayne Aspinall, Orrin Hatch, Jake Garn, and Paul Laxalt. Leading sagebrush rebels at the state level, Dean Rhoads, Calvin Black, and Joe Lane, had remarkably similar backgrounds. Preservation organizations assumed powerful roles in shaping public land policy for the first time during the 1950s and 1960s, so that the characteristics of the Wilderness Society and Sierra Club partially explained the course of events, especially when they are compared to the sagebrush rebel organizations such as the League for the Advancement of States Equal Rights and the Mountain States Legal Defense Fund.

More than any other single person, Howard Zahniser was responsible for the Wilderness Act of 1964. His background did not foretell such a strong interest in wilderness: he grew up as the son of a clergyman in small-town Pennsylvania where his first extensive exposure to nature was through the Junior Audubon Club.[1] His wide-ranging liberal arts education was at Greenville (Illinois) College where he developed writing and editorial skills.[2] With an interest in natural environments and a dedication to preservation, he took employment with the Biological Survey (later to become the U.S. Fish and Wildlife Service) where he was tolerated as a maverick. When Robert Sterling

Yard died in 1945, leaving vacant the executive directorship of the Wilderness Society, Zahniser took over at half the pay from his position with the survey. For nearly twenty years, he used his executive directorship as a platform for the establishment of the Wilderness Act.

Zahniser was an intellectual, an abstract thinker more comfortable with the regular book review column he wrote for *Nature Magazine* than with the rough-and-tumble world of Washington politics.[3] He loved books, art, and music. Art galleries, he said, were like wilderness areas: "available to everyone, actually visited by relatively few, highly valued by many who never visit them."[4] He was no Robert Marshall; wilderness was more of an ideal than an experience. He was hardly up to a wilderness experience in any case, with a sciatic condition in his legs and a poor heart condition. He was a forceful spokesman for wilderness preservation, however, and through his writing, speaking, and political maneuvering he convinced others.

His greatest assets were his sincerity, his ability to compromise, and persistence. Balding, bespeckled, and of friendly demeanor, he worked with and was liked by men as diverse as David Brower of the Sierra Club, and Congressman Wayne Aspinall. His sincerity won over Forest Service skeptics who concluded that he was a "more appropriate role model for the average wilderness lover than the wealthy, backpacking elitist who figured prominently in the demonology created by wilderness opponents."[5] Although he was an idealist, he was pragmatic enough to develop compromises at critical junctures: he proposed the arrangement whereby the Bureau of Reclamation agreed not to violate national park areas along the Colorado River in exchange for support for some dams in the Colorado River Storage Project. His ability to identify and implement compromises in the legislative wars over the Wilderness Act led to its final approval. His persistence was demonstrated by his attendance at every hearing for the bill and his dogged devotion to its passage despite stirrings of doubt even within his own organization.

While Zahniser did not fit the "demonology" views of the wilderness opponents, his associate and excecutive director of the Sierra Club, David Brower, appeared to them to be the archdruid.[6] A combative preservationist who adopted extreme positions and articulated them with skillful, biting effectiveness, Brower, in his role as executive director of the Sierra Club from 1952 to 1969, was the leading public exponent of environmentalism and preservation of the period. Brower was the major force behind successful efforts to prevent the construction of dams in the Grand Canyon and in Dinosaur National Monument,[7] as well as in preserving the North Cascades from logging. With regard to wilderness legislation, he and Zahniser worked as a curious team. Brower was a charismatic debater, while Zahniser quietly pursued the objectives through negotiation. Neither would have been as effective without the cooperation of the other.

Brower grew up in Berkeley, California, where his parents ran a small, dilapidated apartment complex that was a source of embarrassment for him.[8] In order to escape Depression-era problems, the family camped and hiked in the Sierra Nevada, so that throughout his youth, Brower became accustomed to contrasting the refreshing wilderness of the high country with crowding and near-poverty of the city. By the time he reached high school, he had a well-developed inferiority complex based partly on the condition of his home and partly on his physical appearance that included crooked teeth from a childhood accident. He remained extremely shy at the University of California, and finally he quit the university in his sophomore year.

For much of the next decade, he hiked and climbed in the Sierra Nevada. He was the first to climb thirty-three peaks in the range, and he came to know the region and its wilderness experience as few others could. To earn a living, he worked at Yosemite National Park and later as an editor with the University of California Press, but his living eventually became a calling. He developed his use of the written and spoken word to an unusual degree, and he overcame his shyness. When Brower became director of the Sierra Club, he was more than a manager, he was a propagandist.

Brower's preservation philosophy derived form the perspectives of Muir and Thoreau, with a nearly religious reverence for the natural world and wilderness. He was combative and militant to a degree seldom seen in major conservation leaders, a characteristic that turned out to be both an asset and a liability. When arguing against the development of an open pit copper mine on Miner's Ridge in the Glacier Peak Wilderness, he garnered much attention to the issue by indicating that the pit would be visible from the moon. When informed that this statement was incorrect, Brower groused that it would be visible "with a small telescope."[9]

Zahniser and Brower manipulated the public and lawmakers to secure passage of the Wilderness Act, but they succeeded because of the efforts of Clinton Anderson. No one would have referred to Anderson as part of the "demonology" of the preservation movement, but without his legislative skill, dedication to the wilderness ideal, and powerful position in Congress, the Wilderness Act would not have become a reality in the early 1960s.

Anderson grew up at the turn of the century in small towns in South Dakota in a family that made frequent moves to keep his father in a job as storekeeper, sheriff, farmer, and equipment supplier.[10] During his early college education at Dakota Wesleyan University, an English professor, Clyde Tull, interested Anderson in speaking and writing. Enthusiastic about a journalistic career, Anderson applied to the Columbia University School of Journalism, but he was not accepted. He resented the rejection because he thought that the eastern university considered him "too much of a hick to make good in New York."[11] Throughout his life he considered himself an outsider with regard to eastern society and politics, even when he was the most insider of senators.

After his education at the University of Michigan, a stint in South Dakota to care for his disabled father, and some experience as a small-town reporter, Anderson moved to Albuquerque, New Mexico, to cure his tuberculosis.[12] He worked his way up the local Democratic publishing hierarchy; in 1921 he became managing editor and investigative reporter for the *Albuquerque Journal.* His "outsider" image was enhanced when he produced a series of sensational articles outlining the illegal dealings between Sinclair Oil and Secretary of Interior Albert Fall (who owned a New Mexico ranch). Fall confronted Anderson in the newspaper office, asking "Are you the son-of-a-bitch who has been writing those lies about me?" Anderson's reply was typical of him: "I may be a son-of-a-bitch, but I'm not a liar."[13] Anderson's reporting led to the "Teapot Dome" scandal and Fall's conviction. Soon thereafter, local Republicans bought the *Albuquerque Journal,* and Anderson became an insurance executive.

Anderson formed his environmental opinions during his time as newspaper writer and businessman in Albuquerque. Politically, he was a progressive in most matters, and he met a kindred spirit in Aldo Leopold, who at the time was secretary of the Albuquerque Chamber of Commerce.[14] Anderson developed an intense interest in planning and scientific management of natural resources, and he spoke publicly on conservation issues. When Leopold agitated for the protection of roadless areas in the Gila National Forest of New Mexico, Anderson was one of his outspoken allies.

Anderson's conservation orientation was present throughout his political career, first as a New Deal administrator, then as a member of the House of Representatives committee on Indian affairs, irrigation, reclamation, and public lands, as Truman's secretary of agriculture, and finally as a member of the Senate from 1948 to 1972. By the time of the wilderness controversy, he was chair of the Senate Interior Committee and a member of the "inner circle" of the most influential legislators in Washington. A tall man with an open, round face and a friendly public manner, he was gruff and blunt even with friends. He was a master mechanic of the legislative process, a well-read, well-prepared debater who expected others to be equally prepared.

As committee chair, he guided complex discussions without procedural error and without patience for participants who could not keep up. When Senator Ernest Gruening of Alaska asked whether his favored amendment had survived a particularly complex session, Anderson snapped "If you don't know, you don't need it."[15] This pragmatic, nuts-and-bolts approach to the legislative process was important for the Wilderness Act, because Zahniser was the philosopher who drafted a bill that was more a wish list than a viable piece of legislation, and Brower was the publicist who could generate public support but not congressional support.

Anderson's primary nemesis in the congressional battles over the Wilderness Act was Wayne Aspinall, chair of the House Interior Committee.[16] Aspinall was born in Middleburg, Ohio, but spent much of his youth in Western Colorado. He had a lifelong interest in resource development and public land management, in part derived from his early experiences in an area dominated by public land. In 1925, he obtained his law degree from the University of Denver and became a practicing attorney on the "western slope," that part of the Rocky Mountains in Colorado lying on the rural western side of the continental divide.[17] His involvement in the management of the peach orchard industry of western Colorado engendered an interest in water reclamation and irrigation that, in time, made him one of the nation's foremost experts on western water management.

In 1931, Aspinall began his political career as a member of the Colorado House of Representatives. He was continually elected to the House until 1939, when he was elected to the state Senate where he served until 1949. Thereafter, he was a Colorado member of the House of Representatives for eleven terms. His remarkable consistency in the electoral process was a product of diligent representation of his constituents who were farmers, ranchers, miners, and loggers whose livelihoods depended on federal public land and water. His opposition to wilderness preservation was a fair representation of the expressed interests of his constituents until the influx of backpackers, skiers, and young people changed the nature of his district in the late 1960s.

Aspinall was a short man, with sharply defined features and piercing eyes. He was every bit a match for Anderson in the areas of legislative mechanics and debate, and he was adept at exercising his considerable influence in Congress for the benefit of his constituents. His fractious relationship with Anderson was legendary in the capitol because they often disagreed yet had to work together because they chaired the Interior committees of their respective houses. Aspinall remarked that he and Anderson often agreed—Anderson agreed to have their meetings in his Senate office and Aspinall agreed to attend.[18]

By the time of the fourth sagebrush rebellion in the late 1970s, Aspinall was out of office, but three senators continued his philosophy and were the leading exponents of the rebellion in the national legislature. Utah Senators Orrin Hatch and Jake Garn were prominent. Hatch introduced the sagebrush rebellion bill in the Senate in 1979, and for a time he appeared to be the major national spokesman for the rebellion. Though born in Pennsylvania, he spent most of his life in Utah. After receiving his undergraduate degree at Brigham Young University and his law degree from the University of Pittsburgh, he practiced law in Utah with an interest in business. Politically conservative, he garnered such awards as "Mr. Free Enterprise" and "Guardian of Small Business" before being elected to the Senate in 1977.[19] His support of the sagebrush rebellion was partially an outgrowth of his interest in the generic entrepreneur.

Edwin Jacob (Jake) Garn was an unlikely sagebrush rebel in that most of his professional career was focused on cities. Born in Utah and educated at the University of Utah, he had a long history of service with local urban agencies and regional planning boards. Before his election to the Senate, Garn was mayor of Salt Lake City and was active in the state's insurance industry.[20]

Hatch and Garn in part derived their conservative values regarding public land and its management from their religion; both were Mormons, members of the Church of Jesus Christ of Latter-day Saints (LDS). In the LDS religion, there is great emphasis on productivity in all respects, including economic productivity related to the land.[21] Preservation of natural areas without conversion of their resources to economic benefits for humans is largely antithetical to the Mormon philosophy. The religion, which dominates Utah politics, also has a long history of antifederalism dating from the 1850s when federal troops were sent to Salt Lake City to enforce federal demands on the territory, which was then an LDS theocracy.

Hatch made the connection between religion and public land policy explicit in his address to the LASER convention in Salt Lake City. Hatch compared the mission of the sagebrush rebels with the mission of the Mormon pioneer leader Brigham Young. He quoted Young's admonition to "progress, improve, and make beautiful everything around you." Young, Hatch, and most devout Mormons defined that improvement and beauty to mean productive economic development.

The third prominent sagebrush rebel in Washington was Paul Laxalt, senator from Nevada. Born in Reno, with a law degree from the University of Denver, and experience as a district and city attorney, Laxalt continually spoke from a conservative position that was consistent with the sagebrush rebellion. He was lieutenant governor and governor of Nevada before his election to the Senate in 1975,[22] and he had an extensive network among local politicians of the state. That network provided the primary support for the first state sagebrush bill, and used its connection to Laxalt, who was Ronald Reagan's campaign manager in 1976 and 1980, to influence the highest office in the land.

Much of the activity on the part of the sagebrush rebels in the late 1970s was at the state level where individuals could still bend the political machinery. Dean Rhoads, Calvin Black, and Joe Lane were examples. Dean Rhoads was able to move the Nevada state legislature to pass a sagebrush bill and to convince its select committee on public lands to conduct a campaign for the passage of similar bills in other state legislatures.[23] As a rancher who had to deal with the increasingly restrictive Bureau of Land Management, and as head of an organization of stock owners, he could combine personal interest with sincerely held political positions.

In Utah, the leading rebel was Calvin Black, rancher and miner from the rural southeastern part of the state where national parks and potential wilderness areas were restricting grazing. Black had been a state legislator before the rebellion and was intensely involved in county-level politics. He was a positive force for the economic development of the region, pushing for road paving projects to increase access for industry and tourism, proposing irrigation projects, and serving as a focal point for planning processes. Black operated a uranium mine, and with five employees his company was the smallest member of the Utah Mining Association. Nonetheless, he convinced the association to pass a resolution supporting the sagebrush rebellion,[24] lending credibility to the fledgling effort.

In Arizona, the rebellion focused on state legislator Joe Lane. Lane was a rancher born in Roswell, New Mexico, educated at the University of Arizona, and a fixture in the political scene in southeast Arizona where he owned a cattle operation. Like Rhoads and Black, he dealt with increasingly difficult regulations on federal public land and the potential encroachment of wilderness areas on less restricted areas. Whether or not he believed that the federal government actually would transfer lands to the states is debatable, but the rebellion served him well as a means to bring the attention of an increasingly urbanized state to the plight of its long-time rural residents.

A facet of the fourth sagebrush rebellion not seen in previous episodes was the emerging influence of preservation organizations, especially the Wilderness Society and the Sierra Club. The Wilderness Society provided needed focus on the wilderness preservation issue over several decades, slowly growing in membership and financial resources. During the early 1940s the organization subsisted on the dues of fewer than 1,000 members and the income from a portion of the estate of Robert Marshall that produced a total budget of less than $10,000.[25]

When Zahniser assumed the executive directorship of the organization he introduced three important changes. First, the previously insular organization began to actively seek cooperation from other groups on lobbying. Second, the society functions became concentrated under Zahniser's control, increasing (for a time) efficiency. Finally, the new executive director sought to broaden the membership of the organization to define a truly national grass-roots clientele. By 1970 the previously tiny organization had grown tenfold to a membership of 87,000, by 1983 to 100,000, and by 1986 to 160,000.[26]

This rapid growth in size and influence brought internal pressures to the Wilderness Society.[27] After Zahniser's death in 1964, his understudy, Steward Brandborg, became executive director. Brandborg kept the society's effort focused on wilderness issues at a time when the environmental movement was broadening its interests to encompass a wide range of issues, a trend that diluted the emphasis on wilderness preservation in other organizations. Brandborg emphasized participation in the wilderness area review process, but beginning in the early 1970s he gave increasing attention to the Alaska lands issues. This regional emphasis offended members and activists from other regions, resulting in a loss of institutional momentum. The organization began to lose money, and internal disagreements resulted in the resignation or firing of key personnel, including Robert Cutler whose knowledge of forestry and wilderness management was invaluable. Finally in 1976, the council forced Brandborg's own resignation and two years later hired William Turnage to be executive director.

Turnage, son of a State Department official, was influenced toward public service by his parents and William Sloane Coffin, Jr., of Yale University. Turnage became interested in environmental matters while doing graduate work at Oxford, where he spent much time hiking in the Cotswolds. Later, as an employee of the State Department in Washington, D.C., he hiked the Shenandoah and Great Smoky Mountain trail networks, and he concluded that his avocation should be his vocation. He returned to Yale's Forestry School where he had a chance meeting with Ansel Adams.[28] Adams's nature photography and Turnage's interests in the outdoors neatly coincided, and Turnage became the photographer's business manager. Adams, an activist in the Wilderness Society, was instrumental in securing the executive directorship for Turnage, who won the position over 50 other applicants.

Turnage's appointment in 1978 was critical because the fourth sagebrush rebellion was gathering momentum at a time when the Wilderness Society was adrift. Turnage instituted sweeping changes in the society that quickly restored its effectiveness.[29] He hired professionals for the staff, organized the society into functioning departments much like a business, acquired foundation grants to supplement other income sources, and instituted rotating membership on the council rather than reappointing the same individuals. When Gaylord Nelson, the conservation-oriented senator from Wisconsin, was defeated in the 1980 Reagan electoral sweep, Turnage was instrumental in securing his appointment as president of the Wilderness Society. Nelson was the most prominent person associated with the group and lent credibility to its lobbying and publicity efforts during the conclusion of the fourth rebellion.

A most important organizational ally of the Wilderness Society during the fourth rebellion was the Sierra Club. It too had passed through periods of rapid development, internal discord, and finally, smooth-running success. When David Brower assumed control of the organization in 1952 it was still largely focused on West Coast interests and had only 7,000 members. He brought a larger perspective and broadened the activities of the club to include opposition to major federal projects that threatened to invade many potential wilderness areas. Recognizing that mobilization of an awakening electorate was essential to the success of the environmental agenda, Brower instituted a series of photographic and prose exhibit books in 1960, which were part art and part propaganda.[30] Each volume focused on a threatened area and served to marry photographic art with political activism. In nine years, the series generated $10 million in income for the club. Unfortunately, it cost the club even more, and internal discord over the series and Brower's management style resulted in his ouster in 1969.

During the 1970s and 1980s, the club recovered its equilibrium and was the most widely recognized militant group among the many environmental organizations. However, from the standpoint of wilderness preservation, it was ineffective for two reasons. First, the club took on too many environmental issues, which ranged from air pollution to overpopulation to harzardous wastes. Second, during the 1970s much of the club's public-land-related activities focused on Alaska, where immense areas of national park and wilderness land were at stake. A powerful friend of the Wilderness Society, the Sierra Club had become incredibly diverse. But its diversity reflected expansion: by 1987 the Sierra Club had a membership of 426,000 and an annual operating budget of more than $28 million.[31]

The preservation lobby was successful during the wilderness controversy and the fourth rebellion because it adopted the organizational tactics of the commodity groups and refined them. The Wilderness Society and Sierra Club were nothing more than tremendously successful counterparts to associations of livestock growers, miners, or timber producers. The difference was that the preservation groups appealed to a wider audience, commanded greater monetary resources, and were more closely associated with the political opinions of the increasingly urban American electorate. Oddly, the major sagebrush rebellion organization, the Leage for the Advancement of States Equal Rights (LASER) was patterned after the preservation groups.

LASER was incorporated in the state of Utah during summer of 1979. Its stated aims were "charitable, educational, scientific," and the analyses of economic, social, environmental, and legal problems relating to the western states.[32] John L. Harmer was chief executive of the group and was a member of its board of trustees along with Calvin Black (the southeastern Utah politico and mine owner), Alex G. Oblad (a mining professor at the University of Utah), Robert Lunt (a Salt Lake City attorney), and Jonathan E. Johnson (a California attorney and previous associate of Harmer's). LASER represented itself as an organization dedicated to the divestiture of federal public lands, and invited as members individuals and businesses that had similar interests. The group solicited contributions, saying that "it will take hundreds of thousands of dollars for LASER to reach its goals,'' and claiming a tax-exempt status for donations.[33] No financial records of the group were made public, but its short life suggests that the hundreds of thousands of dollars were not forthcoming. According to a local news reporter, some of the expenses of the conference were paid by large corporations with interests in mining: Mountain Fuel Suppliers (coal), Rio Algom (uranium), and Kennecott Minerals (copper). [34]

LASER did not stand on its own, but rather was associated with Harmer's business practice. He, Mark Hurst, and C. McClain (''Big Mac'') Haddow were partners in the firm of Mountain States Advertising,[35] and after the conference, visitors to LASER were directed to their corporate offices.[36] Haddow, a legislator, was influential in the Republican Party in Utah and provided an important congressional connection, because he was a member of Hatch's staff.[37] After the conference, LASER faded from the public scene, abandoned by national figures and bereft of purpose, funds, and members.

More prominent (and longer lasting) than LASER among the sagebrush rebel organizations was the Mountain States Legal Defense Fund. Mountain States was part of a network of probusiness organizations that were counterparts to the environmental groups. The concept originated with the establishment of the Pacific Legal Foundation in 1973 by the California Chamber of Commerce.[38] Joseph Coors, the Colorado brewer, duplicated the concept at the national level by funding the initiation of the National Center for the Public Interest. Clifford Rock, a consultant for Coors, initiated the Mountain States Legal Defense Fund in 1977 with the stated objectives of research, analysis, legal representation, and education.[39] The initial board of directors included Coors, Karl Eller (an Arizona business executive), Leonard J. Theberge (of the National Center for Public Interest), and lobbyists for the Association of Commerce and Industry. Later directors included Wayne Aspinall.

By the early 1980s, the incorporation documents of the group included ammendments to allow it to pursue litigation on behalf of its members. During its first year, the fund had a budget of $194,000, but at the height of the rebellion in 1980, with James Watt as its president and chief legal officer, its income was $1.2 million.[40] The fund successfully manipulated the news media by timely filing of court briefs in cases related to resource development, wilderness issues, and public lands, and it garnered additional attention when Watt was appointed secretary of interior. The organization declined as Watt's national career disintegrated.

Notable by their absence during the move to establish the Wilderness Act and the subsequent sagebrush rebellion were scientists. During the irrigation controversy, the establishment of the national forests, and the grazing debates, geologists, geographers, foresters, and soil scientists played key roles in the formulation of public policy for the federal lands. Although the geologist Robert Currey was active in wilderness issues in the northern Rocky Mountains, and Aldo Leopold's geologist son, Luna, served on the council of the Sierra Club, science was surprisingly absent from the wilderness debate. Individual ecologists, botanists, and zoologists occasionally made public statements about the desirability of wilderness preservation or argued against the divestiture of federal lands, but generally the life sciences did not appear in force in the public arena.

The resulting information void meant that the issues were debated on moral and philosophical grounds without substantive discussion of the scientific value of wilderness to society. Additionally, scientific analyses of the different impacts of federal versus state management of public lands based on demonstrable evidence rarely entered into the sagebrush rebellion debate. This general lack of interest in public policy on the part of the community of natural scientists resulted in an undereducated public being left to sort through the rhetoric with little authoritative guidance. A few individuals and organizations had a free field in influencing public opinion, without need to represent their cases in the light of environmental research.

Chapter 18

On a Clear Day You Can See Four Corners

During the period of the fourth sagebrush rebellion, the American West was a land transformed from a mostly wilderness region inhabited by few people to a mostly developed, integrated part of the larger nation where wilderness was a scarce, mostly western resource (Figure 19). The physical, social, cultural, political, and economic environments of the West underwent drastic changes from the 1960s to the 1980s, changes that were reflected in the unique nature of the most recent rebellion. Perceptions of the western environment also changed, often transmitting a new vision of reality that replaced venerated myths of the past.

The effectiveness of management of grazing on public lands was often couched in terms of protecting the landscape from excessive erosion. The rapid erosion rates observed in the 1920s and 1930s had, by the 1940s, reversed as did the development of gullies and arroyos. Extensive sheet erosion was also much less common in subsequent decades. The reversal of processes that apparently occurred in the early 1940s resulted in widespread surface stability in the interior West through the early 1980s.[1] Despite the overall stability of the river systems, federal public land still was subject to erosion that was unacceptable. In 1973, professional range scientists classified 84 percent of Bureau of Land Management rangeland as in "fair, poor, or bad" condition.[2] The Natural Resources Defense Council used this figure in its suit against the bureau to force the writing of environmental impact statements to accompany management plans for grazing allotments.

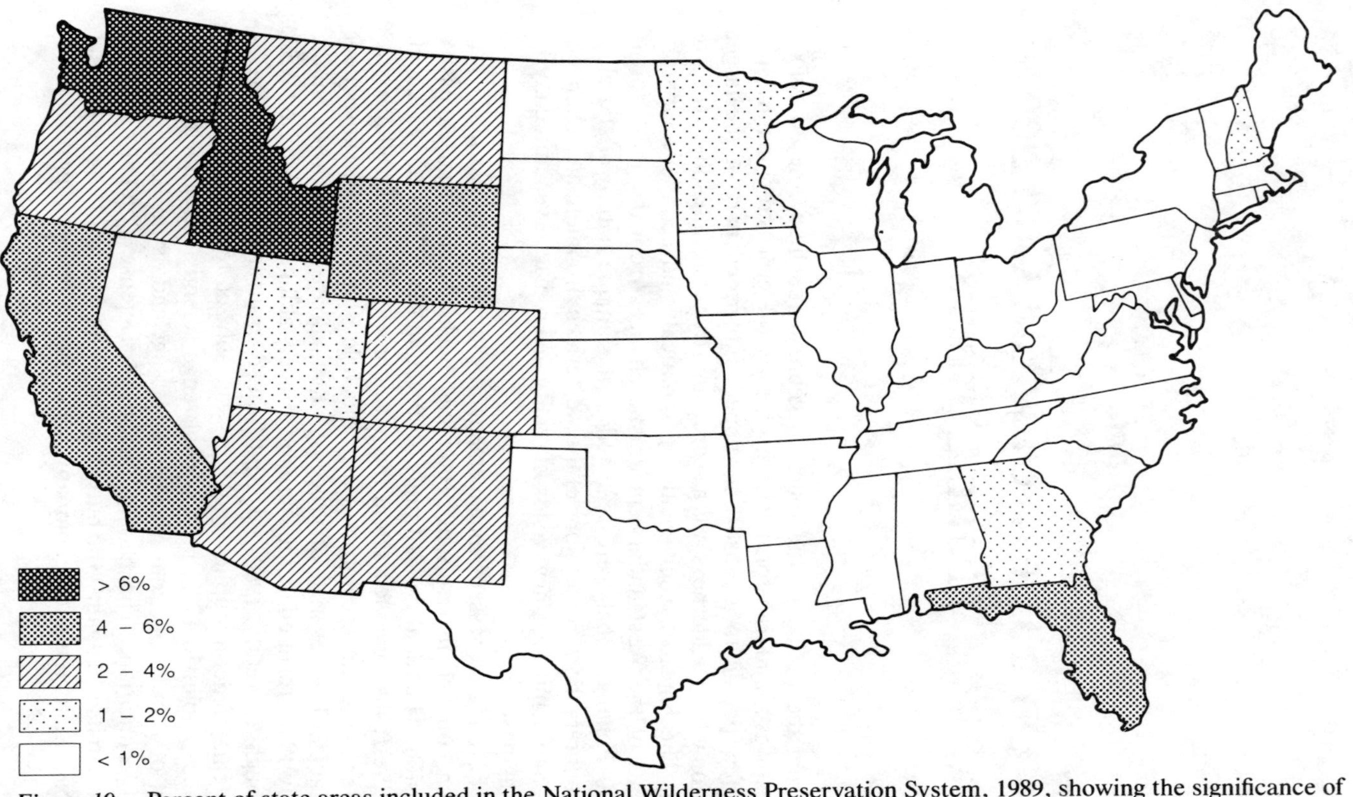

Figure 19. Percent of state areas included in the National Wilderness Preservation System, 1989, showing the significance of western states (Wilderness Society data).

There were three problems with the assessment of the quality of the rangelands. First, there is no proof that the lack of quality forage is due to overgrazing. Most of the lands in question lie in regions with less than fifteen inches of annual rainfall, so that even under entirely undisturbed conditions their productivity is low. Second, the fifteen inches of annual rainfall is an average figure that does not take into account the year-to-year variability that causes actual rainfall amounts to vary by more than 100 percent from one year to the next. Finally, surface materials on much of the bureau's lands do not support good forage: in the Colorado Plateau region there are large areas of bare rock, and in other regions steep slopes or alkaline soils retard the growth of vegetation.[3] These factors confuse the accurate assessment of the success or failure of a grazing management program, and allowed the sagebrush rebels to argue that stock owners were not responsible. In any case, localized overgrazing could be observed in at least some areas, providing an angle of attack for the bureau's opponents.

Trees were coming to cover increasingly large areas of the West during the period of the fourth rebellion. Some rangelands were being invaded by undesirable species, such as juniper, that reduced the grass cover available for forage.[4] As environmental regulations became more strict, range users who had less freedom to remove the offending trees felt that the federal government in Washington simply did not understand the problem. In the western mountains, renewed forest growth after extensive logging resulted in the expansion of the mountain forests at the expense of meadows on valley floors that graziers viewed as desirable pasture. Sites photographed at the turn of the century and then rephotographed in the 1960s and 1970s revealed significant expansion of the mountain forests.[5]

During the period of the fourth sagebrush rebellion, air pollution became a new environmental quality issue in the interior West. Forest fires had produced periodic reductions in visibility even before the incursion of Anglo-Americans,[6] but generally the region's pristine air quality was one of its most notable features. Throughout the early twentieth century, local degradation of atmospheric quality resulted from urban sources of smoke and chemicals, while smelters generated sulfur gases.[7] The development of coal-fired power plants in the southwestern United States during and after the 1960s resulted in substantial declines in visibility, while air pollution regulations and the designation of wilderness areas influenced the regulation of such facilities and added to the regional antagonism toward wilderness policy.

The issue of power generating plants, air quality laws, and the influence of wilderness areas in the enforcement of those laws has been a point of conflict between preservationists and regional economic developers in the Southwest. In part, preservationists were responsible for the development of the plants because of their opposition to the development of hydroelectric power in the canyons of the Colorado River.[8] The Bureau of Reclamation originally had planned to develop hydroelectric dams in the Grand Canyon to provide power for pumping irrigation water to the Central Arizona Project, but during the 1960s the Sierra Club was successful in blocking the construction plans.[9] In order to save the canyon, however, the club had to forego opposition to the development of coal-fired power plants as replacements for the dams. Regional utilities constructed several large generating stations in northeast Arizona and northwest New Mexico, and by the 1970s air quality had declined to an alarming degree.[10] Opposition to further power development by preservationists centered on the interaction of federal laws and wilderness areas.

Federal air pollution laws derived mostly from urban experiences with air pollution that threatened public health.[11] The 1967 Air Quality Act erected a legal framework for the enforcement of standards for industrial and automobile emissions,[12] and subsequent amendments in 1970 expanded the federal role in setting and enforcing standards.[13] Further amendments in 1977 established a geographical component to the laws by designating three classes of areas based on objectives for air quality. Class I areas were located where pristine conditions existed in zones to be preserved (including national parks and wilderness areas).[14] The hearth region of the sagebrush rebellion, the four-corner states of Utah, Colorado, New Mexico, and Arizona, had the highest concentration of Class I areas in the nation.[15] Environmental regulations in California were more strict than in these interior states, the interior states having coal resources to burn (literally), and local sentiment often favored power development as an opportunity for regional economic development. Many power plant sites became unviable because their stack gases would threaten the air quality of distant Class I airsheds in wilderness areas. The result was increased local support for the rebellion.

The social and cultural environment of the western states was becoming more complex during the fourth rebellion, but changes during the 1960-1980 period generally weakened the position of the sagebrush rebels. Migration to the western states from other parts of the nation brought new people into western society who did not share the ranching-mining-lumbering backgrounds of long-term residents.

Because the in-migrants were mostly urban dwellers, the basic demography of the West changed. Although the western states always had significant percentages of their people in cities, by 1980 cities had come to dominate the western life-style. In Arizona, for example, 80 to 90 percent of the population became urban in the period of the fourth rebellion.[16] Because the sagebrush rebellions had always drawn their support from rural populations, this urbanization diluted the political base of the rebels, who could not even command solid support in their home states.

The changing social character of the western population directly impacted the primary rebels—the ranchers—by diminishing their political influence within their own states. Until the late 1960s, ranchers were disproportionately represented in western state legislatures,[17] but redistricting to account for the newly populous cities reduced the representation of the ranchers thereafter. Their decline was accelerated by the dominance of urban communication systems. Major newspapers targeted urban markets and largely ignored the rural areas, in part because timely delivery of daily newspapers to far-flung rural customers was impossible.[18] In most of the interior western states, a few urban television stations (carried by statewide transponders) dominated the news market, and they were more likely to broadcast stories with an environmental slant than they were to present ranching news.

Urban residents quickly learned to deal with the federal land management agencies. Urban preservationists mounted effective, persistent campaigns where once the only voices heard were those of ranchers, miners, and loggers. The preservationists were better organized and more numerous than the rural commodity interests.[19] Urbanites staffed the land management agencies too, as young employees made their way into middle-management positions. In the Bureau of Land Management, the change was particularly noticeable. Middle management previously had been dominated by graduates of western state universities that drew rural dwellers into their agricultural programs. By the 1970s, however, many of those same positions were held by graduates of natural science or management programs in nonwestern universities.[20] Western ranchers had groused about Gifford Pinchot's "college boys" at the turn of the century, and about the "do gooder scientists" of the Soil Conservation Service in the 1930s. From the 1970s on, they had a new group to complain about.

The Park Service posed special problems for the western rancher, and Park Service policies added fuel to the sagebrush rebellion. New national parks (where grazing generally has been prohibited) threatened to remove large areas from the range resource. In northern New Mexico, for example, a proposal for the creation of Valle Grande National Park in the 1960s would have absorbed a large privately owned forest and meadow area. Although Senator Clinton Anderson supported federal purchase of the area, there was strong local opposition, and the idea was abandoned.[21]

In the same period, Canyonlands National Park was created in south-central Utah. The Park Service envisioned a primitive park with emphasis on wilderness preservation in what was mostly a roadless area used only by ranchers. Local boosters dreamed of another Grand Canyon National Park, with roads, lodges, recreation developments, and tourist dollars flowing into the local economy. Twenty years after the creation of the park, and after bitter controversy, more than 90 percent of the area was designated wilderness.[22] Similarly, when President Lyndon Johnson enlarged Capitol Reef National Monument, Utah, from about 40,000 acres to 215,000 acres in 1969, local ranching interests feared the loss of winter ranges. The town board of nearby Boulder, Utah, voted to change the town name to "Johnson's Folly."[23] Eventually, local grazing rights were grandfathered into the management plan of the area by the Park Service.

Antifederal political positions have always played well to the western electorate. Despite the newly urbanized nature of the population, this antagonism continued during the fourth rebellion as a product of issues that were often unique to each sagebrush state. In Colorado, grazing management and the placement of the 1972 Winter Olympics in the Rockies were issues. A statewide referendum turned the Olympics away, and Governor Richard Lamm was coauthor of a book that outlined the history of the grievances of western states against federal land management. He meekly concluded that the sagebrush rebellion was not the answer, however, because it involved a small minority of westerners whose claims had been accepted without reservation by a naive eastern news media.[24]

In Utah, attitudes toward the federal government derived from a lengthy history of Mormon friction with Washington. Eventually the sagebrush rebellion evolved into an effort to trade isolated state land parcels for large, consolidated blocks of federal land. The plan, called Project BOLD,[25] was viewed with cautious optimism by most interested parties, but progress was slow because of questions about values of the lands the state would give up relative to those of lands the federal government would relinquish.

In New Mexico, Spanish and Mexican land grants continued to be a major public land issue. When the United States acquired the Southwest as a result of the 1846 war with Mexico, the federal government promised to honor these grants. Through legal chicanery and dissent among the Hispanic owners, much land-grant property was subdivided, sold, and later absorbed into national forests.[26] Rural Hispanic New Mexicans have continued to regard many of the lands as their birthright, and routinely ignore regulations governing national forests. In 1966, a minor armed revolt developed in northern New Mexico and culminated in a shoot-out at a county court house.[27] Designation of large areas as wilderness throughout the region over the next two decades made access and use of the traditional Hispanic lands even more difficult, and the issue was a continuing source of local dissatisfaction.[28]

In Arizona, increasingly liberal urban populations in Phoenix and Tucson threatened the traditional conservatism of the state, but antifederal attitudes, especially regarding public lands, continued throughout the fourth rebellion. In the congressional and legislative elections of the early 1980s, many Democratic candidates and all Republicans held antifederal views. Donna Carlson-West, a state legislator running for a congressional seat, was one of the many candidates supporting the rebellion. Her position on the issue was so similar to that of her opponents that the primary election was jokingly referred to as a demonstration for a mythical popular recording group to be called "Donna and the Monotones."

Even environmentalists in Arizona were dissatisfied with federal administration because of continued sight-seeing flights by airplanes and helicopters over the wilderness areas of the inner Grand Canyon. The issue simmered for several years until a deadly midair collision converted it from a question of environmental management to one of public safety. Resolution was by an act of Congress.

Nevada, where ranchers agitated against federal land policies throughout the early 1980s, was a major force in the fourth rebellion. Increasingly, the rebels found themselves associated with environmentalists in opposition to defense-related uses of Nevada federal land. The Carter administration had proposed to construct several thousand launch sites for the MX intercontinental missile in central and eastern Nevada and portions of neighboring Utah. The proposal would have connected the sites with a rail network over which a limited number of missiles would be shuttled in the largest shell game ever imagined. The MX issue, with its dependence on western public lands and scarce water supplies, became a major rallying point for the rebellion.[29]

Rural residents of southern Nevada, southern Utah, and northern Arizona were likely recruits for the rebellion because of their ongoing dispute with the federal government concerning the management of the Nevada nuclear weapons test site. Until the Nuclear Test Ban Treaty in 1962, atmospheric detonations of nuclear devices at the test site were common.[30] With the increased environmental awareness of the 1970s and 1980s, residents of the region questioned the impact on public health of fallout from the explosions. A class-action lawsuit against the federal government charged that mismanagement and failure to warn the public of dangerous fallout resulted in cancer-related deaths of rural residents.[31] The suit dragged on unsuccessfully for several years and was a continuing source of bitterness for those involved.

Despite these strongly developed and widely held antifederal opinions, the West continued to be a "child of subsidy."[32] In 1976, the federal government spent $32 per person on defense-related salaries in New York, a paltry sum compared to the figures for western states: $275 in Colorado and $306 in Utah.[33] Almost all Fortune 500 companies maintained their headquarters elsewhere, ensuring that decisions affecting resource development in the West were made elsewhere. Though the region flourished economically, its residents did not control their economic destiny. The myth of western independence continued unabated, even as the ideal diverged from reality to a greater degree than ever before.

During the fourth rebellion, the West was enjoying an economic boom unlike any it had experienced before.[34] The growing urban areas attracted industries associated with the most modern technology. Denver became a banking and finance center; Albuquerque and Tucson developed sophisticated defense-related industries; Salt Lake City and the Wasatch Front entered the space travel business as a producer of booster rockets; Phoenix grew into a major producer of electronic gear; and Las Vegas continued its own unique brand of urban development. The rural West did not share in this general improvement as resource commodity prices, especially for minerals and fuels, declined.

The story of uranium mining in the interior West provides an example of the connection among commodities, public land, and the fourth rebellion. The four-corners area has been a source of radioactive ores for almost a century, though major exploration and mining did not occur until after World War II. In the race to develop domestic supplies of uranium during the Cold War in the late 1940s and early 1950s, the federal government offered financial incentives to individuals who could find and develop deposits on public lands. Many remote areas in the four-corners states administered by the Bureau of Land Management lost their wilderness characteristics as they were penetrated by crude roads for the uranium exploration. For a period, the entire region was caught up in a new form of the gold rush: a stock market dealing exclusively with uranium stocks opened in Salt Lake City with trading often at the price of a penny per share.[35] A few struck it rich, like Charlie Steen, who with his wife and two young sons lived an exciting but impoverished existence until he discovered a huge uranium deposit that converted them into millionaires.

When the federal government determined in 1958 that it had sufficient supplies of uranium for nuclear weapons, the incentives ended and the uranium boom went bust, leaving behind a few large mining companies and a few part-time prospectors who pressed on more for fun than for profit. As with most minerals and fuels, the price of uranium fluctuated widely in response to demands for nuclear reactor fuel and changes in politically unstable foreign sources.[36] In the middle 1970s, uranium prices skyrocketed, fueling the demand for access to potential wilderness areas. Later, the price of uranium plummeted as supplies outstripped demand and construction of new nuclear power stations stagnated (Figure 20).

By the early 1980s, there were no active uranium mines in the four-corners area, where the individual uranium mines had become a favorite character and a new symbol of independence. To the miners, the problems were a stubborn federal government that had mismanaged the nation's resources and the environmental lobby that could not understand the need to develop resources. Calvin Black, the arch Utah rebel, had to close his small mine. Backcountry and wilderness hikers were viewed with suspicion. The local sentiment was adequately expressed by the sign on a closed gate over a mine access road: "Keep out—authorized personnel only. Mining is America's Future. You Environmentalists and Sahara [sic] Clubbers and other commie pinkos are a threat to that future. Keep out."[37] Bumper stickers in southern Utah made the statement more succinctly: "More Nucs, Less Kooks."

Part of the local rural animosity toward outsiders was a function of increased use of the backcountry and wilderness areas by urbanites from all parts of the country. The West no longer belonged only to westerners. The national environmental movement that had begun with Earth Day in 1970 had, by the early 1980s, become a less spectacular but firmly entrenched part of the national culture. A survey by the President's Council on Environmental Quality in 1980 showed

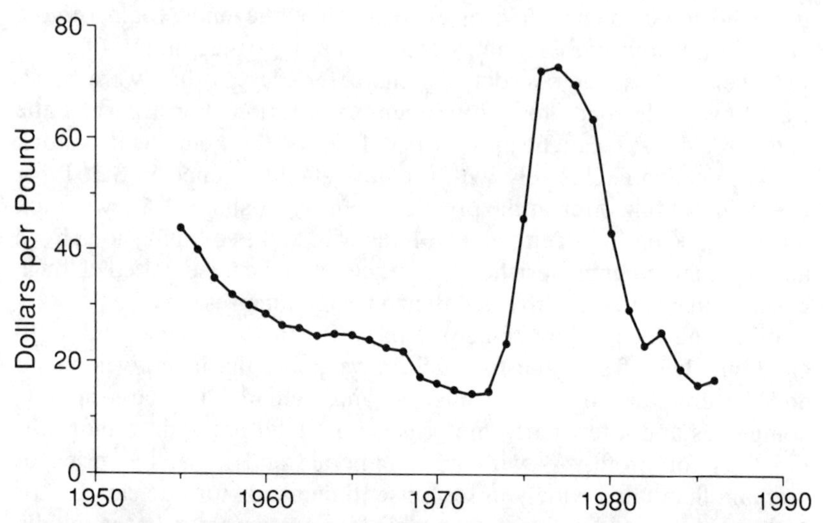

Figure 20. Spot prices for uranium oxide showing the radical increases in the mid- and late 1970s that stimulated miners' interest in public lands at a time when reservations for wilderness were also increasing as shown in Figure 16 (U.S. Department of Energy data).

that an amazing 73 percent of Americans considered themselves "definitely" or "somewhat" environmentalist.[38] The West was the region with the most activists, but also the region with the most opposition to environmentalists.[39] The West was also intricately bound up with the wilderness issue because most wilderness was located in the West and because the perceptions of many Americans of a "wilderness area" included the perception of a Western landscape.

Photography had a strong impact on the American perception of wilderness. The most important photographer in the effort to secure the Wilderness Act was Ansel Adams. Adams grew up in San Francisco and developed a sensitive eye for texture and light in natural scenes, especially in California. His medium was the black-and-white print, which he single-handedly elevated into an art form.[40] From the 1920s to the 1980s, Adams's views of natural landscapes brought a new vision of the West into the American psyche. He was active in the effort to preserve wilderness as early as Robert Marshall's era of the 1930s. Adams circulated his photos of the Sierra Nevada to congressmen and senators in an unsuccessful attempt to have a section of the mountains preserved in a new national park.[41] He collected the photos into a book and sent a copy with a personal plea to Secretary of

Interior Harold Ickes. Ickes was sold, and Kings Canyon National Park became a reality.

Adams was instrumental in the Sierra Club's Exhibit Series as well, and his work appeared as the first volume of the series. He was one of the most powerful members of the club's board, and he influenced the direction of the organization. He was the first to propose David Brower as executive director, and hc was the first to propose Brower's removal from the position.[42] Later in his career, Adams became increasingly active in the Wilderness Society, and his photography continued to be a force in public education about wilderness areas until his death in 1984.

Color photography also was a prominent art form in communicating previously unknown landscapes to a large American audience. Elliott Porter hiked with his camera to remote locations and constructed striking images that created a new mythical West where fragile beauty could be enjoyed in the contemplation of solitude.[43] Many copied Porter's style, and western wilderness areas became images known and admired even by those who had never visited them.

Writers who dealt with the western wilderness in the period of the fourth rebellion were unlike their artistic predecessors. Essayist and novelist Edward Abbey, for example, represented a new and aggressive perspective for the preservationist. Abbey's essays were blistering indictments of development, including the intrusive construction of facilities for automobile-bound tourists in national parks and wilderness areas.[44] His fictional characters were ecoterrorists who, although they avoided personal attacks, were willing to engage in the destruction of property to protect natural areas.[45] An advocate of the destruction of Glen Canyon Dam in order to restore the natural Colorado River upstream, Abbey was an activist who reflected a new vision of the West where heroes were not fur trappers, cowboys, or Indians, but were ordinary Anglo-Americans willing to go to extraordinary lengths to preserve the remaining wilderness experience.[46]

The West of the 1980s was a region radically changed from previous decades. Its physical environments were undergoing rapid change, its culture was in flux, its politics and economics were unstable, and the image it presented to the rest of the nation was a new vision. These changes were taking place against the backdrop of a western geography that included a new phenomenon, units of the National Wilderness System as areas where the truly original West remained undisturbed.

Chapter 19

How Much Is Enough?

In 1982 it appeared that the West was about to undergo a massive transformation with respect to its public lands. With one devoted sagebrush rebel in the White House and another secretary of interior, western states passing rebellion legislation and appropriating funds to support legal offensives, the national Congress in a restive mood, it seemed that the fourth rebellion would succeed where none had before. However, the hoped-for success of the rebellion carried the seeds of its own demise, and the inevitable response finally appeared.

Burgeoning preservation groups formed an umbrella group called "Save Our Public Lands" to deal with the rebellion by coordinating contributions, speakers, and publications. The preservation lobby finally awakened, and opposition to Secretary Watt's proposals grew in Congress. The sagebrush rebellion bills in the various western states collected dust in the archives, and their appropriations for legal challenges expired without court cases. The bills introduced by Senator Hatch and Congressman Santini in the national legislature died quiet deaths. Secretary Watt aroused so much opposition that he became a liability to the Reagan administration, and when he committed the last of a series of speaking gaffes in 1983 by referring to the membership of his coal advisory board as "a black, . . . a woman, two Jews and a cripple," he was forced to resign.[1]

Watt's resignation in late 1983 marked the virtual end of the fourth rebellion. After three years of little attention, there was a renewal of the introduction of state-by-state wilderness bills for Forest Service lands setting aside some land as wilderness, reserving some for further study, and releasing the remaining areas for multiple-use management. In 1988, the same process began for BLM wilderness bills on a state-by-state basis.[2] The rebellion ended because it could not establish its basic legal claim that the public domain belonged to the states. From a more practical perspective it ended because it had become successful enough to stimulate a debilitating preservation reaction at the national level. Graziers and other western sympathizers were outnumbered once the debate became a national one, with eastern congressional members who represented large populations unwilling to grant even the scaled-down demands of the rebels.

The rebellion ended in part because it had achieved some of its goals. Federal agencies managing land in the West became sensitized to the demands of local users, and more care was taken in the process of wilderness review than had been in the recent past. The wilderness reservation process had been slowed nearly to a halt, winning further delays for the rebels. Timber sales, grazing leases, and leases for oil and gas exploration were accelerated, benefiting the local western land users at the expense of the preservationists.

The most important reason that the rebellion ended was its abandonment by its own national leaders. President Reagan did not pursue land disposal policies or relaxation of the provisions of the Federal Land Policy and Management Act, despite his claims of identity as a sagebrush rebel. Despite his success in generating antipreservation media coverage, Secretary Watt backed away from the rebellion almost immediately after taking office by saying that he saw as his major objective the search for ways to "defuse" the rebellion.[3] A leader of the Arizona sagebrush delegation and state representative, Joe Lane, dryly observed that "he is not helping the cause with statements like that."[4]

Congressional representatives who had courted the votes of the rebels in their own states also abandoned the cause once they were in Washington. Orrin Hatch of Utah, primary congressional spokesman for the rebellion, allowed his bill to expire and did not resurrect it. He moved on to other issues including national health care. Jake Garn, prominent participant in the LASER conference in Salt Lake City, took on loftier interests than the rebellion and focused his energies on support of the aerospace program, becoming the first congressman to make a space trip. James Santini pursued his support of the mining industry but did not renew his quest for passage of a sagebrush rebellion bill.

Without continued support from the top, the sagebrush organizations disintegrated. LASER ceased its activities and hosted no more conferences. The organization's dynamic leader, John Harmer, moved on to other issues, and became president of the National Center for Constitution Studies. An occasional point in his public agenda was the problem of the administration of federal public lands.[5] Sagebrush Rebellion, Inc., disappeared from the public eye and also became inactive. Meanwhile, the preservation organizations flourished. The Wilderness Society continued its growth in size and influence, approaching 200,000 members in 1988 and participating in hundreds of separate wilderness area studies and proposals. The more broadly based Sierra Club counted almost half a million members.

The dialogue between opposing sides in the issue of wilderness preservation took on new dimensions with the state-by-state review processes. Led by congressional delegations, commodity users and preservationists were forced to deal with each other early in the reservation process rather than at the end. Negotiations involving concessions by both sides resulted in bills likely to succeed in passing both houses in Congress, and local interests at the state level could participate. Perhaps the most important outcome of the fourth rebellion was this new air of negotiated settlements. Although neither side would be completely satisfied, every party had access to the system. Although it had taken 55 years, Robert Marshall's dream of a National Wilderness System approached 91 million acres, a magnitude that even he might not have suspected (Figure 21).

This American wilderness system is a product of the conflicts of four periods in land policy history. A review of these episodic periods of federal action and western reaction in the form of a sagebrush rebellion show that each episode has remarkable similarities to the others. In every rebellion there was a lack of knowledge about the most fundamental of American resources: the land and its conditions. The irrigation question developed at a time when dam sites were unsurveyed; national forests were set aside without clear knowledge of their ecosystems; the grazing controversy appeared in a time when long-term range surveys were not available; and wilderness areas still have unknown mineral potential.

Environmental change has never been adequately accounted for in public-land decisions. Water, forest, grazing, and recreation resources are susceptable to long-term trends ultimately traceable to climatic changes. Though these changes are profound and directly affect the resources of the public lands, they are poorly understood and ignored. Environmental changes resulting from human activities have brought about a consistent reduction in the areas available for wilderness

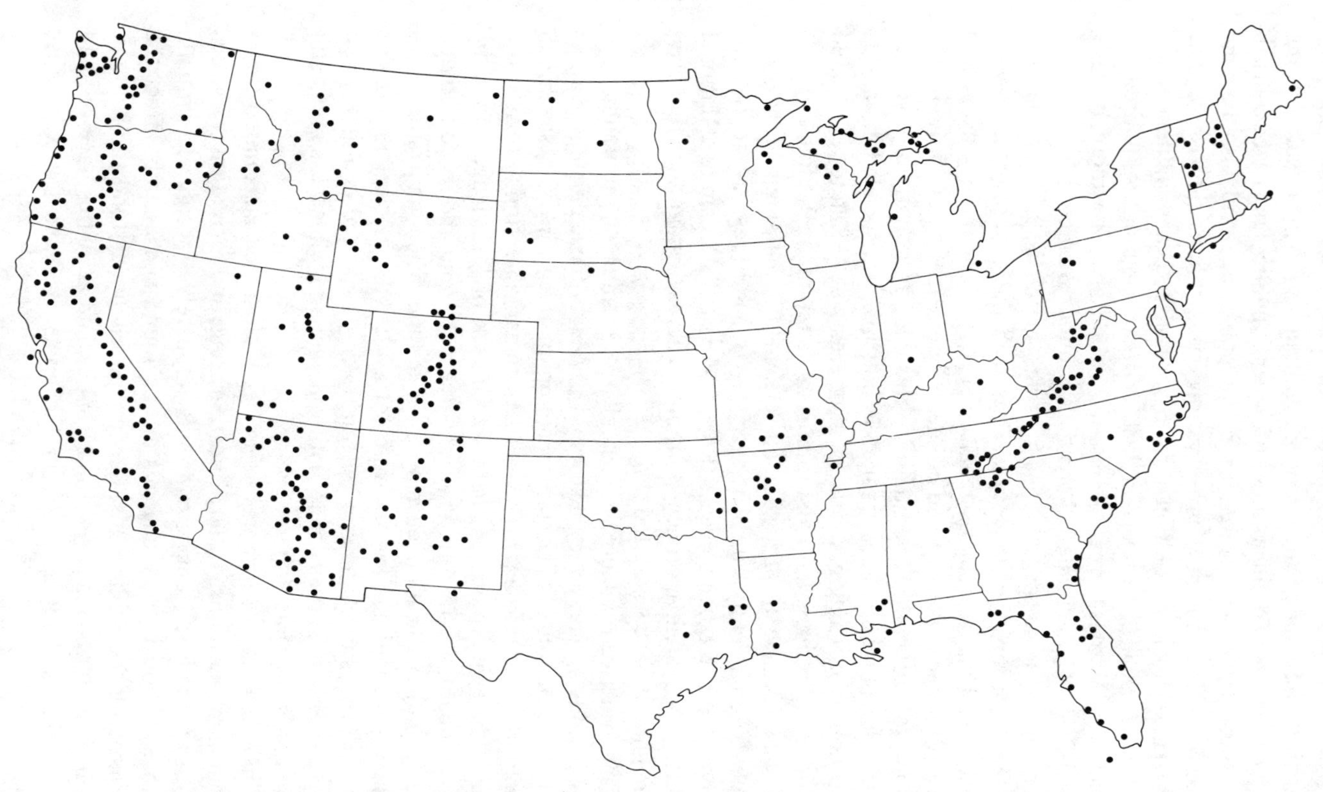

Figure 21. The distribution of units in the National Wilderness Preservation System, 1989, with each unit shown by a dot, irrespective of the size of the unit (Wilderness Society and National Geographic Society data).

preservation throughout all the periods of federal attempts at preservation and the rebellions.

Throughout the history of federal preservation and antifederal rebellions, there have been four aspects of the policy landscape that have driven events. Foremost is greed. Of land, Americans never seem to get enough. From commodity developers to wilderness preservationists, all the participants in policy debates have gone to great lengths to ensure that their particular views of land management are extended to as much of the federal lands as possible. Each group has endangered its credibility and effectiveness when its political reach has exceeded its logical grasp.

A definable cycle has been evident throughout the story of the sagebrush rebellions. A cycle of explosive development, restrictive regulation, and resulting rebellion by western interests has occurred several times. Overall, the federal regulatory role has consistently become more intensive and extensive. With the passage of each rebellion, the authority has become more centralized despite efforts and protestations to the contrary. The centralized management agencies have become increasing distant from the users by cultural as well as physical measures.

The debates about the proper management of the federal public lands in general, and particularly about those lands that ultimately were designed as wilderness, reveal a series of divisions among Americans with respect to these issues. An understanding of these divisions and their conflicts provides a basic understanding of the preservation/sagebrush rebellion process. There are philosophical divisions between the commodity developer who seeks economic gain either for himself or for his local community, and the preservationist who seeks aesthetic gain either for himself or for a larger community. Prodevelopment groups have almost always been small collections of professionals, while preservation groups have been larger collections of amateurs.

Another recurring theme in the public land debate has been the division between the "expert" manager and the user. Commodity developers have a short-term, practical orientation while the manager deals with long-term, large-scale issues. The former often believes that a clear vision of reality depends on scuffed boots and dirty fingernails obtained by labor in the mines, forests, ranges, or trails, and the latter believes in something called "the big picture."

The urban versus rural division regarding public land issues has been especially well developed in all the rebellions. Rural and small town populations have consistently aligned themselves with the sagebrush rebels because they are most directly affected by changes in the administration of public lands. Urban populations with less direct economic ties to the public lands have been hotbeds of support for the preservation perspective. At the national level, the political division has been East versus West, but this simplification has begun to disintegrate with the migration of population to western cities. The once unified western voting block that could be counted on by sagebrush rebels is crumbling quickly.

The issue of states' rights has usually been a stalking horse for the sagebrush rebels seeking increased influence in the management of federal lands. It is a strongly developed undercurrent in the politically conservative states of the interior West. Preservationists have seen little hope of success in influencing this western political machinery and so have become federalists by default.

In the largest sense, the preservation/sagebrush rebellion processes outlined in this story are driven by three basic components of American culture: land ownership, independence, and individualism. The most basic point of disagreement between sagebrush rebels and protectionists is "who owns the land?" The preservationist groups, representing diverse constituencies, consider the public lands as belonging to everyone and administered on their behalf by the federal government. The career federal managers agree. Western commodity interests point out that many of these so-called owners not only have never seen "their" lands, they are not even aware of their ownership. To the resource developer and many rural westerners, the land belongs to the user.

Regional independence in the West is a cornerstone upon which all the sagebrush rebellions have constructed elaborate, anti-Washington crusades. The rancher, miner, logger, and water developer sought to be free of federal bureaucratic entanglements in their quests for independent resource development. Yet, at the same time, they sought from the federal government farm subsidies, mineral price supports, below-cost timber contracts, and reclamation projects. In 1980, for every federal tax dollar sent to Washington, the West received $1.20 in return. The mythical quality of western independence does not diminish its influence, however, and as long as citizens of the region believe in the myth, it guides their political actions as though it were reality.

Finally, individualism has been the hidden mainspring of the sagebrush rebellions and the preservation ethic. For the western rebels the frontier heritage of individualism is expressed in the economic arena. The owner of a small ranch with leased federal land for grazing, or the operator of a small mining operation on public lands, expresses individualism as success in resource development, and he views restrictive federal regulation as a threat to individual opportunity. The wilderness preservationist sees in wilderness the last opportunity for modern Americans to obtain an experience of self-reliance and individual contact with the natural world. For the urban American, economic individualism offers little promise in a world increasingly dominated by big business, big government, and even big preservation organizations.

As the twentieth century draws to a close, federal land management in the United States continues to be dominated by the conflict between wilderness preservation and the sagebrush rebellions. Although the conflict may change in appearance and rhetoric, it will continue as the underlying tension that has produced one of the most marvelous paradoxes of Western civilization. A nation energized by economic opportunism has seen fit to preserve for everyone the opportunity to recreate one of the country's most basic historic and geographic experiences: to be alone in the wilderness.

Notes

Introduction

1. R. Nash, *Wilderness and the American Mind*, 2d ed. (New Haven, Conn.: Yale University Press), 1973.
2. W. Zelinski, ''North America's Vernacular Regions,'' *Annals of the Association of American Geographers* 70(1980): 1–16.
3. *U.S. Census*, 1980.

Chapter 1
Surveying the West

1. W. C. Darrah, *Powell of the Colorado* (Princeton: Princeton University Press, 1951).
2. *Utah Historical Quarterly* 15, 16, 17 (1947).
3. W. Stegner, *Beyond the Hundredth Meridian: John Wesley Powell and the Second Opening of the West* (Boston: Houghton Mifflin Company, 1953), 10–20.
4. M. C. Rabbitt, *John Wesley Powell: Pioneer Statesman of Federal Science*, U.S. Geological Survey Professional Paper 669-A (Washingon, D.C.: Government Printing Office, 1969), 1–21.
5. R. A. Bartlett, *Great Surveys of the American West* (Norman: University of Oklahoma Press, 1962), 373–77.
6. H. Adams, *The Education of Henry Adams* (Boston: Houghton Mifflin Company, 1918), 322.
7. J. W. Powell, *Report on the Lands of the Arid Region of the United States, with a More Detailed Account of the Lands of Utah* (Washington, D.C.: U.S. Geographical and Geological Survey of the Rocky Mountain Region, 1878).
8. M. C. Rabbitt, *Minerals, Lands, and Geology for the Common Defense and General Welfare*. Vol. 1. *Before 1879* (Washington, D.C.: U.S. Geological Survey, 1979).
9. Stegner, *Beyond the Hundredth Meridian*, 233.

10. J. W. Powell, "The Non-irrigable Lands of the Arid Region," *Century Illustrated Magazine* 39(1890): 915–22.

11. J. W. Powell, "Ownership of Lands in the Arid Region," *Irrigation Age* 6(1894): 143–49.

12. Darrah, *Powell of the Colorado,* 221–36.

13. Stegner, *Beyond the Hundredth Meridian*, 235–42.

14. M. Clawson, *Uncle Sam's Acres* (New York: Mead and Company, 1951).

15. S. V. Connor, and O. B. Faulk, *North America Divided: The Mexican War, 1846–1848* (New York: Oxford University Press, 1971).

16. R. W. van Alstyne, "International Rivalries in the Pacific Northwest," *Oregon Historical Quarterly* 46(1945): 221–50.

17. P. N. Garber, *The Gadsden Treaty* (Philadelphia: University of Pennsylvania Press, 1923).

18. R. Robbins, *Our Landed Heritage: The Public Domain, 1776–1936* (Princeton, N.J.: Princeton University Press, 1942).

19. B. H. Hibbard, *A History of Public Land Policies* (New York: Macmillan, 1939).

20. M. Conover, *The General Land Office: Its History, Activities, and Organization* (Baltimore: Johns Hopkins University Press, 1923).

21. Hibbard, *Public Land Policies*, 245.

22. T. H. Watkins, and C. S. Watson, Jr., *The Lands No One Knows* (San Francisco: Sierra Club Books, 1975), 42–47.

23. P. W. Gates, "The Homestead Law in an Incongruous Land System," *American Historical Review* 41(1936): 652-81.

24. Bureau of Land Management, *Public Land Statistics* (Washington, D.C.: Department of Interior, 1978), 6.

25. T. Donaldson, *The Public Domain* (Washington, D.C.: U.S. Department of Interior, 1884), 355.

26. F. A. Shannon, "The Homestead Act and the Labor Surplus," *American Historical Review* 14(1936): 637–51.

27. L. Cazier, *Surveys and Surveyors of the Public Domain, 1785–1975* (Washington, D.C.: Government Printing Office, 1976), 47–53.

28. Bureau of Land Management, *Public Land Statistics*, 3, 6.

29. Arizona State Land Department, *Annual Report, 1979-1980* (Phoenix: State of Arizona, 1980), 23.

30. W. P. Webb, *The Great Plains* (New York: Grosset & Dunlap, 1931), 17–27.

31. Bureau of Land Management, *Public Land Statistics,* 3.

32. S. A. Puter, *Looters of the Public Domain* (Portland, Oreg.: Portland Printing House, 1908).

33. C. B. McIntosh, "Use and Abuse of the Timber Culture Act," *Annals of the Association of American Geographers* 65(1975):347-62.

34. L. Atherton, *The Cattle Kings* (Lincoln: University of Nebraska Press, 1961).

35. W. E. Hollon, *The Great American Desert: Then and Now* (Lincoln: University of Nebraska Press, 1975), 138.

36. A. Sampson, *The Arms Bazaar: From Lebanon to Lockheed* (New York: Viking Press, 1977).

37. H. A. Dunham, "Some Crucial Years of the General Land Office, 1875–1890," *Agricultural History* 11(1937): 117–41.

38. W. A. J. Sparks, *Annual Report, General Land Office* (Washington, D.C.: Government Printing Office, 1885), 3–4.

39. M. C. Rabbitt, *Minerals, Lands, and Geology for the Common Defense and General Welfare,* Vol. 2, *1879–1904* (Washington, D.C.: U.S. Geological Survey, 1980), 36–41.

40. Bartlett, *Great Surveys of the American West*, 90–122.

41. Rabbitt, *Minerals, Lands, and Geology,* Vol. 1, 196-204.

42. W. T. Jackson, "The Creation of Yellowstone National Park," *Mississippi Valley Historical Review* 29(1942): 187–206.

43. C. P. Russell, *100 Years in Yosemite* (Yosemite National Park: Yosemite Natural History Association, 1968).

44. H. Huth, "Yosemite: The Story of an Idea," *Sierra Club Bulletin* 33(1948): 47–48.

45. C. Goodrich, *Government Promotion of American Canals and Railroads, 1800–1890* (New York: Columbia University Press, 1960).

46. F. Norris, *The Octopus, A Story of California* (New York: Doubleday, Page, and Company, 1901).

47. G. Catlin, *North American Indians: Being Letters and Notes on Their Manners, Customs, and Conditions, Written During Eight Years Travel Amongst the Wildest Tribes of Indians in North America*, 2 vols. (Edinburg: J. Grant, 1841).

48. L. L. Noble, *The Life and Works of Thomas Cole* (Cambridge: Oxford University Press, 1964).

49. R. Nash, *Wilderness and the American Mind*, 2d ed. (New Haven, Conn.: Yale University Press, 1973), 101.

50. S. Fox, *John Muir and His Legacy, The American Conservation Movement* (Boston: Little, Brown, and Company, 1981).

51. Darrah, *Powell of the Colorado*, 400–420.

52. Powell, *Report on the Lands of the Arid Region*, map in pocket.

53. R. P. Teele, *Irrigation in the United States: A Discussion of Its Legal, Economic and Financial Aspects* (New York: D. Appleton and Company, 1915).

54. Conover, *The General Land Office*.

55. Dunham, "General Land Office, 1875–1890."

56. E. W. Sterling, "The Powell Irrigation Survey, 1888–1893," *Mississippi Valley Historical Review* 27(1940): 421-34.

57. *Congressional Record, VIII*, Pt. 3, 45th Cong., 3d sess., 1879.

58. J. W. Powell, *Tenth Annual Report of the Director of the United States Geological Survey,* Part 2, *Irrigation* (Washington, D.C.: Department of Interior, 1889).

59. H. H. Smith, "Rain Follows the Plow: The Notion of Increased Rainfall for the Great Plains, 1844–1880," *Huntington Library Quarterly* 10(1947): 169–93.

60. General Land Office, *Annual Report* (Washington, D.C.: Department of Interior, 1888).

61. *New York Times*, 19 December 1885.

62. Dunham, "General Land Office, 1875–1890," 117–41.

63. R. S. Bradley, *Precipitation History of the Rocky Mountain States* (Boulder, Colo.: Westview Press, 1976).
64. R. H. Mattison, "The Hard Winter of the Range Cattle Business, Montana," *The Magazine of History* 1(1950): 5–20.
65. General Land Office, *Annual Report* (Washington, D.C.: Department of Interior, 1885, 1886).
66. Powell, *Tenth Annual Report*, letter by Stockslager, 20 February 1888.
67. *Congressional Record*, 50th Cong., 1st sess., 7031, 1888.

Chapter 2
The Big Thinkers

1. W. H. Stewart, *Reminiscences* (Washington, D.C.: Neale Publishing, 1908).
2. *Congressional Record*, 51st Cong., 1st sess., 1890, 6301.
3. E. W. Sterling, "The Powell Irrigation Survey, 1888–1893," *Mississippi Valley Historical Review* 27(1940): 421–34.
4. E. Ellis, *Henry Moore Teller, Defender of the West* (Caldwell, Idaho: Caxton Press, 1941).
5. M. C. Rabbitt, *Minerals, Lands, and Geology for the Common Defense and General Welfare,* Vol. 2, *1879–1904* (Washington, D.C.: U.S. Geological Survey, 1980).
6. H. A. Dunham, "Some Crucial Years of the General Land Office, 1875-1890," *Agricultural History* 11(1937): 117-41.
7. J. W. Powell, *Eleventh Annual Report,* Part 2, *Irrigation* (Washington, D.C.: U.S. Geological Survey, 1891), attached map.
8. U.S. Congress. House. H. Rept. 2407, 1890.
9. W. Stegner, *Beyond the Hundredth Meridian: John Wesley Powell and the Second Opening of the West* (Boston: Houghton, Mifflin and Company, 1954), 331.
10. W. E. Connelley, *Life of Preston B. Plumb* (Chicago: University of Chicago Press, 1913).
11. *Congressional Record*, 50th Cong., 1st sess., 1889, 7031.
12. Durham, "Crucial Years of the General Land Office," 137–40.
13. H. A. Dunham, *Government Handout: A Study in the Administration of the Public Lands, 1875-1891* (Ann Arbor, Mich.: Edwards Litho-Printers, 1941).
14. J. W. Powell, *Canyons of the Colorado* (Meadville, Pa.: Flood and Vincent, 1895).
15. W. C. Darrah, *Powell of the Colorado* (Princeton, N.J.: Princeton University Press, 1951).
16. J. W. Powell, *Eleventh Annual Report,* Part 2, *Irrigation* (Washington, D.C.: U.S. Geological Survey, 1891), 289.
17. C. V. Easum, *The Americanization of Carl Schurz* (Chicago: University of Chicago Press, 1929), 10–20.
18. H. George, *Progress and Poverty: An Inquiry into the Cause of Indus-*

trial Depressions and of Increase of Want with Increase of Wealth (New York: Appleton, 1881).

19. Dunham, *Government Handout*, 234.

20. M. C. Rabbitt, *Minerals, Lands, and Geology for the Common Defense and General Welfare,* Vol. 1, *Before 1879* (Washington, D.C.: U.S. Geological Survey, 1979), 288.

21. C. Schurz, *The Reminiscences of Carl Schurz*, 2 vols. (New York: McClure Company, 1907).

22. Dunham, "Crucial Years of the General Land Office," 136.

23. *Wyoming Daily Sun*, 1 February 1887.

24. A. Nevins, *Grover Cleveland: A Study in Courage* (New York: Dodd, Mead and Company, 1933).

25. L. Q. C. Lamar, *Annual Report* (Washington, D.C.: U.S. Department of Interior, 1887), 7.

26. A. Nevins, *Abraham S. Hewitt, with Some Account of Peter Cooper* (New York: Harper and Row, 1935).

27. H. B. Hammett, *Hilary Abner Herbert, A Southerner Returns to the Union* (Philadelphia: American Philosophical Society, 1976), Memoir 110.

28. Dunham, "Crucial Years of the General Land Office," 129.

29. Dunham, *Government Handout*, 212.

30. Rabbitt, Vol. 1, *Before 1879*, 1.

31. M. C. Rabbitt, *John Wesley Powell: Pioneer Statesman of Federal Science* (Washington, D.C.: U.S. Geological Survey, 1969), Professional Paper 669-A, 1–21.

32. G. K. Gilbert, *Lake Bonneville* (Washington, D.C.: U.S. Geological Survey, 1890), Monograph 1.

33. S. J. Pyne, *Grove Karl Gilbert, Great Engine of Research* (Austin: University of Texas Press, 1980), 220.

34. J. S. Diller, "Clarence E. Dutton," *Geological Society of America Bulletin* 24(1913): 10–18.

35. *Congressional Record*, 51st Cong., 1st sess., 1892, 6301.

36. S. Fox, *John Muir and His Legacy: The American Conservation Movement* (Boston: Little, Brown and Company, 1981).

37. R. U. Johnson, *Remembered Yesterdays* (Boston: Little, Brown and Company, 1923).

38. H. R. Jones, *John Muir and the Sierra Club* (San Francisco: Sierra Club Books, 1965).

39. Pyne, *Grove Karl Gilbert*, 204.

40. D. Lowenthal, *George Perkins Marsh: Versatile Vermonter* (New York: Columbia University Press, 1958).

41. G. P. Marsh, *Man and Nature* (New York: Charles Scribner, 1864).

42. P. Brooks, *Speaking for Nature* (Boston: Houghton Mifflin Company, 1980), 90.

43. G. P. Marsh, *The Earth as Modified by Human Action* (New York: Arno, 1874), 327.

44. H. Huth, *Nature and the American: Three Centuries of Changing Attitudes* (Lincoln: University of Nebraska Press), 1957.

Chapter 3
The Mythical Land of Gilpin

1. H. H. Bancroft, *History of the Life of William Gilpin: A Character Study* (San Francisco: A. L. Bancroft and Company, 1889).
2. B. DeVoto, "Geopolitics with the Dew on It," *Harper's Magazine* 187(1944): 313–23.
3. S. Adams, *Colorado River Expeditions of Samual Adams*, U.S. House of Representatives, Miscellaneous Document No. 37, 42d Cong., 1st sess., 1882.
4. P. Brooks, *Roadless Area* (New York: Ballantine, 1964), 123–41.
5. W. E. Hollon, *The Great American Desert: Then and Now* (Lincoln: University of Nebraska Press, 1975), 9–20.
6. C. D. Wilber, *The Great Valleys and Prairies of Nebraska and the Northwest* (Omaha, Nebr.: Daily Republican Printer, 1881), 69.
7. G. K. Gilbert, *Lake Bonneville* (Washington, D.C.: U.S. Geological Survey, 1890), Monograph 1.
8. H. N. Smith, "Rain Follows the Plow: The Notion of Increased Rainfall for the Great Plains, 1844–1880," *Huntington Library Quarterly* 10(1947): 169–93.
9. Wilber, *Nebraska and the Northwest*, 1.
10. W. B. Hazen, "The Great Middle Region of the United States and Its Limited Space of Arrable Land," *North American Review* 120(1875): 1–34.
11. H. C. Fritts, *Tree Rings and Climate* (London: Academic Press, 1976).
12. M. P. Lawson, and C. W. Stockton, "Desert Myth and Climatic Reality," *Annals of the Association of American Geographers* 71(1981): 527–35.
13. M. P. Lawson, and C. W. Stockton, "The Desert Myth Evaluated in the Context of Climatic Change," in Smith, C. D., and Parry, M., eds., *Consequences of Climatic Change* (Nottingham: University of Nottingham, 1981), 106–18.
14. R. Claiborne, *Climate, Man, and History* (New York: W. W. Norton and Company, 1970).
15. J. Chang, *Atmospheric Circulation Systems and Climatic Systems* (Honolulu: Oriental Publishing Company, 1972).
16. R. G. Barry, and R. J. Chorley, *Atmosphere, Weather, and Climate* (New York: Holt, Rinehart and Winston, 1970).
17. R. S. Bradley, *Precipitation History of the Rocky Mountain States* (Boulder, Colo.: Westview Press, 1976), 172.
18. J. Borchert, "The Dust Bowl in the 1970s," *Annals of the Association of American Geographers* 61(1971): 1–22.
19. J. W. Powell, *Tenth Annual Report,* Part 2, *Irrigation (Washington, D.C.: U. S. Geological Survey, 1889), 87.*
20. O. B. Faulk, Arizona: A Short History (Norman: University of Oklahoma Press, 1970), 168–72.
21. O. E. Young, Jr., *Western Mining* (Norman: University of Oklahoma Press), 1970.
22. G. W. James, *Reclaiming the Arid West: The Story of the United States Reclamation Service* (New York: Appleton, 1917).

23. M. E. Carlson, "William E. Smyth: Irrigation Crusader," *Journal of the West* 7(1968): 41–47.

24. W. E. Smyth, *The Conquest of Arid America* (New York: Harper and Brothers, 1889).

25. W. C. Darrah, *Powell of the Colorado* (Princeton, N.J.: Princeton University Press, 1951), 312.

26. U.S. Department of Agriculture, *Agricultural Statistics* (Washington, D.C.: Government Printing Office, 1980), 421.

27. A. Mills, *My Story* (Washington, D.C.: C. H. Claudy, 1918).

28. J. W. Powell, *Thirteenth Annual Report* (Washington, D.C.: U.S. Geological Survey, 1893).

29. J. M. Hunter, *The Trail Drivers of Texas* (New York: Argosy-Antiquarian, 1963).

30. C. A. Perkins, M. G. Nielson, and L. B. Jones, *Saga of San Juan* (Monticello, Utah: San Juan County Daughters of Utah Pioneers, 1968).

31. F. A. Shannon, *The Centennial Years, A Political and Economic History of America From the Late 1870s to the Early 1890s* (Garden City, N.Y.: Doubleday, 1967).

32. L. Atherton, *The Cattle Kings* (Lincoln: University of Nebraska Press, 1961).

33. Ibid., 168.

34. W. M. Denevan, "Livestock Numbers in Nineteenth-Century New Mexico, and the Problem of Gullying in the Southwest," *Annals of the Association of American Geographers* 57(1967): 691–703.

35. R. U. Cooke, and R. W. Reeves, *Arroyo Development and Environmental Change in the American South-West* (Oxford: Clarendon Press, 1976).

36. R. A. Bartlett, *Great Surveys of the American West* (Norman: University of Oklahoma Press, 1962).

37. H. A. Dunham, "Some Crucial Years of the General Land Office, 1875–1890," *Agricultural History* 11(1937): 117–41.

38. G. M. Butter, *Some Facts About Ore Deposits* (Tucson: Arizona Bureau of Mines, 1935), Bulletin 139.

Chapter 4
Resolution and the Seeds of Discontent

1. E. W. Sterling, "The Powell Irrigation Survey, 1888–1893," *Mississippi Historical Review* 27(1940): 421-34.

2. M. C. Rabbitt, *Minerals, Lands, and Geology for the Common Defense and General Welfare,* Vol. 2, *1879–1904* (Washington, D.C.: U.S. Geological Survey, 1980).

3. *Congressional Record*, 51st Cong., 1st sess., 1890, 7320.

4. *Washington Star*, 28 May 1890.

5. *Congressional Record*, 51st Cong., 1st sess., 1890, 5419.

6. T. H. Watkins, and C. S. Watson, *The Lands No One Knows* (San Francisco: Sierra Club Books, 1975).

7. U.S. Bureau of the Census, *Compendium of the Eleventh Census, 1890,*

Part I, *Population* (Washington, D.C.: Government Printing Office, 1890), xlviii.

8. W. C. Barnes, rev. by B. H. Granger, *Arizona Place Names* (Tucson: University of Arizona Press, 1960), 110.

9. W. C. Darrah, *Powell of the Colorado* (Princeton, N.J.: Princeton University Press, 1951), 400.

Chapter 5
Forests Without Trees

1. R. A. Bartlett, *Great Surveys of the American West* (Norman: University of Oklahoma Press, 1962), 184.

2. J. W. Powell, *Report on the Lands of the Arid Region of the United States, with a More Detailed Account of the Lands of Utah* (Washington, D.C.: U.S. Geographical and Geological Survey of the Rocky Mountain Region, 1878), Plate I.

3. H. K. Steen, *The U.S. Forest Service: A History* (Seattle: University of Washington Press, 1976), 10.

4. J. Ise, *The United States Forest Policy* (New Haven, Conn.: Yale University Press, 1920), 37.

5. E. L. Jacobsen, "Franklin B. Hough: A Pioneer in Scientific Forestry in America," *New York History* 15(1934): 311-18.

6. Ise, *Forest Policy*, 31–38.

7. U.S. Congress. House. H.R. 1310, 44th Cong., 1st sess., 1876.

8. Steen, *U.S. Forest Service*, 76.

9. F. B. Hough, *Report Upon Forestry,* Vol. 1 (Washington, D.C.: Government Printing Office, 1878).

10. Ibid., 16.

11. F. B. Hough, *Report Upon Forestry,* Vol. 2 (Washington, D.C.: Government Printing Office, 1880).

12. F. B. Hough, *Report Upon Forestry,* Vol. 3 (Washington, D.C.: Government Printing Office, 1882).

13. R. W. Raymond, *Report of the General Land Office*, House Executive Document 207, 41st Cong., 2d sess., 1877, 373.

14. S. S. Burdett, *Report of the General Land Office* (Washington, D.C.: U.S. Department of Interior, 1884), xvi, 6.

15. Steen, *U. S. Forest Service*, 19.

16. H. Clepper, "Crusade for Conservation: The Centennial History of the American Forestry Association," *American Forests* 81(1975): 37–44.

17. H. Clepper, "Forestry's Uncertain Beginning," *Journal of Forestry* 67(1969): 218–21.

18. Steen, *U.S. Forest Service*, 18.

19. A. D. Rogers, *Bernhard Edward Fernow: A Story of North American Forestry* (Princeton, N.J.: Princeton University Press, 1951).

20. O. M. Butler, "70 Years of Campaigning for American Forestry," *American Forests* 52(1946): 456–59, 512.

21. A. H. Carhart, *The National Forests* (New York: Alfred A. Knopf, 1959).
22. W. B. Greeley, *Forests and Men* (New York: Doubleday and Company, 1951).
23. U.S. Forest Service, *Highlights in the History of Forest Conservation* (Washington, D.C.: Government Printing Office, 1952), Agricultural Bulletin No. 83.
24. S. P. Hays, *Conservation and the Gospel of Efficiency: The Progressive Conservation Movement, 1890–1920* (Cambridge, Mass.: Harvard University Press, 1960).
25. C. L. Fenton, and M. A. Fenton, *The Story of the Great Geologists* (Garden City, N.Y.: Doubleday and Company, 1945).
26. Rodgers, *Bernhard Edward Fernow*, 154–55.
27. 26 U.S. Stat. 1095.
28. Clepper, "Forestry's Uncertain Beginning."
29. R. U. Johnson, *Remembered Yesterdays* (Boston: Little, Brown and Company, 1923), 293-94.
30. R. Pettigrew, *Triumphant Plutocracy: The Story of American Public Life from 1870 to 1920* (New York: Academy Press, 1921), 11–12.
31. Ise, *Forest Policy*, 112–14.
32. S. T. Dana, *Forest and Range Policy* (New York: McGraw-Hill, 1956).
33. B. Frank, *Our National Forests* (Norman: University of Oklahoma Press, 1955).
34. S. A. D. Puter, and H. Stevens, *Looters of the Public Domain* (Portland, Oreg.: Portland Printing House, 1908).
35. C. Goodrich, *Government Promotion of American Canals and Railroads, 1800–1890* (New York: Doubleday and Company, 1960).
36. Ise, *Forest Policy*, 84.
37. T. A. Richard, *A History of American Mining* (New York: Appleton, 1932).
38. W. L. Graf, "Mining and Channel Response," *Annals of the Association of American Geographers* 69(1979): 262–75.
39. P. H. Roberts, *Hoof Prints on Forest Ranges: The Early Years of National Forest Range Administration* (San Antonio: Naylor Company, 1963).
40. C. E. Winter, *Four Hundred Million Acres: The Public Lands and Resources: History, Acquisition, Disposition, Proposals, Memorials, Brief, States* (Casper, Wyo.: Overland Publishing Company, 1932).
41. L. Wooley, "Federal vs. State Control of the National Forests of the United States." Bachelor's thesis, University of Utah, Salt Lake City, Utah, 1911.
42. *Congressional Record*, 51st Cong., 2d sess., 1891, 3894.
43. Ise, *Forest Policy*, 70.
44. U.S. House of Representatives Resolution 119, 1891.
45. E. A. Sherman, "Thirty-Five Years of National Forest Growth," *Journal of Forestry* 24(1926): 129-35.
46. G. Cleveland, "Second Annual Message," in *Compilation of the Messages and Papers of the Presidents* (New York: Bureau of National Literature, 1894), vol. 13, 5975.

47. S. B. Sutton, *Charles Sprague Sargent and the Arnold Arboretum* (Cambridge, Mass.: Harvard University Press, 1970), 159.
48. G. Pinchot, *Breaking New Ground* (Seattle: University of Washington Press, 1947, reprinted, 1972).
49. Johnson, *Remembered Yesterdays*, 190-93.
50. Clepper, "Forestry's Uncertain Beginning," 220.
51. National Academy of Sciences, *The Report of the National Academy of Sciences Upon the Inauguration of a Forest Policy for the Forested Lands of the United States* (Washington, D.C.: National Academy of Sciences, 1897).
52. *Congressional Record*, 6 May 1897, 909.
53. Ibid., 910.
54. Ibid., 912.
55. F. P. Dunne, *Mr. Dooley: First in the Hearts of His Countrymen* (Boston: Small, Maynard and Company, 1899), 147.
56. E. Richardson, *The Politics of Conservation, Crusades and Controversies, 1897–1913* (Berkeley: University of California Press, 1962), 4–7.
57. Richardson, *The Politics of Conservation*, 5.
58. *Laramie, Wyoming Boomerang*, 4 January 1894.
59. Richardson, *The Politics of Conservation*, 9.
60. *Congressional Record*, 6 May 1897, 909–12.
61. W. B. Stevens, "When a Missourian Forced a Special Session of Congress," *Missouri Historical Review* 23(1928): 46–47.
62. U.S. Senate Document No. 68, 55th Cong., 1st sess., 1897.
63. G. Pinchot, "How the National Forests Were Won," *American Forests and Forest Life* 36(1930): 615-19.
64. Richardson, *The Politics of Conservation*, 6.
65. Clepper, "Forestry's Uncertain Beginning," 221.
66. U.S. Senate Document No. 189, 55th Cong., 2d sess., 1898.
67. H. Gannett, *Nineteenth Annual Report*, Part 5, *Forests* (Washington, D.C.: U.S. Geological Survey, 1899).
68. G. E. Mowry, *The Era of Theodore Roosevelt and the Birth of Modern America, 1900–1912* (New York: Harper, 1962), 106.
69. Sherman, "National Forest Growth," 130.

Chapter 6
The Cowboy and the Dude in Washington

1. T. Roosevelt, "First Annual Message," in *A Compilation of the Messages and Papers of the Presidents* (New York: Bureau of National Literature, 1901), vol. 15, 6641–80.
2. G. Pinchot, "How the National Forests Were Won," *American Forests and Forest Life* 36(1930):618.
3. F. Roth, "Administration of U.S. Forest Reserves," *Forestry and Irrigation* 9(1902): 191–93, 241–44.
4. W. D. Langwill, "Mostly Division 'R' Days," *Oregon Historical Quarterly* 57(1956): 301–13.

5. G. Pinchot, *Breaking New Ground* (Seattle: University of Washington Press, 1947, reprinted ed. 1972).

6. H. K. Steen, *The U.S. Forest Service: A History* (Seattle: University of Washington Press, 1976), 75–80.

7. *Congressional Record*, 11 February 1901, 2247.

8. *Congressional Record*, 9 June 1902, 6509–6526.

9. 43 U.S. Stat. 371.

10. J. P. Degnan, "The Desert Shall Rejoice, and Be Made to Blossom as the Rose," *The Living Wilderness* 46(Summer 1982): 14–26.

11. E. R. Richardson, *The Politics of Conservation, Crusades and Controversies, 1897–1913* (Berkeley: University of California Press, 1962), 10.

12. *Annual Report* (Washington, D.C.: U.S. Department of Interior, 1903), 12.

13. S. A. D. Puter, and H. Stevens, *Looters of the Public Domain* (Portland, Oreg.: Portland Publishing House, 1908).

14. J. Ise, *The United States Forest Policy* (New Haven, Conn.: Yale University Press, 1920), 182.

15. *Congressional Record*, 10 May 1897, 965.

16. U.S. Department of Agriculture, *Agricultural Statistics* (Washington, D.C.: Government Printing Office, 1980), 324.

17. P. H. Roberts, *Hoof Prints on Forest Ranges: The Early Years of National Forest Range Administration* (San Antonio: Naylor Company, 1963).

18. Richardson, *The Politics of Conservation*, 28.

19. G. Hardin, "The Tragedy of the Commons," *Science* 162(1968): 1243.

20. Ise, *Forest Policy*, 173.

21. Pinchot, *Breaking New Ground*, 200.

22. Roberts, *Hoof Prints on Forest Ranges*, 112.

23. H. T. Pinkett, *Gifford Pinchot: Private and Public Forester* (Urbana: University of Illinois Press), 1970.

24. Roberts, *Hoof Prints on Forest Ranges*, 28–31.

25. Richards, *Politics of Conservation*, 28–30.

26. Ibid., 31.

27. Pinchot, *Breaking New Ground*, 211.

28. Ise, *Forest Policy*, 168–70.

29. U.S. Congress. Senate. S. Doc. 189, 58th Cong., 3d sess., 1905, Append. Pt. 5, 8–24.

30. M. C. Rabbitt, *Minerals, Lands, and Geology for the Common Defence and General Welfare,* Vol. 2, *1879–1904* (Washington D.C.: U.S. Geological Survey, 1980), 356.

31. Richards, *Politics of Conservation*, 22.

32. U.S. Congress. Senate. S. Doc. No. 189, 58th Cong., 3d sess., 1905.

33. Ise, *Forest Policy*, 151.

34. Rabbitt, *Minerals, Lands, and Geology,* Vol. 2, 343–45.

35. 33 U.S. Stat. 628.

36. E. A. Sherman, "Thirty-Five Years of National Forest Growth," *Journal of Forestry* 24(1926): 129–35.

37. Steen, *U.S. Forest Service*, 84.

38. M. Frome, *Whose Woods These Are: The Story of the National Forests* (Garden City, N.Y.: Doubleday and Company, 1962), 57.

39. Pinchot, *Breaking New Ground*, 248.
40. Steen, *U.S. Forest Service*, 99.
41. Ise, *Forest Policy*, 197–200.
42. *Congressional Record*, 21 February 1907, 352.
43. 34 U.S. Stat. 1269.
44. Steen, *U.S. Forest Service*, 99.
45. Pinchot, "How the National Forests Were Won," 615–19.
46. Pinkett, *Gifford Pinchot*, 270.
47. T. Roosevelt, *Autobiography* (New York: Charles Scribner's Sons, 1913), 404–45.

Chapter 7
Conformers and Reformers

1. J. Ise, *The United States Forest Policy* (New Haven, Conn.: Yale University Press, 1920), 82.
2. *Annual Report* (Washington, D.C.: U.S. Department of Interior, 1885), 234.
3. Editorial, *Conservation* 2(1908): 579–84.
4. Ise, *Forest Policy*, 179.
5. *Annual Report of the General Land Office* (Washington, D.C.: Government Printing Office, 1901), 327.
6. *Annual Report* (Washington, D.C.: U.S. Department of Interior, 1905), 327.
7. R. L. Williams, *The Loggers* (Alexandria, Va.: Time-Life Books, 1976), 195.
8. R. W. Hidy, F. E. Hill, and A. Nevins, *Timber and Men: The Weyerhaeuser Story* (New York: Macmillan, 1963).
9. F. K. Weyerhaeuser, *Trees and Men* (Princeton, N.J.: Newcomen, 1951).
10. G. Myers, *History of the Great American Fortunes* (New York: Modern Library, 1936).
11. W. A. Swanberg, *Citizen Hearst: A Biography of William Randolph Hearst* (New York: Charles Scribner's Sons, 1961), 36.
12. H. O'Connor, *The Guggenheims: The Making of an American Dynasty* (New York: Covici, Friede, 1937).
13. C. B. Glasscock, *The War of the Copper Kings: Builders of Butte and Wolves of Wall Street* (New York: Bobbs-Merrill Company, 1935).
14. W. K. Wyant, *Westward in Eden: The Public Lands and the Conservation Movement* (Berkeley: University of California Press, 1982), 177.
15. W. D. Mangam, *The Clarks: An American Phenomenon* (New York: Silver Bow Press, 1941), 69.
16. L. Atherton, *The Cattle Kings* (Lincoln: University of Nebraska Press, 1961), 267–69.
17. F. M. Tanner, *The Far Country: A Regional History of Moab and La Sal, Utah* (Salt Lake City: Olympus Publishing Company, 1976), 173–74.
18. P. H. Roberts, *Hoof Prints on Forest Ranges: The Early Years of*

National Forest Range Administration (San Antonio: Naylor Company, 1963), 35.

19. W. T. Jackson, "The Wyoming Stock Growers Association, Political Power in Wyoming Territory, 1873–1900," *Missouri Valley Historical Review* 33(1947): 571–94.

20. This case is discussed in the previous chapter.

21. E. R. Richardson, *The Politics of Conservation, Crusades and Controversies, 1897–1913* (Berkeley: University of California Press, 1962), 8.

22. Ibid., 18–20.

23. Ibid., 20–24.

24. Wyant, *Westward in Eden*, 306.

25. T. A. Larson, *The History of Wyoming* (Lincoln: University of Nebraska Press, 1965), 382.

26. H. Clepper, "Forestry's Uncertain Beginning," *Journal of Forestry* 67(1969): 218–21.

27. E. L. Jacobson, "Franklin B. Hough: A Pioneer in Scientific Forestry in America," *New York History* 15(1934): 311–18.

28. D. Lowenthal, *George Perkins Marsh: Versatile Vermonter* (New York: Columbia University Press, 1958), 268.

29. G. Pinchot, "How the National Forests Were Won," *American Forests and Forest Life* 36(1930): 616.

30. H. K. Steen, *The U.S. Forest Service: A History* (Seattle: University of Washington Press, 1976), 37–38.

31. S. Fox, *John Muir and His Legacy: The American Conservation Movement* (Boston: Little, Brown and Company, 1981), 113.

32. M. N. McGeary, *Gifford Pinchot: Forester-Politician* (Princeton, N.J.: Princeton University Press, 1960), 37.

33. Clepper, "Forestry's Uncertain Beginnings," 221.

34. E. Morris, *The Rise of Theodore Roosevelt* (New York: Random House, 1979).

35. L. Lang, *Ranching with Roosevelt* (New York: Lippincott, 1926).

36. H. Hagedorn, *Roosevelt in the Bad Lands* (New York: Houghton Mifflin, 1921), 208.

37. Morris, *Theodore Roosevelt*, 283–84.

38. H. T. Pinkett, *Gifford Pinchot: Private and Public Forester* (Urbana: University of Illinois Press), 1970.

39. McGeary, *Gifford Pinchot*, 15.

40. Steen, *U.S. Forest Service*, 45–46.

41. C. A. Schenck, *The Biltmore Story: Recollection of the Beginnings of Forestry in the United States* (St. Paul, Minn.: Minnesota Historical Society), 1955.

42. M. Frome, *Whose Woods These Are: The Story of the National Forests* (New York: Doubleday and Company, 1962), 51.

43. G. Pinchot, *Breaking New Ground* (Seattle: University of Washington Press, 1947, reprinted ed. 1972), 150–52.

44. McGeary, *Gifford Pinchot*, 20–30.

45. Richardson, *Politics of Conservation*, 26–30.

46. Roberts, *Hoof Prints on Forest Ranges*, 38–39.

47. C. DuBois, *Trail Blazers* (Stonington, Conn.: Stonington Publishing Company, 1957), 23.
48. Frome, *Whose Woods These Are*, 64.
49. Roberts, *Hoof Prints on Forest Ranges*, 4.
50. Ibid., 67.
51. Ibid., 68–69.
52. J. Riis, *Ranger Trails* (Richmond, Va.: Dietz Press, 1937).
53. Tanner, *The Far Country*, 195–201.
54. L. H. LeFevre, and N. G. Woolsey, *The Boulder Country and Its People* (Springville, Utah: Art City Publishing, 1973), 187–208.
55. T. Roosevelt, *Hunting Trips of a Ranchman* (New York: Putnam, 1885).
56. P. R. Cutright, *Theodore Roosevelt the Naturalist* (New York: Harper's, 1956).
57. P. Brooks, *Speaking for Nature: How Literary Naturalists from Henry David Thoreau to Rachel Carson Have Shaped America* (Boston: Houghton Mifflin Company, 1980), 114–15.
58. W. T. Hornaday, *Our Vanishing Wildlife* (Washington, D.C.: National Audubon Society, 1913).
59. J. C. Van Dyke, *The Desert: Further Studies in Natural Appearances* (New York: Charles Scribner's Sons, 1901).
60. F. Bergon, ed., *The Wilderness Reader* (New York: New American Library, 1980), 273–74.
61. Brooks, *Speaking for Nature*, 183-86.
62. Ibid., 186.
63. M. Austin, *The Land of Little Rain* (Boston: Houghton Mifflin Company, 1903).
64. H. R. Jones, *John Muir and the Sierra Club* (San Francisco: Sierra Club Books, 1965).
65. Fox, *John Muir and His Legacy*, 120.
66. Ibid., 130–34.

Chapter 8
The Hole in the Donut

1. Briefing, Utah State Representatives, Department of Energy, 15 February 1977.
2. J. Garreau, *The Nine Nations of North America* (Boston: Houghton Mifflin Company, 1981), 287–89.
3. U.S. Department of Commerce, *Historical Statistics of the United States, Colonial Times to 1970* (Washington, D.C.: Government Printing Office, 1975), vol. 1, 428–29.
4. W. C. Everhart, *The National Park Service* (New York: Praeger Publishers, 1975), 252–55.
5. R. D. Sale, and E. D. Karn, *American Expansion: A Book of Maps* (Lincoln: University of Nebraska Press, 1962), 25.
6. H. E. Gregory, and R. C. Moore, *The Kaiparowits Region: A Geo-*

graphic and Geologic Reconnaissance of Parts of Utah and Arizona (Washington, D.C.: U.S. Geological Survey, 1931), Professional Paper 134.

7. J. C. Hendee, G. H. Stankey, and R. C. Lucas, *Wilderness Management* (Washington, D.C.: U.S. Department of Agriculture, 1978), U.S. Forest Service Misc. Publ. No. 1365, 68.

8. R. Nash, *Wilderness and the American Mind*, 2d ed. (New Haven, Conn.: Yale University Press, 1973), 42–49.

9. W. J. Bryan, *The First Battle: A Story of the Campaign of 1896* (Chicago: W. B. Conkey Company, 1896), 199–206.

10. F. E. Haynes, "The New Sectionalism," *Quarterly Journal of Economics* 10(1896): 269–70, 280–95.

11. N. de Gennaro, *Arizona Statistical Abstract: A 1979 Data Handbook* (Flagstaff, Ariz.: Northland Press, 1979), 277–87.

12. B. DeVoto, *1846, The Year of Decision* (Boston: Little, Brown and Company, 1943).

13. R. D. Britton, "The B. H. Roberts Case of 1898-1900," *Utah Historical Quarterly* 25(1957): 27–46.

14. J. W. Garner, "The Case of Senator Smoot and the Mormon Church," *North American Review* 184(1907): 46.

15. L. J. Arrington, *Great Basin Kingdom: Economic History of the Latter-Day Saints, 1830–1900* (Lincoln: University of Nebraska Press, 1958), 406.

16. G. M. Greesley, ed., *Bostonians and Bullion: The Journal of Robert Livermore, 1892-1915* (Lincoln: University of Nebraska Press, 1968).

17. C. F. Park, Jr., *Minerals and the Political Economy* (San Francisco: Freeman, Cooper and Company, 1968).

18. Nash, *Wilderness and the American Mind*, 42.

19. R. A. Billington, *Westward Expansion: A History of the American Frontier* (New York: Macmillan, 1974), 632.

20. R. Higgs, "Railroad Rates and the Populist Uprising," *Agricultural History* 44: 197.

21. C. H. Dunning, and E. H. Peplow, *Rock to Riches: The Story of Arizona Mines and Mining* (Pasadena, Calif.: Hicks Publishing, 1966).

22. C. B. Glasscock, *The War of the Copper Kings: Builders of Butte and Wolves and Wall Street* (New York: Bobbs-Merrill Company, 1935).

23. W. S. Greever, *The Bonanza West: The Story of the Western Mining Rushes, 1848–1900* (Norman: University of Oklahoma Press, 1963).

24. Nash, *Wilderness and the American Mind*, 30–31.

25. R. W. Hidy, F. E. Hill, and A. Nevins, *Timber and Men: The Weyerhaeuser Story* (New York: Macmillan, 1963).

26. G. Pinchot, *Breaking New Ground* (Seattle: University of Washington Press, 1947).

27. R. U. Cooke, and R. W. Reeves, *Arroyos and Environmental Change in the American South-West* (Oxford: Clarendon Press, 1976).

28. W. L. Graf, "The Arroyo Problem: Paleohydrology and Paleohydraulics in the Short Term," in Gregroy, K. J., ed., *Perspectives on Paleohydrology* (London: John Wiley & Sons, 1983), 262–303.

29. W. L. Graf, "Late Holocene Sediment Storage in Canyons of the Colorado Plateau," *Geological Society of America Bulletin* 99(1987): 261–71.

30. T. T. Swift, "Date of Channel Trenching in the Southwest," *Science* 63(1925): 70–71.
31. W. L. Graf, "Downstream Changes in Stream Power, Henry Mountains, Utah," *Annals of the Association of American Geographers* 73(1983): 373-87.
32. H. E. Gregory, *The San Juan Country* (Washington, D.C.: U.S. Geological Survey, 1938).
33. D. E. Burkam, *Channel Changes of the Gila River in Safford Valley, Arizona* (Washington, D.C.: U.S. Geological Survey, 1972), Professional Paper 655-G, 3–6.
34. F. Waters, *The Colorado* (New York: Holt, Rinehart and Winston, 1946), 300–310.
35. F. M. Tanner, *The Far Country: A Regional History of Moab and La Sal, Utah* (Salt Lake City, Utah: Olympus Publishing Company, 1976), 196–97.
36. *Grand Valley Times*, 1 December 1905.
37. F. A. Branson, G. F. Gifford, K. G. Renard, and R. F. Hadley, *Rangeland Hydrology* (Dubuque, Iowa: Kendall/Hunt Publishers, 1981).
38. V. T. Chow, ed., *Handbook of Hydrology* (New York: McGraw-Hill, 1964).
39. R. Geiger, *The Climate Near the Ground* (Cambridge, Mass.: Harvard University Press, 1966).
40. R. G. Barry, and R. J. Chorley, *Atmosphere, Weather, and Climate* (New York: Holt, Rinehart and Winston, 1970).
41. W. E. K. Middleton, *A History of the Theories of Rain and Other Forms of Precipitation* (New York: F. W. Watts, 1965), 153.
42. D. E. Brown, and C. H. Lowe, *Biotic Communities of the Southwest* (Washington, D.C.: U.S. Department of Agriculture, 1980), U.S. Forest Service General Technical Report RM-78.
43. S. J. Pyne, *Grove Karl Gilbert: A Great Engine of Research* (Austin: University of Texas Press, 1980), 204–5.
44. H. Swan, *History of Music in the Southwest, 1825–1950* (San Marino, Calif.: Huntington Library, 1952).
45. R. L. Davis, *A History of Opera in the American West* (Englewood Cliffs, N.J.: Prentice-Hall, 1965).
46. E. R. Hart, ed., *That Awesome Space: Human Interaction with the Intermountain Landscape* (Salt Lake City, Utah: Westwater Press, 1981).

Chapter 9
Resolution . . . Again

1. J. A. Shaforth, "Real Conservation and Federal Ownership," *Continental Magazine* (May 1912): 112–13.
2. G. L. Knapp, "The Other Side of Conservation," *North American Review* 191(1910): 465–81.
3. Editorial, "Conservation from a Colorado Viewpoint—By a Critic of the Government," *Collier's Weekly* 50(1913): 24.

4. W. B. Greeley, "Shall the National Forests Be Abolished?" *Mining Congress Journal* 13(1927): 594–97.
5. E. R. Richardson, *The Politics of Conservation, Crusades and Controversies, 1897–1915* (Berkeley: University of California Press, 1962), 35.
6. J. Ise, *The United States Forest Policy* (New Haven, Conn.: Yale University Press, 1920), 174–76.
7. *Rocky Mountain News*, 17 June 1907.
8. Richardson, *The Politics of Conservation*, 38.
9. F. C. Johnson, *Proceedings of the Public Lands Convention Held at Denver, Colorado, June 18, 19, 20, 1907, by the States and Territories Containing Public Lands of the United States and Lying West of the Missouri River* (Denver: Public Lands Conference, 1907), 168.
10. Richardson, *The Politics of Conservation*, 39.
11. Ibid., 40.
12. S. T. Dana, *Forest and Range Policy* (New York: McGraw-Hill, 1956), 146.
13. G. C. Coogins, and C. F. Wilkinson, *Federal Public Land and Resources Law* (Mineola, N.Y.: Foundation Press, 1981), 122–24.
14. 22 U.S. 523.
15. 137 U.S. 160.
16. P. P. Wells, "Phillip P. Wells in the Forest Service Law Office," *Forest History* 16(1972): 22–29.
17. G. Pinchot, *Breaking New Ground* (Seattle: University of Washington Press, 1947), 319–26.
18. U.S. House. Congress. H. Doc. 1425, 60th Cong., 2 sess., 1908.
19. Pinchot, *Breaking New Ground*, 326–33.
20. H. K. Steen, *The U.S. Forest Service: A History* (Seattle: University of Washington Press, 1976), 8–9.
21. Richardson, *The Politics of Conservation*, 46.
22. U.S. Congress. Senate. S. Doc. 676, 60th Cong., 2d sess., 1908.
23. Richardson, *The Politics of Conservation*, 48.
24. Ibid., 63–64.
25. D. F. Anderson, *William Howard Taft: A Conservative's Conception of the Presidency* (Ithaca, N.Y.: Cornell University Press, 1965), 72–76.
26. T. T. Munger, "Recollections of My Thirty-Eight Years in the Forest Service, 1908–1946," *Timberline* 16(1962): 13–14.
27. U.S. Congress. Senate. S. Doc. 719, 61st Cong., 2d sess., 1911.
28. Richardson, *The Politics of Conservation*, 80.
29. Munger, "Recollections," 14.
30. Dana, *Forest and Range Policy*, 157.
31. H. Clepper, *Professional Forestry in the United States* (Baltimore: Johns Hopkins University Press), 1971.
32. Ise, *Forest Policy*, 180.
33. Richardson, *Politics of Conservation*, 89.
34. *Proceedings of the Twelfth Trans-Mississippi Commercial Congress Held at Denver, Colorado, August 16–21, 1909* (Denver: Commercial Congress, 1909).
35. 137 U.S. 160.
36. Richardson, *Politics of Conservation*, 90–91.

37. *Bellingham Washington Herald*, 11 September 1910.
38. F. W. Harris, and A. P. Johnson, *The Neglected West* (Seattle, Wash.: City of Seattle, 1911), Municipal Monograph No. 1.
39. Richardson, *Politics of Conservation*, 99.
40. *Salt Lake City Tribune*, 7 August 1910.
41. Taft, W. H., "Appendix to Second Annual Message, Address to the National Conservation Congress in St. Paul, Minnesota, September 5, 1910," *A Compilation of the Messages and Papers of the Presidents* (New York: Bureau of National Literature, 1910), Vol. 16, 7555–74.
42. H. T. Pinkett, *Gifford Pinchot: Private and Public Forester* (Urbana: University of Illinois Press, 1970).
43. D. Brandis, "The Late Franklin B. Hough," *Indiana Forester* 11(1885): 429.
44. B. E. Fernow, *Economics of Forestry* (New York: Thomas Y. Crowell Publishers, 1902).
45. A. D. Rodgers III, *Bernhard Edward Fernow: A Story of North American Forestry* (Princeton, N.J.: Princeton University Press, 1951).
46. Steen, *U.S. Forest Service*, 46.
47. S. P. Hays, *Conservation and the Gospel of Efficiency: The Progressive Conservation Movement, 1890–1920* (Cambridge, Mass.: Harvard University Press, 1960).
48. Richardson, *The Politics of Conservation*, 94.
49. Bureau of Land Management, *Public Land Statistics* (Washington, D.C.: U.S. Department of Interior, 1980), 33.
50. G. O. Robinson, *The Forest Service: A Study in Public Land Management* (Baltimore: Johns Hopkins University Press, 1975).
51. 137 U.S. 160.
52. 179 U.S. 243.
53. 220 U.S. 523.
54. R. Nash, *Wilderness and the American Mind*, 2d ed. (New Haven, Conn.: Yale University Press, 1973).
55. L. Urdang, *The Timetable of American History* (New York: Simon and Schuster, 1981), 260–81.

Chapter 10
Sacred Cows and the Pastures of Heaven

1. *American Sheep Breeder and Wool Grower*, 37 (January 1917):19.
2. H. K. Steen, *The U.S. Forest Service: A History* (Seattle: University of Washington Press, 1976), 163.
3. "Convention Notes," *American Sheep Breeder and Wool Grower*, 36 (January 1916): 26.
4. P. O. Foss, *The Politics of Grass: The Administration of Grazing on the Public Domain* (Seattle: University of Washington Press, 1960), 43.
5. Ibid., 44.
6. Ibid., 45.
7. E. L. Peffer, *The Closing of the Public Domain: Disposal and Reserva-*

tion Policies, 1900–1950 (Stanford, Calif.: Stanford University Press, 1951), 180.

8. Senate Resolution 4076, 1927.

9. C. E. Rachford, *Range Appraisal Report* (Washington, D.C.: U.S. Forest Service, 1924), 1.

10. W. L. Dutton, "History of Forest Service Grazing Fees," *Journal of Range Management* 6 (1953): 393–98.

11. *Congressional Record*, 68th Cong., 2d sess., 1924–1925, 5505.

12. Steen, *Forest Service*, 166.

13. W. Voigt, Jr., *Public Grazing Lands: Use and Misuse by Industry and Government* (New Brunswick, N.J.: Rutgers University Press, 1976), 61.

14. Hearings on U.S. Congress. Senate. 2584, 69th Cong., 1st sess., 1925–1926, 853.

15. Ibid., 907.

16. Peffer, *Public Domain*, 188–92.

17. U.S. Congress. Senate. S. 2584, 69th Cong., 1st sess, 1925, 500.

18. Steen, *Forest Service*, 167.

19. Editorial, *San Francisco Examiner*, 11 March 1928.

20. Dutton, "History of Forest Service Grazing Fees," 395.

21. *Congressional Record*, 71st Cong., 1st sess., 1929, 3570–71.

22. R. Robbins, *Our Landed Heritage: The Public Domain 1776–1936* (Lincoln: University of Nebraska Press, 1942), 413.

23. *Congressional Record*, 71st Cong., 1st sess., 1929, 3572.

24. R. L. Wilbur, and A. M. Hyde, *The Hoover Policies* (New York: Charles Scribner's Sons, 1937), 230.

25. *Congressional Record*, 71st Cong., 2d sess., 1929–1930, 408.

26. Wilbur and Hyde, *Hoover Policies*, 231–32.

27. Peffer, *Public Domain*, 204.

28. U.S. Congress. House.: *Grazing on the Public Domain, Hearings on H.R. 11816*. 72d Congr., 1st sess., 1932.

29. H. Hoover, *State Papers* (Washington, D.C.: Government Printing Office, 1934), 2:117–18.

30. H. B. Graves, "The Public Domain," *The Nation* 81(1930): 147–49.

31. W. Shephard, "The Handout Magnificent," *Harper's Monthly* 163(1931): 594–603.

32. Editorial, "Flinging Away an Empire," *The New Republic* 70(24 February 1932): 32.

33. Peffer, *Public Domain*, 233.

34. Editorial, *Billings Montana Gazette*, 23 November 1929.

35. Editorial, "The Public Lands," *Outlook and Independent* 153(11 September 1929): 54.

36. G. D. Nash, *The American West in the Twentieth Century: A Short History of an Urban Oasis* (Albuquerque: University of New Mexico Press, 1977), 141–43.

37. *Historical Statistics of the United States, Colonial Times to 1970* (Washington, D.C.: Department of Commerce, Bureau of the Census, 1975), 1:519.

38. Foss, *Politics of Grass*, 48–49.

39. *U.S. Statutes*, 45:380.

40. *A National Plan for American Forestry*. U.S. S. Doc. 12, 73d Cong., 1st sess., 1933.

41. *Congressional Record, Report on H. R. 11816*, 72d Cong., 1st sess., 1932, 7.

42. *Congressional Record*, 63d Cong., 2d sess., 1913–1914, 13680.

43. *Congressional Record, Hearings on H. R.* 6462. 73d Cong., 2d sess., 1934, 44.

44. Peffer, *Public Domain*, 217.

45. U.S. Congress. House. Committee on Public Lands. *To Provide for the Orderly Use, Improvement, and Development of the Public Range,'' Hearings on H. R. 2835, Congressional Record*, 73d Cong., 1st sess., 1933, 171–72.

46. T. H. Watkins, and C. S. Watson, Jr., *The Lands No One Knows* (San Francisco: Sierra Club, 1975), 112–14.

49. U.S. Congress. House. Committee on Public Lands. *To Provide for the Orderly Use, Improvement, and Development of the Public Range, Hearings on H. R. 2835. Congressional Record*, 73d Cong., 1st sess., 1933, 55.

50. Foss, *Politics of Grass*, 58.

51. Ibid., 58–59.

53. *Congressional Record*, 74th Cong., 1st sess., 1935, 6013.

54. 48 USCA 1269.

55. *Annual Report of the Department of Interior* (Washington, D.C.: U.S. Department of Interior, 1934), 61.

56. Peffer, *Public Domain*, 224.

57. W. Calef, *Private Grazing and Public Lands: Studies of the Local Management of the Taylor Grazing Act* (Chicago: University of Chicago Press, 1960), 53.

Chapter 11
The Wild West Show

1. R. A. Foresta, *America's National Parks and Their Keepers* (Washington, D.C.: Resources for the Future, 1984), 20.

2. R. N. Searle, *Saving Quetico-Superior: A Land Set Apart* (St. Paul, Minn.: Minnesota Historical Society Press, 1977), 19.

3. J. C. Hendee, G. H. Stankey, and R. C. Lucas, *Wilderness Management* (Washington, D.C.: U.S. Department of Agriculture, Forest Service, Miscellaneous Publication 1365), 64.

4. W. C. Everhart, *The National Park Service* (New York: Praeger Publishers, 1972), 24–25.

5. R. E. McArdle, ''Wilderness Politics: Legislation and Forest Service Policy,'' *Forest History* 19(1975): 166–79.

6. C. W. Allin, *The Politics of Wilderness Preservation* (Westport, Conn.: Greenwood Press, 1982), 68.

7. D. W. Baldwin, *The Quiet Revolution: Grass Roots of Today's Wilderness Preservation Movement* (Boulder, Colo.: Prescott, 1972), 36–42.

8. S. L. Flader, *Thinking Like a Mountain: Aldo Leopold and the Evolution*

of an Ecological Attitude Toward Deer, Wolves and Forests (Lincoln, Nebr.: University of Nebraska Press, 1974), 76–121.

9. A. Leopold, "The Wilderness and Its Place in Forest Recreational Policy," *Journal of Forestry* 29(1921): 718–21.

10. Ibid., 720.

11. J. P. Gilligan, "The Development of Policy and Administration of Forest Service Primitive and Wilderness Areas in the Western United States." Ph.D. diss., University of Michigan, 1953, 114.

12. Baldwin, *The Quiet Revolution*, 65.

13. J. D. Wellman, *Wildland Recreation Policy: An Introduction* (New York: John Wiley & Sons, 1987), 134–35.

14. "Recreational Resources of Federal Lands" (Washington, D.C.: National Conference on Outdoor Recreation, 1928), 86–103.

15. Hendee et al., *Wilderness Management*, 62.

16. R. Marshall, "The Problem of Wilderness," *The Scientific Monthly* 30 (1930): 142–43.

17. U.S. Department of Agriculture, *Forest Service Manual*, section G, "Wilderness Areas" (Washington, D.C.: U.S. Department of Agriculture, Forest Service, 1955), 23.

18. Hendee et al., *Wilderness Management*, 62–63.

19. S. Fox, *John Muir and His Legacy: The American Conservation Movement* (Boston: Little, Brown, 1981), 191.

20. T. H. Beck, J. N. Darling, and A. Leopold, *Report of the President's Committee on Wildlife Restoration* (Washington, D.C.: Government Printing Office, 1934), 7.

21. Fox, *John Muir*, 193.

22. Ibid., 194.

23. E. B. Nixon, ed., *Franklin D. Roosevelt and Conservation, 1911–1945* (Hyde Park, N.Y.: General Services Administration, 1957), 2: 405–6.

24. Bureau of Land Management, *Public Land Statistics* (Washington, D.C.: Department of Interior, Bureau of Land Management, 1978), 12.

25. U.S. Congress. Senate. *A National Plan for American Forestry*. S. Doc. 12. 73d Cong., 1st sess., 1933.

26. H. K. Steen, *The U. S. Forest Service: A History* (Seattle: University of Washington Press, 1976), 206.

27. W. Voigt, Jr., *Public Grazing Lands: Use and Misuse by Industry* (New Brunswick, N.J.: Rutgers University Press, 1976), 67.

28. W. L. Dutton, "History of Forest Service Grazing Fees," *Journal of Range Management* 6 (1953): 393–98.

29. Voigt, *Public Grazing Lands*, 68–69.

30. F. E. Mollin, "On Western Grazing," *Successful Farming* 42 (October, 1937), 4.

31. F. E. Mollin, *If and When It Rains: The Stockman's View of the Range Question* (Denver: American National Livestock Association, 1938), iii.

32. "Changing the Name of the Interior Department, Hearings on S. 2665," *Congressional Record*, 74th Cong., 1st sess., 1935.

33. *Congressional Record*, 76th Cong., 3d sess., 1940–1941, Appendix, 533, 574, 1572–73.

34. E. L. Peffer, *The Closing of the Public Domain: Disposal and Reserva-*

tion Policies 1900–1950 (Stanford, Calif.: Stanford University Press, 1951), 236.

35. *Government Organization, Report to Accompany S. 2970*. Report from the Select Committee on Government Organization, *Congressional Record*, 75th Cong., 1st sess., 1937.

36. T. H. Watkins, "The Terrible-Tempered Mr. Ickes," *Audubon* 82 (1984): 93–111.

37. T. E. Conklin, "Public Land Law Review Commission Revisited—A Potpourri of Memories," *Denver Law Journal* 54 (1977): 445–47.

38. P. O. Foss, *Politics and Grass: The Administration of Grazing on the Public Domain* (Seattle: University of Washington Press, 1960), 130–60.

39. *Dewar v. Brooks*, 60 Nev. 219.

40. *Dewar v. Brooks*, P. 2d 755.

41. Bureau of Land Management, "A Comparison of Grazing Fees on National Forest and Grazing Districts," unpublished paper, National Advisory Board Council, Bureau of Land Management, 1950, 14.

42. U.S. Department of Interior, *Annual Report* (Washington, D.C.: Government Printing Office, 1944), 222.

43. *Congressional Record*, 76th Cong., 3d sess., 1940, 2593.

44. *Congressional Record*, 78th Cong., 2d sess., 1944, 9558–59.

45. *Congressional Record*, 79th Cong., 2d sess., 1946, 20–21.

46. *Congressional Record*. 79th Cong., 1st sess., 1946, 463.

47. *Congressional Record*, 79th Cong., 2d sess., 1946, 4634.

48. B. Shanks, *This Land Is Your Land: The Struggle to Save America's Public Lands* (San Francisco: Sierra Club, 1984), 182.

49. Voigt, *Public Grazing Lands*, 92.

50. Editorial, *American Cattle Producer* 23 (1947): 8.

51. *Congressional Record*, 79th Cong., 2d sess., 1945, S. 1945.

52. Voigt, *Public Grazing Lands*, 93.

53. B. DeVoto, "The Anxious West," *Harper's Monthly* 93 (December, 1946): 481–91.

54. B. DeVoto, "The West Against Itself," *Harper's Monthly* 94 (January, 1947): 231–56.

55. W. Stegner, *The Uneasy Chair: A Biography of Bernard DeVoto* (New York: Doubleday, 1973), 301.

56. Editorial, *Denver Post*, 2 February 1947.

57. Voigt, *Public Grazing Lands*, 94–110.

58. Stegner, *The Uneasy Chair*, 304.

59. Voigt, *Public Grazing Lands*, 115.

60. "Hearings Pursuant to H.R.93," *Congressional Record*, 80th Cong., 1st sess., 1947.

61. A. D. Brownfield, and S. C. Hyatt, "A Proposal for Handling the Public Lands," *American Cattle Producer* 28 (1946): 8.

Chapter 12
Cowboys, Columnists, and Conservationists

1. B. H. Granger, *Will C. Barnes' Arizona Place Names* (Tucson: University of Arizona Press, 1979), 139.

2. *National Cyclopedia of American Biography* (New York: James T. White Company, 1916–1981), 42:127–28.
3. Ibid., 415–16.
4. *New York Times*, 16 April 1945, 23.
5. *National Cyclopedia*, supp. C:133.
6. H. K. Steen, *The U.S. Forest Service: A History* (Seattle: University of Washington Press, 1976), 167.
7. D. Morris, and I. Morris, *Who Was Who in American Politics* (New York: Hawthorn Books, 1974), 554.
8. *Congressional Record*, 63d Cong., 2d sess., 13680.
9. M. Block, ed., *Current Biography* (New York: H. W. Wilson, 1941), 852.
10. *National Cyclopedia*, supp. D:214.
11. *Dictionary of American Biography* (New York: Charles Scribner's Sons, 1981), supp. 5:443–45.
12. *New York Herald Tribune*, 29 September 1954, 26.
13. W. Voigt, Jr., *Public Grazing Lands: Use and Misuse by Industry and Government* (New Brunswick, N.J.: Rutgers University Press, 1976), 94.
14. Voigt, *Public Grazing Lands*, 94–95.
15. *Dictionary of American Biography*, supp. 7:30–32.
16. Morris and Morris, *Who Was Who*, 53.
17. H. Hoover, *The Memoirs of Herbert Hoover* (New York: Macmillan, 1952–1953), 3 vols.
18. *National Cyclopedia*, 56:295–302.
19. R. L. Wilbur, and A. M. Hyde, *The Hoover Policies* (New York: Charles Scribner's Sons, 1937), 100–120.
20. N. Hundley, Jr., *Dividing the Waters: A Century of Controversy Between the United States and Mexico* (Berkeley: University of California Press, 1966), 48–63.
21. *National Cyclopedia*, supp. C:12.
22. R. L. Wilbur, and A. M. Hyde, *The Hoover Policies* (New York: Charles Scribner's Sons, 1937), ix–x.
23. E. B. Nixon, ed., *Franklin D. Roosevelt and Conservation, 1911–1945* (Hyde Park, N.Y.: General Services Administration, 1957).
24. O. Dows, *Franklin Roosevelt at Hyde Park* (New York: American Artists' Group, 1949).
25. *Dictionary of American Biography*, supp. 3:641–67.
26. *National Cyclopedia*, 40:4–6; supp. F:23–26.
27. T. H. Watkins, "The Terrible-Tempered Mr. Ickes," *Audubon* 82(1984): 93–111.
28. H. L. Ickes, *The Autobiography of a Curmudgeon* (New York: Reynal and Hitchcock, 1943).
29. R. Polenberg, *Reorganizing Roosevelt's Government: The Controversy Over Executive Reorganization, 1936–1939* (Cambridge, Mass.: Harvard University Press, 1966).
30. Steen, *U.S. Forest Service*, 240.
31. L. T. Lear, *Harold L. Ickes: The Aggressive Progressive, 1874–1933* (New York: Garland Publishers, 1940).

32. J. B. Trefethen, *Crusade for Wildlife: Highlights in Conservation Progress* (Harrisburg: Stackpole Company, 1961).
33. A. Leopold, "The Green Lagoons," *American Forests* 51(1945): 414.
34. S. L. Flader, *Thinking Like a Mountain: Aldo Leopold and the Evolution of an Ecological Attitude Toward Deer, Wolves and Forests* (Lincoln: University of Nebraska Press, 1974).
35. A. Leopold, *Game Management* (New York: Charles Scribner's Sons, 1933).
36. P. L. Errington, "In Appreciation of Aldo Leopold," *Journal of Wildlife Management* 12(1948): 341–50.
37. A. Leopold, *A Sand County Almanac* (Oxford: Oxford University Press, 1949).
38. J. D. Wellman, *Wildland Recreation Policy: An Introduction* (New York: John Wiley & Sons, 1987), 147–53.
39. P. Brooks, *Speaking for Nature: How Literary Naturalists from Henry Thoreau to Rachel Carson Have Shaped America* (Boston: Houghton Mifflin, 1980), 263–69.
40. Wellman, *Wildland Recreation Policy*, 154.
41. Brooks, *Speaking for Nature*, 267.
42. S. Fox, *John Muir and His Legacy: The American Conservation Movement* (Boston: Little, Brown, 1981), 175–77.
43. *Time Magazine*, 3 November 1930, 23.
44. Fox, *John Muir*, 178.
45. F. P. Jaques, *Birds Across the Sky* (April, 1942), 182.
46. Fox, *John Muir*, 226.
47. Editorial, *Outdoor America* 12(April 1947): 7.
48. Editorial, *Outdoor America* 10(April 1945): 16.
49. O. Sawey, *Bernard DeVoto* (New York: Twayne Publishers, 1969).
50. W. Stegner, *The Uneasy Chair: A Biography of Bernard DeVoto* (Garden City, N.Y.: Doubleday, 1974).
51. B. DeVoto, "The West: A Plundered Province," *Harper's Magazine* 164(August 1934): 355–64.

Chapter 13
The West in Transition

1. Bureau of the Census, *Historical Statistics of the United States: Colonial Times to 1970* (Washington, D.C.: U.S. Department of Commerce, 1975), 519.
2. V. Carstensen, ed., *The Public Lands: Studies in the History of the Public Domain* (Madison: University of Wisconsin Press, 1962), 481.
3. S. A. Mosk, "Land Policy and Stock Raising in the Western United States," in *The Public Lands: Studies in the History of the Public Domain*, ed., V. Carstensen (Madison: University of Wisconsin Press, 1962), 411–34.
4. Arizona State Land Commission, *Report of the Arizona State Land Commission, June 6, 1912—December 1, 1914* (Phoenix: State of Arizona, 1914), 22–23.

5. F. V. Coville, *Bulletin 62* (Washington, D.C.: U.S. Forest Service, 1905), 55.

6. P. H. Roberts, *Hoof Prints on Forest Ranges* (San Antonio: Naylor Company, 1963), 102.

7. W. K. Wyant, *Westward in Eden: The Public Lands and the Conservation Movement* (Berkeley: University of California Press, 1982), 309.

8. W. Calef, *Private Grazing and Public Lands* (Chicago: University of Chicago Press, 1960), 43.

9. G. D. Nash, *The American West Transformed: The Impact of the Second World War* (Bloomington: Indiana University Press, 1985), 17–34.

10. A. S. Lott, *A Long Line of Ships* (Annapolis: U.S. Naval Academy, 1954), 212–14.

11. U.S. Bureau of the Census, *Statistical Abstracts* (Washington, D.C.: U.S. Department of Commerce, 1953), 805.

12. C. M. Wiltse, *Aluminum Policies of the War Production Board and Predecessor Agencies* (Washington, D.C.: War Production Board, 1945), 173–75.

13. L. J. Arrington, and G. Jensen, *The Defense Industry of Utah* (Logan: Utah State University Press, 1965), 13–43.

14. Nash, *The West Transformed*, 28–29.

15. D. Hornbeck, review of *The West Transformed*, by G. R. Nash, *Professional Geographer* 38(1986): 448.

16. W. C. Palmer, *Meteorological Drought* (Washington, D.C.: National Oceanic and Atmospheric Administration, Research Paper 45, 1965), 4–10.

17. T. R. Karl, and R. W. Knight, *Atlas of Monthly Palmer Drought Severity Indices for the Contiguous United States* (Washington, D.C.: National Oceanic and Atmospheric Administration, 1985), 2 vols.

18. R. W. Bailey, ''Epicycles of Erosion in the Valleys of the Colorado Plateau Province,''*Journal of Geology* 43(1935):337–55.

19. H. C. Lockett, and M. Snow, *Along the Beale Trail: A Photographic Account of Wasted Range Land* (Washington, D.C.: U.S. Office of Indian Affairs, 1938), 2–3.

20. R. U. Cooke, and R. W. Reeves, *Arroyos and Environmental Change in the American South-West* (Oxford: Clarendon Press, 1976), 1.

21. H. D. Miser, *The San Juan Canyon, Southeastern Utah* (Washington, D.C., U.S. Geological Survey Water-Supply Paper 538, 1924), 55–71.

22. W. O. Smith, C. P. Vetter, and G. B. Cummings, *Comprehensive Survey of Sedimentation in Lake Mead, 1948–1949* (Washington, D.C.: U.S. Geological Survey Professional Paper 295, 1960), 1–3.

23. V. A. Vanoni, ed., *Sedimentation Engineering* (New York: American Society of Civil Engineers, 1975), 653–63.

24. W. D. Sellers, and R. H. Hill, *Arizona Climate 1931–1972* (Tucson, Ariz.: University of Arizona Press, 1974), 63.

25. C. H. Reitan, ''Trends in the Frequencies of Cyclone Activity Over North America,'' *Monthly Weather Review* 107(1980): 1684–88.

26. W. L. Graf, ''Fluvial Erosion and Federal Public Policy in the Navajo Nation,'' *Physical Geography* 7(1986):97–115.

27. L. B. Leopold, ''Reversal of Erosion Cycle and Climatic Change,'' *Quaternary Research* 6(1976): 557–62.

28. R. Hereford, "Climate and Ephemeral-Stream Processes: Twentieth Century Geomorphology and Alluvial Stratigraphy of the Little Colorado River, Arizona," *Geological Society of America Bulletin* 95(1984): 654–68.

29. R. Hereford, "Modern Alluvial History of the Paria River Drainage Basin," *Quaternary Research* 25(1986): 293–311.

30. W. L. Graf, "Late Holocene Sediment Storage in Canyons of the Colorado Plateau," *Geological Society of America Bulletin* 99(1987): 261–71.

31. J. N. Oka, *Sedimentation Study, Glen Canyon Dam, Colorado River Storage Project* (Salt Lake City: U.S. Bureau of Reclamation, 1962), 7–8.

32. N. Hundley, *Dividing the Waters* (Berkeley: University of California Press, 1966), 60.

33. M. C. Robinson, *Water for the West: The Bureau of Reclamation* (Chicago: Public Works Historical Society, 1979), 121.

34. R. Price, *History of Forest Service Research in the Central and Southern Rocky Mountain Regions 1908–1975* (Ft. Collins, Colo.: U.S. Forest Service General Technical Reort RM–27, 1976), 1–19.

35. C. W. Thornthwaite, *Climate and Accelerated Erosion in the Arid and Semi-Arid Southwest, with Special Reference to the Polacca Wash Drainage Basin, Arizona* (Washington, D.C.: U.S. Department of Agriculture Technical Bulletin 808, 1942), 125–29.

36. C. F. S. Sharpe, *What Is Soil Erosion?* (Washington, D.C.: U.S. Department of Agriculture Miscellaneous Publication 286, 1938), 1–2.

37. G. C. Lusby, V. H. Reid, and O. D. Knipe, *Effects of Grazing on the Hydrology and Biology of the Badger Wash Basin in Western Colorado, 1953–1966* (Washington, D.C.: U.S. Geological Survey Water-Supply Paper 1532-D, 1971), 62–64.

38. W. L. Graf, "Temporal Variation of Sediment Yield in the Upper Colorado River Basin," abst., *Program Abstracts, Association of American Geographers 1985 Annual Meeting*, 1985, 193.

39. S. W. Trimble, *Man-Induced Soil Erosion on the Southern Piedmont 1700–1900* (Ankeny, Iowa: Soil Conservation Society of America, 1974), 101.

40. W. K. Everson, *A Pictorial History of the Western Film* (New York: Citadel Press, 1969).

41. R. G. Athern, *The Mythic West in Twentieth-Century America* (Lawrence: University of Kansas Press, 1986), 138.

42. N. Gillespie, "The Spirit of Place," *Southwest Profile* (February 1988): 26–32.

43. W. H. Goetzmann, and W. N. Goetzmann, *The West of the Imagination* (New York: W. W. Norton & Company, 1986), 357.

44. L. P. Rudnick, *Mabel Dodge Luhan: New Woman, New World* (Albuquerque: University of New Mexico Press, 1984), 143.

45. L. Lyle, *Portrait of an Artist: A Biography of Georgia O'Keeffe* (New York: New York Graphic Society, 1980), 105.

46. A. Adams, *Ansel Adams: An Autobiography* (Boston: Little, Brown and Company, 1985), 121–38.

47. L. R. Borne, *Dude Ranching: A Complete History* (Albuquerque: University of New Mexico Press, 1983), 10–20.

48. H. J. Pererson, "Wyoming," in T. C. Donnelly, ed., *Rocky Mountain Politics* (Albuquerque: University of New Mexico Press, 1940), 122.

49. Athern, *The Mythic West*, 139.

Chapter 14
The Law of the Wilderness

1. M. McCloskey, "The Wilderness Act of 1964: Its Background and Meaning," *Oregon Law Review* 45(1966): 288–321.

2. R. Marshall, *The People's Forests* (New York: Harrison Smith and Robert Haas, 1933), 177.

3. M. Frome, *The Battle for the Wilderness* (New York: Praeger Publishers, 1974), 105.

4. H. H. Chapman, "National Parks, National Forests, and Wilderness Areas," *Journal of Forestry* 36(1938): 469-74.

5. S. Fox, *John Muir and His Legacy: The American Conservation Movement* (Boston: Little, Brown and Company, 1980), 287.

6. U.S. Senate, Committee on Interior and Insular Affairs, *Natural Resources Policy*, January 31–February 7, 1949, 260–70.

7. R. Leonard, and E. L. Sumner, "Protecting Mountain Meadows," *Sierra Club Bulletin* 32, no. 5 (1947): 53–62.

8. R. A. Baker, *Conservation Politics: The Senate Career of Clinton P. Anderson* (Albuquerque: University of New Mexico Press, 1985), 100–102.

9. H. Clepper, "Conservation's Grand Lodge," *American Forests* 73, no. 10 (1967): 22–27, 58–61.

10. J. C. Hendee, G. H. Stankey, and R. C. Lucas, *Wilderness Management*, Forest Service Miscellaneous Publication 1365 (Washington, D.C.: U.S. Department of Agriculture, 1978), 38.

11. C. F. Keyser, *The Preservation of Wilderness Areas—An Analysis of Opinion on the Problem* (Washington, D.C.: Legislative Reference Service, Library of Congress, 1949), 114.

12. List of Officers of the Wilderness Society, *Living Wilderness* (Summer 1949), i.

13. H. C. Zahniser, "How Much Wilderness Can We Afford to Lose?" in D. R. Brower, ed., *Wildlands in Our Civilization* (San Francisco: Sierra Club Books, 1964), 33–39.

14. J. B. Craig, "A Circle That Took Us In," *American Forests* 70, no. 3 (1964): 8–9.

15. Fox, *John Muir*, 289.

16. *Congressional Record* 101 (1 June 1955).

17. J. P. Gilligan, "The Development of Policy and Administration of the Forest Service Primitive and Wilderness Areas in the Western United States." Ph.D. diss., University of Michigan, 1954.

18. *Congressional Record*, 102 (11 July 1956), 12314–16.

19. C. W. Allin, *The Politics of Wilderness Preservation* (Westport, Conn.: Greenwood Press, 1982), 105.

20. U.S. Congress. Senate. Committee on Interior and Insular Affairs. *National Wilderness Preservation Act: Hearings on S. 1176*. June 19–20, 1957, 198.

21. *Congressional Record*, 102 (7 June 1956), 9778.

22. *Congressional Record*, 102 (11 June 1956), 11838.

23. Frome, *Battle for the Wilderness*, 27–104.

24. U.S. Congress. Senate. Committee on Interior and Insular Affairs. *National Wilderness Preservation Act: Hearings on S. 1176*. June 19–20, 1957, 331.

25. G. W. Barr, *Recovering Rainfall, Part 1: Arizona Watershed Program* (Phoenix: Arizona State Land Department, Salt River Valley Water Users Association, and University of Arizona, 1956), 1.

26. A. R. Hibbert, "Increases in Streamflow After Converting Chaparral to Grass," *Water Resources Research* 7(1971): 71–80.

27. U.S. Congress. Senate. Committee on Interior and Insular Affairs. *National Wilderness Preservation Act*, November 1958, 953.

28. Baker, *Clinton Anderson*, 115.

29. Ibid., 120.

30. U.S. Congress. Senate. Committee on Interior and Insular Affairs. *National Wilderness Preservation Act: Hearings on S. 1176*. June 19–20, 1957, 398.

31. ——— *National Wilderness Preservation Act*, November 1958, 926.

32. National Wildlife Federation, *Conservation News*, 1 March 1959, 1.

33. M. E. McCloskey, and J. P. Gilligan, eds., *Wilderness and the Quality of Life* (San Francisco: Sierra Club Books, 1969), 245–47.

34. U.S. Congress. Senate. Committee on Interior and Insular Affairs. *Hearings on S. 1176*. June 19–20, 1957, 379.

35. McCloskey, "Wilderness Act," 289-90.

36. U.S. Congress. Senate. Committee on Interior and Insular Affairs. *Hearings on S. 1176*. June 19–20, 1957, 258.

37. ——— *Hearings on S. 1176*. June 19–20, 1957, 331.

38. ——— *National Wilderness Preservation Act: Hearings on S. 4028*. November 7, 10, 12, and 14 1958, 2–6.

39. ——— *National Wilderness Preservation Act: Hearings on S. 4028*. November 7, 10, 12, and 14, 1958, 1.

40. *Congressional Record*, 107 (5 September 61), 18100.

41. Baker, *Clinton Anderson*, 120.

42. R. A. Foresta, *America's National Parks and Their Keepers* (Washington, D.C.: Resources for the Future, 1984), 62.

43. S. T. Dana, and S. K. Fairfax, *Forest and Range Policy*, 2d ed. (New York: McGraw-Hill, 1980), 200–210.

44. G. C. Cogins, and C. F. Wilkinson, *Federal Public Land and Resources Law* (Mineola, N.Y.: Foundation Press, 1981), 479–81.

45. E. C. Crafts, "Saga of a Law," *American Forests* 76(6, 1970): 18–19.

46. *Congressional Record*, 106 (2 July 1960), 15564.

47. *New York Times*, 24 February 1961, 12.

48. Allin, *Politics*, 106–12.

49. Editorial, *Washington Star*, 26 February 1961.

50. U.S. Congress. Senate. Committee on Interior and Insular Affairs. *Hearings on S. 174*. February 27, 28, 1961, 1–2.

51. ——— *Hearings on S. 174*. February 27, 28, 1961, 28.

52. C. P. Anderson, *Outsider in the Senate* (New York: World Publishing Company, 1970), 233.

53. J. L. Sundquist, *Politics and Policy: The Eisenhower, Kennedy, and Johnson Years* (Washington, D.C.: Brookings Institute, 1968), 355–61, 489–93.

54. Outdoor Recreation Resources Review Commission, *Wilderness and Recreation: A Report on Resources, Values, and Problems* (Washington, D.C.: Government Printing Office), 1962.
55. Baker, *Clinton Anderson*, 58.
56. U.S. Congress. House. Subcommittee on Public Lands. *Hearings on S. 174*. May 7–11, 1962, 1062–79.
57. *Congressional Quarterly, Weekly Report* 22 (17 August 1962), 1369.
58. Allin, *Politics*, 128–29.
59. Baker, *Clinton Anderson*, 199.
60. *Congressional Record* 109 (9 April 1963), 5887–5932.
61. *Congressional Record* 109 (7 November 1963), 18606–607.
62. U.S. Congress. House. *Establishment of Public Land Law Review Commission*. H. Rept. 88-1008, December 7, 1963, 1–4.
63. Baker, *Clinton Anderson*, 7.
64. P. Brooks, "Congressman Aspinall vs. the People of the United States," *Harper's Magazine* (March 1963), 59–63.
65. Baker, *Clinton Anderson*, 209.
66. U.S. Congress. House. *National Wilderness Preservation System*. Rept. 88-1538, July 2, 1964, 1–6.
67. *New York Times*, 31 July 1964, 1.
68. E. H. Hanks, A. D. Tarlock, and J. L. Hanks, *Environmental Law and Policy* (St. Paul, Minn.: West Publishing Company, 1975), 267–72.

Chapter 15
Winning the Wilderness Game

1. M. E. McCloskey, and J. P. Gilligan, eds., *Wilderness and the Quality of Life* (San Francisco: Sierra Club Books, 1969), 245–47.
2. J. P. Foote, "Wilderness—A Question of Purity," *Environmental Law* 3(1973): 259–60.
3. M. Frome, *The Battle for the Wilderness* (New York: Praeger Publishers, 1974), 177.
4. 16 U. S. C. A., sec. 1271–1287.
5. A. D. Tarlock, and R. Tippy, "The Wild and Scenic Rivers Act of 1968," *Cornell Law Review* 55(1970): 707–739.
6. Office of the President, Council on Environmental Quality, *Environmental Quality 1984* (Washington, D.C.: Government Printing Office, 1984), 652.
7. C. W. Allin, *The Politics of Wilderness Preservation* (Westport, Conn.: Greenwood Press, 1982), 143–46.
8. McCloskey and Gilligan, *Wilderness*, 11.
9. G. O. Robinson, *The Forest Service: A Study in Public Land Management* (Baltimore: Johns Hopkins University Press and Resources for the Future, 1975), 181.
10. E. H. Hanks, A. D. Tarlock, and J. Hanks, *Environmental Law and Policy: Cases and Materials* (St. Paul, Minn.: West Publishing Company, 1975), 282–88.

11. 309 F. Supp. 593, affirmed 448 F. 2d 793 (10th Cir. 1971), cert. denied, 405 U.S. 989, 92 S. Ct. 1252, 31 L. Ed. 455 (1972).
12. Note, "*Parker v. United States*: The Forest Service Role in Wilderness Preservation," *Ecology Law Quarterly* 3(1973): 145–50.
13. 541 F. 2d 1292, cert. denied, 430 U.S. 922, 97 S. Ct. 1340, 51 L.Ed. 2d 601.
14. 92 Stat. 1649.
15. R. J. Costley, "An Enduring Resource," *American Forests* 78(June 1972): 11.
16. J. Shepard, *The Forest Killers—The Destruction of the American Wilderness* (New York: Weybright and Talley, 1975), 252.
17. U.S. Department of Agriculture, Forest Service, *Roadless Area Review and Evaluation: Final Report* (Washington, D.C.: Government Printing Office, 1973), 1.
18. 3 ELR 20071 (N. D. Cal. 1972).
19. Office of the President, Council on Environmental Quality, *Environmental Quality 1974* (Washington, D.C.: Government Printing Office, 1974), 200.
20. J. D. Leshy, *The Mining Law: A Study in Perpetual Motion* (Washington, D.C.: Resources for the Future, 1987), 233–34.
21. "A Copper Company vs. the North Cascades: Should the Future of a National Treasure Depend on How a Giant Corporation Defines Its Responsibilities to the Public?" *Harper's Magazine* 235 (September 1967): 48–50.
22. M. McCloskey, "Can Recreational Conservationists Provide for a Mining Industry?" *Proceedings of the Rocky Mountain Mineral Law Institute* 13(1967): 80–83.
23. R. W. Findley, and D. A. Farber, *Environmental Law: Cases and Materials* (St. Paul, Minn.: West Publishing Company, 1981), 661–64.
24. 353 F. Supp. 698; rev'd., 497 F. 2d 849 (8th Cir. 1974).
25. 42 U. S. C. A., sec. 4331–4335, 4341–4342, 4344.
26. U.S. Department of Interior, Bureau of Land Management, *Draft Environmental Impact Statement, Livestock Grazing Management on National Resource Lands* (Washington, D.C.: Government Printing Office, 1974), 1.
27. 388 F. Supp. 829.
28. G. D. Libecap, *Locking Up the Range: Federal Land Controls and Grazing* (Cambridge, Mass.: Ballinger Publishing Company, 1981), 70–80.
29. *Congressional Record* 121(30 January 1975), 1231.
30. 90 Stat. 2743.
31. U.S. Congress. Senate. Committee on Interior and Insular Affairs. S. Rept. 94-583. 18 December 1975, 43–45.
32. Libecap, *Locking Up the Range*, 76.
33. *Congressional Quarterly Weekly Report*, 35(28 May 1977), 1064.
34. U.S. Forest Service, *Rare II: Final Report* (Washington, D.C.: Government Printing Office, 1979), 5.
35. *Congressional Quarterly Weekly Report*, 37(20 January 1979), 96.
36. Clusen and Scott, 1977, 8.

Chapter 16
Sagebrush Rebellion, Inc.

1. League for the Advancement of States Equal Rights, *Agenda for the '80s: A New Federal Land Policy* (Salt Lake City: LASER), 1980.

2. *Phoenix Gazette*, 1 November 1980.
3. Chap. 633, 1979 Nev. Stat. 1362.
4. Interview, Clifford Young, President National Wildlife Federation, 4 April 1981.
5. *Phoenix Gazette*, 1 November 1980.
6. C. Callison, "Sagebrush Rebellion," *National Parks and Conservation Magazine* (March 1980): 10–13.
7. U.S. Congress. Senate. S. 1680. *Congressional Record* 126 (1979), S. 11657.
8. Callison, "Sagebrush Rebellion," 12.
9. U.S. Congress. House. H.R. 7837. *Congressional Record* 126 (1980), H. 6534.
10. Callison, "Sagebrush Rebellion," 13.
11. *Arizona Republic*, 23 November 1980.
12. 90 Stat. 2744, 43 U. S. C. 1701.
13. *Pollard v. Hagan* 44 U.S. (3 How.) 212 (1846).
14. J. D. Leshy, "Unraveling the Sagebrush Rebellion: Law, Politics and Federal Lands," *U. C. Davis Law Review* 40 (1980): 317–55.
15. 13 Stat. 44.
16. League for the Advancement of States Rights, *Agenda*.
17. S. K. Fairfax, "Beyond the Sagebrush Rebellion: The BLM as Neighbor and Manager in the Western States," in Francis, J. G., and Ganzel, R., eds., *Western Public Lands* (Totowa, N.J.: Rowman and Allenheld, 1984), 79–91.
18. W. C. Patric, *Trust Administration in the Western States* (Denver: Public Lands Institute, 1981), viii.
19. Wyoming Stat. sec. 36-12-10 (Supp. 1980).
20. Utah Code Ann. sec 65-11-9 (Supp. 1980).
21. New Mexico Stat. Ann. sec. 19-15-9 (Supp. 1980).
22. Arizona Rev. Stat. Ann. sec. 37-901 to 909 (Supp. 1980).
23. Washington Rev. Code sec 79.80.020 (Supp. 1980).
24. J. G. Francis, "Environmental Values, Intergovernmental Politics, and the Sagebrush Rebellion," in Francis, J. G., and Ganzel, R., eds., *Western Public Lands: The Management of Natural Resources in a Time of Declining Federalism* (Totowa, N.J.: Rowman and Allenheld, 1984), 33.
25. Francis, "Environmental Values," 37–39.
26. *Arizona Republic*, 1 December 1980.
27. League for the Advancement of States Equal Right, *Agenda*.
28. U.S. Congress. House. *Granting Remaining Unreserved Public Lands to the States: Hearings on H.R. 5840 Before the House Committee on the Public Lands*. 72d Cong., 1st sess. (1932), 15.
29. 426 U.S. 529 (1976).
30. Interview, Clifford Young, president, National Wildlife Federation, 4 April 1981.
31. *Phoenix Gazette*, 21 November 1980.
32. Office of the President, Council on Environmental Quality, *Environmental Quality 1984* (Washington, D.C.: Government Printing Office, 1984), 660.
33. *Phoenix Gazette*, 20 February 1981.
34. *Arizona Republic*, 14 June 1981.
35. *Philadelphia Inquirer*, 26 July 1982.

36. *Arizona Republic*, 6 February 1982.
37. *Arizona Republic*, 25 November 1982.
38. *Arizona Republic*, 29 December 1982.

Chapter 17
Druids and Rebels

1. S. Fox, "We Want No Straddlers," *Wilderness* 48(1985): 5–19.
2. *National Cyclopedia of American Biography* (New York: James T. White Company, 1968), 50: 553–54.
3. R. A. Baker, *Conservation Politics: The Senate Career of Clinton P. Anderson* (Albuquerque: University of New Mexico Press, 1985), 98.
4. Fox, "We Want No Straddlers," 12.
5. Baker, *Conservation Politics*, 99.
6. J. McPhee, *Encounters with the Archdruid* (New York: Farrar, Straus and Giroux), 1971.
7. R. Nash, *Wilderness and the American Mind*, rev. ed. (New Haven, Conn.: Yale University Press, 1973), 227–36.
8. McPhee, *Encounters with the Archdruid*, 28.
9. Ibid., 33.
10. Baker, *Conservation Politics*, 11.
11. Ibid., 12.
12. C. P. Anderson, *Outsider in the Senate* (New York: World Publishing Company), 1970.
13. Baker, *Conservation Politics*, 15.
14. S. L. Flader, *Thinking Like a Mountain: Aldo Leopold and the Evolution of an Ecological Attitude Toward Deer, Wolves and Forests* (Lincoln: University of Nebraska Press, 1974), 15.
15. Baker, *Conservation Politics*, 35.
16. P. Brooks, "Congressman Aspinall vs. the People of the United States," *Harper's Magazine* (March 1963):59–63.
17. *Who Was Who in America with World Notables* (Chicago: Marquis Who's Who, 1985), 13:14.
18. Executive Branch, *U.S. White House Conference on Conservation, Proceedings* (Washington, D.C.: Government Printing Office, 1962), 57–61.
19. *Who's Who in American Politics, 1987–1988*, 11th ed. (New York: R. R. Bowker Company, 1987), 1482.
20. Ibid., 1481.
21. L. J. Arrington, *Great Basin Kingdom: Economic History of the Latter-Day Saints, 1830–1900* (Lincoln: University of Nebraska Press, 1958), 5–6.
22. *Who's Who in American Politics, 1987–1988*, 919.
23. Nevada Legislature, Select Committee on Public Lands, "Sagebrush Rebellion Presentation," *Agenda for the '80s: A New Federal Land Policy* (Salt Lake City, Utah: League for the Advancement of States Equal Rights, 1980), 1–10.
24. Interview, Jack E. Christensen, Executive Vice President, Utah Mining Association, 9 February 1981.

25. Fox, "We Want No Straddlers," 14.
26. "Wilderness Society, Annual Report, 1986," *Wilderness* 50(1987): insert, 30–31.
27. Fox, "We Want No Straddlers," 16.
28 A. Adams, *Ansel Adams: An Autobiography* (New York: New York Graphic Society, 1985), 156–57.
29. Fox, "We Want No Straddlers," 17.
30. McPhee, *Encounters with the Archdruid*, 14.
31. "Sierra Club Financial Report," *Sierra* 73, no. 2(1988) 77–79.
32. Corporate Registration Files, Secretary of State Office, State of Utah, Salt Lake City.
33. *LASER*, Brochure, undated.
34. Interview, Niel Robinson, United Press International, 15 February 1981.
35. Corporate Registration Files, Secretary of State Office, State of Utah, Salt Lake City.
36. Attempt to locate by author during visit to Salt Lake City, February, 1981.
37. Interview, Niel Robinson, United Press International, 15 February 1981.
38. R. Wolf, "New Voice in the Wilderness," *Rocky Mountain Magazine* (March/April 1981):29–34.
39. Corporate Registration Files, Secretary of State Office, State of Colorado, Denver.
40. Wolf, "New Voice in the Wilderness," 33.

Chapter 18
On a Clear Day You Can See Four Corners

1. W. L. Graf, "Late Holocene Sediment Storage in Canyons of the Colorado Plateau," *Geological Society of America Bulletin* 99(1987): 261–71.
2. G. C. Coggins, and C. F. Wilkinson, *Federal Public Land and Resources Law* (Mineola, N.Y.: Foundation Press, 1981), 574.
3. C. B. Hunt, *Natural Regions of the United States and Canada* (San Francisco: Freeman, 1974), 373.
4. D. R. Harris, "Recent Plant Invasions in the Arid and Semiarid Southwest of the United States," *Annals of the Association of American Geographers* 56(1966): 408–422.
5. R. L. Brown, *Ghost Towns of the Colorado Rockies* (Caldwell, Idaho: Caxton Printers), 1968.
6. R. A. Bartlett, *Great Surveys of the American West* (Norman: University of Oklahoma Press, 1962), 87.
7. J. Trijonis, and K. Yuan, *Visibility in the Southwest: An Exploration of the Historical Data Base* (Research Triangle Park, N.C.: U.S. Environmental Protection Agency, 1978).
8. R. Johnson, *The Central Arizona Project, 1918–1968* (Tucson: University of Arizona Press, 1977), 215.
9. R. Nash, *Wilderness and the American Mind*, rev. ed., (New Haven, Conn.: Yale University Press, 1973), 226–36.

10. Trijonis and Yuan, *Visibility in the Southwest*, 1–5.

11. A. V. Kneese, and F. L. Brown, *The Southwest Under Stress: National Resource Development Issues in a Regional Setting* (Baltimore: Johns Hopkins University Press and Resources for the Future, 1981), 106–107.

12. P. L. 90-148.

13. P. L. 91-604.

14. Kneese and Brown, *The Southwest Under Stress*, 134, 153.

15. Ibid., 112.

16. P. Gober, and K. McHugh, "The Urban Growth Process," *Report to the Arizona State Legislature on Urban Growth in Arizona* (Tempe: Morrison Institute, Arizona State University, 1988), 8–9.

17.D. Mann, "The Political Implications of the Migration to the Arid Land States of the U.S.A.," *Natural Resources Journal* (1969): 212–27.

18. Interview, Claire Accord, executive secretary, Utah Wool Growers Association, 11 February 1981.

19. Interview, Gerry E. Magneson, chief planner, Bureau of Land Management, Salt Lake City Office, 9 February 1981.

20. Interview, Claire Accord, executive secretary, Utah Wool Growers Association, 11 February 1981.

21. R. A. Baker, *Conservation Politics: The Senate Career of Clinton P. Anderson* (Albuquerque: University of New Mexico Press, 1985), 170.

22. P. Brooks, *Roadless Area* (New York: Ballantine Books, 1971), 123–41.

23. L. H. LeFevre, *The Boulder Country and Its People* (Springville, Utah: Art City Publishing, 1973), 291.

24. R. D. Lamm, and M. McCarthy, *The Angry West: A Vulnerable Land and Its Future* (Boston: Houghton Mifflin Company, 1982), 323–24.

25. Utah Department of Natural Resources and Energy, *Project BOLD: Alternatives for Utah Land Consolidation and Exchange* (Salt Lake City: State of Utah, 1982), 1.

26. J. R. Chavez, *The Lost Land: The Chicano Image of the Southwest* (Albuquerque: University of New Mexico Press, 1984), 137–39.

27. T. Hillerman, *The Great Taos Bank Robbery and Other Indian Country Affairs* (Albuquerque: University of New Mexico Press, 1973), 111-33.

28. W. deBuys, *Enchantment and Exploitation: The Life and Hard Times of a New Mexico Mountain Range* (Albuquerque: University of New Mexico Press, 1985), 284–300.

29. P. Wiley, and R. Gottlieb, *Empires in the Sun: The Rise of the New American West* (New York: G. P. Putnam's Sons, 1982), 73.

30. Massachusetts Institute of Technology, *The Nuclear Almanac: Confronting the Atom in War and Peace* (Reading, Mass.: Addison-Wesley, 1984), 302–304.

31. J. C. Fuller, *The Day We Bombed Utah: America's Most Lethal Secret* (New York: New American Library, 1984), 237–50.

32. R. C. Athearn, *The Mythic West in Twentieth Century America* (Lawrence: University of Kansas Press, 1986), 206.

33. M. J. Greenwood, *Migration and Economic Growth in the United States: National, Regional, and Metropolitan Perspectives* (New York: Academic Press, 1981), 30–31.

34. G. D. Nash, *The American West in the Twentieth Century: A Short*

History of an Urban Oasis (Albuquerque: University of New Mexico Press, 1973), 228–34.
35. R. W. Taylor, and S. W. Taylor, *Uranium Fever, or No Talk Under $1 Million* (New York: Macmillan, 1970), 240–41.
36. E. Gray, *The Great Uranium Cartel* (Toronto: McClelland and Stewart, 1982), 60–85.
37. *San Juan Record*, 28 May 1981.
38. Council on Environmental Quality, *Public Opinion on Environmental Issues: Results of a National Opinion Survey* (Washington, D.C.: Executive Office of the President, 1980), 2–5.
39. Council on Environmental Quality, *Public Opinion*, 45.
40. A. Adams, *Ansel Adams: An Autobiography* (Boston: Little, Brown and Company, 1985), 126.
41. A. Gray, "Wilderness and the Camera's Eye," *Wilderness* 49, no. 170(1985): 12–17.
42. D. Brower, "A Tribute to Ansel Adams," *Sierra* 69, no. 4(1988): 32–35.
43. W. H. Goetzmann, and W. N. Goetzmann, *The West of the Imagination* (New York: W. W. Norton, 1986), 330.
44. E. Abbey, *Desert Solitaire: A Season in the Wilderness* (New York: Ballantine Books, 1968).
45. E. Abbey, *The Monkey Wrench Gang* (New York: Avon Books), 1975.
46. A. Ronald, *The New West of Edward Abbey* (Albuquerque: University of New Mexico Press, 1982), 4.

Chapter 19
How Much Is Enough?

1. *Arizona Republic*, 10 October 1983.
2. J. Norton, "Field Notes: Southwest," *Wilderness* 51(Summer 1988):55.
3. *Phoenix Gazette*, 26 February 1981.
4. *Phoenix Gazette*, 5 March 1981.
5. *Desert News*, 30 October 1988.

Bibliography

"A Copper Company vs. the North Cascades: Should the Future of a National Treasure Depend on How a Giant Corporation Defines Its Responsibilities to the Public?" *Harper's Magazine* 235(September 1967): 48–50.

Abbey, E. *Desert Solitaire: A Season in the Wilderness*. New York: Ballantine Books, 1968.

———. *The Monkey Wrench Gang*. New York: Avon Books, 1975.

Adams, A. *Ansel Adams: An Autobiography*. Boston: Little, Brown and Company, 1985.

Adams, H. *The Education of Henry Adams*. Boston: Houghton Mifflin Company, 1918.

Adams, S. *Colorado River Expeditions of Samual Adams*. U. S. House of Representatives, Miscellaneous Document 37. 42nd Cong., 1st sess., 1882.

Allin, C. W. *The Politics of Wilderness Preservation*. Westport, Conn.: Greenwood Press, 1982.

Anderson, C. P. *Outsider in the Senate*. New York: World Publishing Company, 1970.

Anderson, D. F. *William Howard Taft: A Conservative's Conception of the Presidency*. Ithaca, N. Y.: Cornell University Press, 1965.

Arizona State Land Commission. *Report of the Arizona State Land Commission, June 6, 1912–December 1, 1914*. Phoenix: State of Arizona, 1914.

Arizona State Land Department. *Annual Report, 1979–1980*. Phoenix: State of Arizona, 1980.

Arrington, L. J. *Great Basin Kingdom: Economic History of the Latter-Day Saints, 1830–1900*. Lincoln: University of Nebraska Press, 1958.

Arrington, L. J., and Jensen, G. *The Defense Industry of Utah*. Logan: Utah State University Press, 1965.

Athearn, R. C. *The Mythic West in Twentieth Century America*. Lawrence: University of Kansas Press, 1986.

Atherton, L. *The Cattle Kings*. Lincoln: University of Nebraska Press, 1961.

Austin, M. *The Land of Little Rain*. Boston: Houghton Mifflin Company, 1903.

Bailey, R. W. "Epicycles of Erosion in the Valleys of the Colorado Plateau Province." *Journal of Geology* 43(1935): 337–55.

Baker, R. A. *Conservation Politics: The Senate Career of Clinton P. Anderson*. Albuquerque: University of New Mexico Press, 1985.

Baldwin, D. W. *The Quiet Revolution: Grass Roots of Today's Wilderness Preservation Movement*. Boulder, Colo.: Prescott, 1972.

Bancroft, H. H. *History of the Life of William Gilpin: A Character Study*. San Francisco: A. L. Bandroft and Company, 1889.

Barnes, W. C., rev. by B. H. Granger. *Arizona Place Names*. Tucson: University of Arizona Press, 1960.

Barr, G. W. *Recovering Rainfall: Part 1: Arizona Watershed Program*. Phoenix: Arizona State Land Department, Salt River Valley Water Users Association, and University of Arizona, 1956.

Barry, R. G., and Chorley, R. J. *Atmosphere, Weather, and Climate*. New York: Holt, Rinehart and Winston, 1970.

Bartlett, R. A. *Great Surveys of the American West*. Norman: University of Oklahoma Press, 1962.

Bergon, F., ed. *The Wilderness Reader*. New York: New American Library, 1980.

Billington, R. A. *Westward Expansion: A History of the American Frontier*. New York: Macmillan, 1974.

Block, M., ed. *Current Biography*. New York: H. W. Wilson, 1941.

Borchert, J. "The Dust Bowl in the 1970s." *Annals of the Association of American Geographers* 61(1971): 1–22.

Borne, L. R. *Dude Ranching: A Complete History*. Albuquerque: University of New Mexico Press, 1983.

Bradley, R. S. *Precipitation History of the Rocky Mountain States*. Boulder, Colo.: Westview Press, 1976.

Brandis, D. "The Late Franklin B. Hough." *Indiana Forester* 11(1885): 429.

Branson, F. A., Gifford, G. F., Renard, K. G., and Hadley, R. F. *Rangeland Hydrology*. Dubuque, Iowa: Kendal/Hunt Publishers, 1981.

Britton, R. D. "The B. H. Roberts Case of 1898–1900." *Utah Historical Quarterly* 25(1957): 27–46.

Brooks, P. "Congressman Aspinall vs. the People of the United States." *Harper's Magazine* 231 (March 1963): 59–63.

———. *Roadless Area*. New York: Ballantine Books, 1964.

———. *Speaking for Nature: How Literary Naturalists from Henry David Thoreau to Rachel Carson Have Shaped America*. Boston: Houghton Mifflin Company, 1980.

Brower, D. "A Tribute to Ansel Adams." *Sierra* 69, no. 4(1988): 32–35.

Brown, D. E., and Lowe, C. H. *Biotic Communities of the Southwest*.

Washington, D. C.: U. S. Department of Agriculture, U. S. Forest Service General Technical Report RM–78, 1980.

Brown, R. L. *Ghost Towns of the Colorado Rockies*. Caldwell, Idaho: Caxton Printers, 1968.

Brownfield, A. D., and Hyatt, S. C. "A Proposal for Handling the Public Lands." *American Cattle Producer* 28(1946): 8.

Bryan, W. J. *The First Battle: A Story of the Campaign of 1896*. Chicago: W. B. Conkey Company, 1896.

Burdett, S. S. *Report of the General Land Office*. Washington, D. C.: U. S. Department of Interior, 1884.

Burkham, D. E. *Channel Changes of the Gila River in Safford Valley, Arizona*. Washington, D. C.: U. S. Geological Survey, Professional Paper 655-G, 1972.

Butler, O. M. "70 Years of Campaigning for American Forestry." *American Forests* 52(1946): 456–59.

Butter, G. M. *Some Facts About Ore Deposits*. Tucson: Arizona Bureau of Mines, Bulletin 139, 1935.

Calef, W. *Private Grazing and Public Lands: Studies of the Local Management of the Taylor Grazing Act*. Chicago: University of Chicago Press, 1960.

Callison, C. "Sagebrush Rebellion." *National Parks and Conservation Magazine* (March 1980): 10–13.

Carhart, A. H. *The National Forests*. New York: Alfred A. Knopf, 1959.

Carlson, M. E. "William E. Smyth: Irrigation Crusader." *Journal of the West* 7(1968): 41–7.

Carstensen, V., ed. *The Public Lands: Studies in the History of the Public Domain*. Madison: University of Wisconsin Press, 1962.

Catlin, G. *North American Indians: Being Letters and Notes on Their Manners, Customs, and Conditions, Written During Eight Years Travel Amongst the Wildest Tribes of Indians in North America*. 2 vols. Edinburg: J. Grant, 1841.

Cazier, L. *Surveys and Surveyors of the Public Domain, 1785–1975*. Washington, D.C.: Government Printing Office, 1976.

Chang, J. *Atmospheric Circulation Systems and Climatic Systems*. Honolulu: Oriental Publishing Company, 1972.

Chapman, H. H. "National Parks, National Forests, and Wilderness Areas." *Journal of Forestry* 36(1938): 469–74.

Chavez, J. R. *The Lost Land: The Chicano Image of the Southwest*. Albuquerque: University of New Mexico Press, 1984.

Chow, V. T., ed. *Handbook of Hydrology*. New York: McGraw-Hill, 1964.

Claiborne, R. *Climate, Man, and History*. New York: W. W. Norton & Company, 1970.

Clawson, M. *Uncle Sam's Acres*. New York: Mead and Company, 1951.

Clepper, H. Conservation's Grand Lodge." *American Forests* 73, no. 10 (1967): 22–27, 58–61.

———. "Forestry's Uncertain Beginning." *Journal of Forestry* 67(1969): 218–21.

———. *Professional Forestry in the United States*. Baltimore: Johns Hopkins University Press, 1971.

———. "Crusade for Conservation: The Centennial History of the American Forestry Association." *American Forests* 81(1975): 37–44.

Cleveland, G. "Second Annual Message." In *Compilation of the Messages and Papers of the Presidents*, vol. 13. New York: Bureau of National Literature, 1894.

Conklin, T. E. "Public Land Law Review Commission Revisited—A Potpourri of Memories." *Denver Law Journal* 54(1977): 445–47.

Connelley, W. E. *Life of Preston B. Plumb*. Chicago: University of Chicago Press, 1913.

Connor, S. V., and Faulk, O. B. *North America Divided: The Mexican War, 1846–1848*. New York: Oxford University Press, 1971.

Conover, M. *The General Land Office: Its History, Activities, and Organization*. Baltimore: Johns Hopkins University Press, 1923.

"Convention Notes." *American Sheep Breeder and Wool Grower* 36(1916): 26.

Coogins, G. C., and Wilkinson, C. F. *Federal Public Land and Resources Law*. Mineola, N. Y.: Foundation Press, 1981.

Cooke, R. U., and Reeves, R. W. *Arroyo Development and Environmental Change in the American South-West*. Oxford: Clarendon Press, 1976.

Costley, R. J. "An Enduring Resource." *American Forests* 78(June 1972): 11–20.

Council on Environmental Quality. *Environmental Quality 1974*. Washington, D. C.: Government Printing Office, 1974.

———. *Public Opinion on Environmental Issues: Results of a National Opinion Survey*. Washington, D. C.: Executive Office of the President, 1980.

———. *Environmental Quality 1984*. Washington, D. C.: Government Printing Office, 1984.

Coville, F. V. *Bulletin 62*. Washington, D. C.: U. S. Forest Service, 1905.

Crafts, E. C. "Saga of a Law." *American Forests* 76, no. 6 (1970): 18–19.

Craig, J. B. "A Circle That Took Us In." *American Forests* 70, no. 3 (1964): 8–9.

Cutright, P. R. *Theodore Roosevelt the Naturalist*. New York: Harper's, 1956.

Dana, S. T., and Fairfax, S. K. *Forest and Range Policy*. 2d ed. New York: McGraw-Hill, 1980.

Dana, S. T. *Forest and Range Policy*. New York: McGraw-Hill, 1956.

Darrah, W. C. *Powell of the Colorado*. Princeton, N.J.: Princeton University Press, 1951.

Davis, R. L. *A History of Opera in the American West*. Englewood Cliffs, N. J.: Prentice-Hall, 1965.

de Gennaro, N. *Arizona Statistical Abstract: A 1979 Data Handbook*. Flagstaff, Ariz.: Northland Press, 1979.

deBuys, W. *Enchantment and Exploitation: The Life and Hard Times of a New Mexico Mountain Range*. Albuquerque: University of New Mexico Press, 1985.

Degnan, J. P. "The Desert Shall Rejoice, and Be Made to Blossom as the Rose." *The Living Wilderness* 46(1982): 14–26.

Denevan, W. M. "Livestock Numbers in Nineteenth-Century New Mexico, and the Problem of Gullying in the Southwest." *Annals of the Association of American Geographers* 57(1967): 691–703.

DeVoto, B. "The West: A Plundered Province." *Harper's Magazine* 164(1934): 355–64.

———. *1846, The Year of Decision*. Boston: Little, Brown and Company, 1943.

———. "Geopolitics with the Dew on It." *Harper's Magazine* 187(1944): 313–23.

———. "The Anxious West." *Harper's Monthly* 93(1946): 481–91.

Dictionary of American Biography. New York: Charles Scribner's Sons, 1981.

Diller, J. S. "Clarence E. Dutton." *Geological Society of America Bulletin* 24(1913): 10–8.

Donaldson, T. *The Public Domain*. Washington, D. C.: U. S. Department of Interior, 1884.

Dows, O. *Franklin Roosevelt at Hyde Park*. New York: American Artists' Group, 1949.

Dubois, C. *Trail Blazers*. Stonington, Conn.: Stonington Publishing Company, 1957.

Dunham, H. A. *Government Handout: A Study in the Administration of the Public Lands, 1875–1891*. Ann Arbor, Mich.: Edwards Litho-Printers, 1941.

Dunham, H. A. "Some Crucial Years of the General Land Office, 1875–1890." *Agricultural History* 11(1937): 117–41.

Dunne, F. P. *Mr. Dooley: First in the Hearts of His Countrymen*. Boston: Small, Maynard and Company, 1899.

Dunning, C. H., and Peplow, E. H. *Rock to Riches: The Story of Arizona Mines and Mining*. Pasadena, Calif.: Hicks Publishing, 1966.

Dutton, W. L. "History of Forest Service Grazing Fees." *Journal of Range Management* 6(1953): 393–98.

Easum, C. V. *The Americanization of Carl Schurz*. Chicago: University of Chicago Press, 1929.

Editorial. "Conservation from a Colorado Viewpoint—By a Critic of the Government." *Collier's Weekly* 50(1913): 24.

Editorial. *American Sheep Breeder and Wool Grower* 37(1917): 19.

Editorial. "The Public Lands." *Outlook and Independent* 153(1929): 54.

Editorial. "Flinging Away an Empire." *The New Republic* 70(1932): 32.

Ellis, E. *Henry Moore Teller, Defender of the West*. Caldwell, Idaho: Caxton Press, 1941.

Errington, P. L. "In Appreciation of Aldo Leopold." *Journal of Wildlife Management* 12(1948): 341–50.

Everhart, W. C. *The National Park Service*. New York: Praeger Publishers, 1975.

Everson, W. K. *A Pictorial History of the Western Film*. New York: Citadel Press, 1969.

Executive Branch, Office of the President. *U. S. White House Conference on Conservation, Proceedings*. Washington, D. C.: Government Printing Office, 1962.

Fairfax, S. K. "Beyond the Sagebrush Rebellion: The BLM as Neighbor and Manager in the Western States." In *Western Public Lands*, ed. by J. G. Francis and R. Ganzel, 79–91. Totowa, N. J.: Rowman and Allenheld, 1984.

Faulk, O. B. *Arizona: A Short History*. Norman: University of Oklahoma Press, 1970.

Fenton, C. L., and Fenton, M. A. *The Story of the Great Geologists*. Garden City, N. Y.: Doubleday and Company, 1945.

Fernow, B. E. *Economics of Forestry*. New York: Thomas Y. Crowell Publishers, 1902.

Findley, R. W., and Farber, D. A. *Environmental Law: Cases and Materials*. St. Paul, Minn.: West Publishing Company, 1981.

Flader, S. L. *Thinking Like a Mountain: Aldo Leopold and the Evolution of an Ecological Attitude Toward Deer, Wolves and Forests*. Lincoln: University of Nebraska Press, 1974.

Foote, J. P. "Wilderness—A Question of Purity." *Environmental Law* 3(1973): 259–60.

Foresta, R. A. *America's National Parks and Their Keepers*. Washington, D. C.: Resources for the Future, 1984.

Foss, P. O. *The Politics of Grass: The Administration of Grazing on the Public Domain*. Seattle: University of Washington Press, 1960.

Fox, S. *John Muir and His Legacy: The American Conservation Movement*. Boston: Little, Brown and Company, 1981.

———. "We Want No Straddlers." *Wilderness* 48(Spring 1985): 5–19.

Francis, J. G. "Environmental Values, Intergovernmental Politics, and the Sagebrush Rebellion." In *Western Public Lands*, ed. by J. G. Francis and R. Ganzel, 29–46. Totowa, N.J.: Rowman and Allenheld, 1984.

Frank, B. *Our National Forests*. Norman: University of Oklahoma Press, 1955.

Fritts, H. C. *Tree Rings and Climate*. London: Academic Press, 1976.

Frome, M. *The Battle for the Wilderness*. New York: Praeger Publishers, 1974.

———. *Whose Woods These Are: The Story of the National Forests*. Garden City, N. Y.: Doubleday and Company, 1962.

Fuller, J. C. *The Day We Bombed Utah: America's Most Lethal Secret*. New York: New American Library, 1984.

Gannett, H. *Nineteenth Annual Report: Part 5: Forests*. Washington, D. C.: U. S. Geological Survey, 1899.

Garber, P. N. *The Gadsden Treaty*. Philadelphia: University of Pennsylvania Press, 1923.

Garner, J. W. "The Case of Senator Smoot and the Mormon Church." *North American Review* 184(1907): 46.

Garreau, J. *The Nine Nations of North America*. Boston: Houghton Mifflin Company, 1981.

Gates, P. W. "The Homestead Law in an Incongruous Land System." *American Historical Review* 41(1936): 652–81.

Geiger, R. *The Climate Near the Ground*. Cambridge, Mass.: Harvard University Press, 1966.

General Land Office. *Annual Report*. Washington, D. C.: U. S. Department of Interior, 1888.

George, H. *Progress and Poverty: An Inquiry into the Cause of Industrial Depressions and of Increase of Want with Increase of Wealth*. New York: Appleton, 1881.

Gilbert, G. K. *Lake Bonneville*. Washington, D. C.: U. S. Geological Survey, Monograph 1, 1890.

Gillespie, N. "The Spirit of Place." *Southwest Profile* (Feb. 1988): 26–32.

Gilligan, J. P. "The Development of Policy and Administration of Forest Service Primitive and Wilderness Areas in the Western United States." Ph.D. diss., University of Michigan, 1953.

Glasscock, C. B. *The War of the Copper Kings: Builders of Butte and Wolves of Wall Street*. New York: Bobbs-Merrill Company, 1935.

Gober, P., and McHugh, K. "The Urban Growth Process." In *Report to the Arizona State Legislature on Urban Growth in Arizona*. Tempe: Morrison Institute, Arizona State University, 1988.

Goetzmann, W. H., and Goetzmann, W. N. *The West of the Imagination*. New York: W. W. Norton & Company, 1986.

Goodrich, C. *Government Promotion of American Canals and Railroads, 1800–1890*. New York: Columbia University Press, 1960.

Graf, W. L. "Mining and Channel Response." *Annals of the Association of American Geographers* 69(1979): 262–75.

———. "Downstream Changes in Stream Power, Henry Mountains, Utah." *Annals of the Association of American Geographers* 73(1983): 373–87.

———. "The Arroyo Problem: Paleohydrology and Paleohydraulics in the Short Term." In *Perspectives on Paleohydrology,* ed. by K. J. Gregory, 262–303. London: John Wiley & Sons, 1983.

———. "Temporal Variation of Sediment Yield in the Upper Colorado River Basin." Abst. *Program Abstracts, Association of American Geographers 1985 Annual Meeting* (1985): 193.

———. "Fluvial Erosion and Federal Public Policy in the Navajo Nation." *Physical Geography* 7(1986): 97–115.

———. "Late Holocene Sediment Storage in Canyons of the Colorado Plateau."*Geological Society of America Bulletin* 99(1987): 261–71.

Granger, B. H. *Will C. Barnes' Arizona Place Names.* Tucson: University of Arizona Press, 1979.

Graves, H. B. "The Public Domain." *The Nation* 81(1930): 147–49.

Gray, A. "Wilderness and the Camera's Eye." *Wilderness* 49(Summer 1985): 12–17.

Gray, E. *The Great Uranium Cartel.* Toronto: McClelland and Stewart, 1982.

Greeley, W. B. "Shall the National Forests Be Abolished?" *Mining Congress Journal* 13(1927): 594–97.

———. *Forests and Men.* New York: Doubleday and Company, 1951.

Greenwood, M. J. *Migration and Economic Growth in the United States: National, Regional, and Metropolitan Perspectives.* New York: Academic Press, 1981.

Greesley, G. M., ed. *Bostonians and Bullion: The Journal of Robert Livermore, 1892–1915.* Lincoln: University of Nebraska Press, 1968.

Greever, W. S. *The Bonanza West: The Story of the Western Mining Rushes, 1848–1900.* Norman: University of Oklahoma Press, 1963.

Gregory, H. E., and Moore, R. C. *The Kaiparowits Region: A Geographic and Geologic Reconnaissance of Parts of Utah and Arizona.* Washington, D. C.: U. S. Geological Survey, Professional Paper 134, 1931.

Gregory, H. E. *The San Juan Country: A Geographic and Geologic Reconnaissance of Southeastern Utah.* Washington, D. C.: U. S. Geological Survey, Professional Paper 188, 1938.

Hagedorn, H. *Roosevelt in the Bad Lands.* New York: Houghton Mifflin, 1921.

Hammett, H. B. *Hilary Abner Herbert: A Southerner Returns to the Union.* Philadelphia: American Philosophical Society, Memoir 110, 1976.

Hanks, E. H., Tarlock, A. D., and Hanks, J. L. *Environmental Law and Policy: Cases and Materials.* St. Paul, Minn.: West Publishing Company, 1975.

Hardin, G. "The Tragedy of the Commons." *Science* 162(1968): 1243.

Harris, D. R. "Recent Plant Invasions in the Arid and Semiarid Southwest of the United States." *Annals of the Association of American Geographers* 56(1966): 408–422.

Harris, F. W., and Johnson, A. P. *The Neglected West*. Seattle, Wash.: City of Seattle, Municipal Monograph 1, 1911.

Hart, E. R., ed. *That Awesome Space: Human Interaction with the Intermountain Landscape*. Salt Lake City, Utah: Westwater Press, 1981.

Haynes, F. E. "The New Sectionalism." *Quarterly Journal of Economics* 10(1896): 269–70, 280–95.

Hays, S. P. *Conservation and the Gospel of Efficiency: The Progressive Conservation Movement, 1890–1920*. Cambridge, Mass.: Harvard University Press, 1960.

Hazen, W. B. "The Great Middle Region of the United States and Its Limited Space of Arrable Land." *North American Review* 120(1875): 1–34.

Hendee, J. C., Stankey, G. H., and Lucas, R. C. *Wilderness Management*. Washington, D. C.: U. S. Forest Service, Miscellaneous Publication 1365, 1978.

Hereford, R. "Climate and Ephemeral-Stream Processes: Twentieth Century Geomorphology and Alluvial Stratigraphy of the Little Colorado River, Arizona."*Geological Society of America Bulletin* 95(1984): 654–68.

Hereford, R. "Modern Alluvial History of the Paria River Drainage Basin." *Quaternary Research* 25(1986): 293–311.

Hibbard, B. H. *A History of Public Land Policies*. New York: Macmillan, 1939.

Hibbert, A. R. "Increases in Streamflow After Converting Chaparral to Grass." *Water Resources Research* 7(1971): 71–80.

Hidy, R. W., Hill, F. E., and Nevins, A. *Timber and Men: The Weyerhaeuser Story*. New York: Macmillan, 1963.

Higgs, R. "Railroad Rates and the Populist Uprising." *Agricultural History* 44: 197–213.

Hillerman, T. *The Great Taos Bank Robbery and Other Indian Country Affairs*. Albuquerque: University of New Mexico Press, 1973.

Hollon, W. E. *The Great American Desert: Then and Now*. Lincoln: University of Nebraska Press, 1975.

Hoover, H. *State Papers*. Washington, D.C.: Government Printing Office, 1934.

———. *The Memoirs of Herbert Hoover*. 3 vols. New York: Macmillan, 1952–1953.

Hornaday, W. T. *Our Vanishing Wildlife*. Washington, D. C.: National Audubon Society, 1913.

Hornbeck, D. Review of *The West Transformed*, by G. D. Nash. *Professional Geographer* 38(1986): 448–49.

Hough, F. B. *Report Upon Forestry: Volume I*. Washington, D. C.: Government Printing Office, 1878.

———. *Report Upon Forestry: Volume II*. Washington, D. C.: Government Printing Office, 1880.

———. *Report Upon Forestry: Volume III*. Washington, D. C.: Government Printing Office, 1882.

Hundley, N. *Dividing the Waters*. Berkeley: University of California Press, 1966.

Hundley, N., Jr., *Dividing the Waters: A Century of Controversy Between the United States and Mexico*. Berkeley: University of California Press, 1966.

Hunt, C. B. *Natural Regions of the United States and Canada*. San Francisco: Freeman, 1974.

Hunter, J. M. *The Trail Drivers of Texas*. New York: Argosy-Antiquarian, 1963.

Huth, H. "Yosemite: The Story of an Idea." *Sierra Club Bulletin* 33(1948): 47–8.

———. *Nature and the American: Three Centuries of Changing Attitudes*. Lincoln: University of Nebraska Press, 1957.

Ickes, H. L. *The Autobiography of a Curmudgeon*. New York: Reynal and Hitchcock, 1943.

Ise, J. *The United States Forest Policy*. New Haven, Conn.: Yale University Press, 1920.

Jackson, W. T. "The Creation of Yellowstone National Park." *Mississippi Valley Historical Review* 29 (1942): 187–206.

———. "The Wyoming Stock Growers Association, Political Power in Wyoming Territory, 1873–1900." *Missouri Valley Historical Review* 33(1947): 571–94.

Jacobsen, E. L. "Franklin B. Hough: A Pioneer in Scientific Forestry in America." *New York History* 15(1934): 311–18.

James, G. W. *Reclaiming the Arid West: The Story of the United States Reclamation Service*. New York: Appleton, 1917.

Johnson, F. C. *Proceedings of the Public Lands Convention Held at Denver, Colorado, June 18, 19, 20, 1907, by the States and Territories Containing Public Lands of the United States*. Denver: Public Lands Conference, 1907.

Johnson, R. *The Central Arizona Project, 1918–1968*. Tucson: University of Arizona Press, 1977.

Johnson, R. U. *Remembered Yesterdays*. Boston: Little, Brown and Company, 1923.

Jones, H. R. *John Muir and the Sierra Club*. San Francisco: Sierra Club Books, 1965.

Karl, T. R., and Knight, R. W. *Atlas of Monthly Palmer Drought Severity Indices for the Contiguous United States*. 2 vols. Washington, D. C.: National Oceanic and Atmospheric Adminstration, 1985.

Keyser, C. F. *The Preservation of Wilderness Areas—An Analysis of Opinion on the Problem*. Washington, D. C.: Legislative Reference Service, Library of Congress, 1949.

Knapp, G. L. "The Other Side of Conservation." *North American Review* 191(1910): 465–81.

Kneese, A. V., and Brown, F. L. *The Southwest Under Stress: National Resource Development Issues in a Regional Setting*. Baltimore: Johns Hopkins University Press and Resources for the Future, 1981.

Lamar, L. Q. C. *Annual Report*. Washington, D. C.: U. S. Department of Interior, 1887.

Lamm, R. D., and McCarthy, M. *The Angry West: A Vulnerable Land and Its Future*. Boston: Houghton Mifflin Company, 1982.

Lang, L. *Ranching with Roosevelt*. New York: Lippincott, 1926.

Langwill, W. D. "Mostly Division 'R' Days." *Oregon Historical Quarterly* 57(1956): 301–13.

Larson, T. A. *The History of Wyoming*. Lincoln: University of Nebraska Press, 1965.

Lawson, M. P., and Stockton, C. W. "Desert Myth and Climatic Reality." *Annals of the Association of American Geographers* 71(1981): 527–35.

———. "The Desert Myth Evaluated in the Context of Climatic Change." In *Consequences of Climatic Change*, ed. by C. D. Smith and M. Parry. Nottingham: University of Nottingham, 1981.

League for the Advancement of States Equal Rights. *Agenda for the '80s: A New Federal Land Policy*. Salt Lake City, Utah: LASER, 1980.

Lear, L. T. *Harold L. Ickes: The Aggressive Progressive, 1874–1933*. New York: Garland Publishers, 1940.

LeFevre, L. H., and Woolsey, N. G. *The Boulder Country and Its People*. Springville, Utah: Art City Publishing, 1973.

Leonard, R., and Sumner, E. L. "Protecting Mountain Meadows."*Sierra Club Bulletin* 32, no. 5 (1947): 53–62.

Leopold, A. *A Sand County Almanac*. Oxford: Oxford University Press, 1949.

———. "The Wilderness and Its Place in Forest Recreational Policy." *Journal of Forestry* 29(1921): 718–21.

———. *Game Management*. New York: Charles Scribner's Sons, 1933.

———. "The Green Lagoons." *American Forests* 51(1945): 414.

Leopold, L. B. "Reversal of Erosion Cycle and Climatic Change." *Quaternary Research* 6(1976): 557-62.

Leshy, J. D. "Unraveling the Sagebrush Rebellion: Law, Politics and Federal Lands." *U. C. Davis Law Review* 40(1980): 317–55.

Leshy, J. D. *The Mining Law: A Study in Perpetual Motion*. Washington, D. C.: Resources for the Future, 1987.

Libecap, G. D. *Locking Up the Range: Federal Land Controls and Grazing*. Cambridge, Mass.: Ballinger Publishing Company, 1981.

"List of Officers of the Wilderness Society." *Living Wilderness* (Summer 1949): i.

Lockett, H. C., and Snow, M. *Along the Beale Trail: A Photographic Account of Wasted Range Land.* Washington, D. C.: U. S. Office of Indian Affairs, 1938.

Lott, A. S. *A Long Line of Ships.* Annapolis: U. S. Naval Academy, 1954.

Lowenthal, D. *George Perkins Marsh: Versatile Vermonter.* New York: Columbia University Press, 1958.

Lyle, L. *Portrait of an Artist: A Biography of Georgia O'Keeffe.* New York: New York Graphic Society, 1980.

Mangam, W. D. *The Clarks: An American Phenomena.* New York: Silver Bow Press, 1941.

Mann, D. "The Political Implications of the Migration to the Arid Lands of the States of the U. S. A." *Natural Resources Journal* (1969): 212–27.

Marsh, G. P. *Man and Nature.* New York: Charles Scribner, 1864.

———. *The Earth as Modified by Human Action.* New York: Arno, 1874.

Marshall, R. *The People's Forests.* New York: Harrison Smith and Robert Haas, 1933.

———. "The Problem of Wilderness." *The Scientific Monthly* 30(1930): 142–43.

Massachusetts Institute of Technology. *The Nuclear Almanac: Confronting the Atom in War and Peace.* Reading, Mass.: Addison-Wesley, 1984.

Mattison, R. H. "The Hard Winter of the Range Cattle Business, Montana."*The Magazine of Western History* 1(1950): 5–20.

McArdle, R. E. "Wilderness Politics: Legislation and Forest Service." *Forest History* 19(1975): 166–79.

McCloskey, M. "Can Recreational Conservationists Provide for a Mining Industry?" *Proceedings of the Rocky Mountain Mineral Law Institute* 13(1967): 80–83.

———. "The Wilderness Act of 1964: Its Background and Meaning." *Oregon Law Review* 45(1966): 288–321.

McCloskey, M. E., and Gilligan, J. P., eds. *Wilderness and the Quality of Life.* San Francisco: Sierra Club Books, 1969.

McGeary, M. N. *Gifford Pinchot: Forester-Politician.* Princeton, N. J.: Princeton University Press, 1960.

McIntosh, C. B. "Use and Abuse of the Timber Culture Act." *Annals of the Association of American Geographers* 65(1975): 347–62.

McPhee, J. *Encounters with the Archdruid.* New York: Farrar, Straus and Giroux, 1971.

Middleton, W. E. K. *A History of the Theories of Rain and Other Forms of Precipitation.* New York: F. W. Watts, 1965.

Mills, A. *My Story.* Washington, D. C.: C. H. Claudy, 1918.

Miser, H. D. *The San Juan Canyon, Southeastern Utah.* Washington, D. C.: U. S. Geological Survey, Water-Supply Paper 538, 1924.

Mollin, F. E. "On Western Grazing." *Successful Farming* 42(1937): 4.

———. *If and When It Rains: The Stockman's View of the Range Question*. Denver: American National Livestock Association, 1938.

Morris, D., and Morris, I. *Who Was Who in American Politics*. New York: Hawthorn Books, 1974.

Morris, E. *The Rise of Theodore Roosevelt*. New York: Random House, 1979.

Mosk, S. A. "Land Policy and Stock Raising in the Western United States." In *The Public Lands: Studies in the History of the Public Domain*, ed. by V. Carstensen, 411–34. Madison: University of Wisconsin Press, 1962.

Mowry, G. E. *The Era of Theodore Roosevelt and the Birth of Modern America, 1900–1912*. New York: Harper, 1962.

Munger, T. T. "Recollections of My Thirty-Eight Years in the Forest Service, 1908–1946." *Timberline* 16(1962): 13–14.

Myers, G. *History of the Great American Fortunes*. New York: Modern Library, 1936.

Nash, G. D. *The American West in the Twentieth Century: A Short History of an Urban Oasis*. Albuquerque: University of New Mexico Press, 1977.

———. *The American West Transformed: The Impact of the Second World War*. Bloomington: Indiana University Press, 1985.

Nash, R. *Wilderness and the American Mind*. New Haven, Conn.: Yale University Press, 1973.

National Academy of Sciences. *Report of the National Academy of Sciences Upon the Inauguration of a Forest Policy for the Forested Lands of the United States*. Washington, D. C.: National Academy of Sciences, 1897.

National Cyclopedia of American Biography. New York: James T. White Company, 1916–1981.

Nevada Legislature, Select Committee on Public Lands. "Sagebrush Rebellion Presentation." In *Agenda for the '80s: A New Federal Land Policy*. Salt Lake City, Utah: League for the Advancement of States Equal Rights, 1980.

Nevins, A. *Grover Cleveland: A Study in Courage*. New York: Dodd, Mead and Company, 1933.

———. *Abraham S. Hewitt, with Some Account of Peter Cooper*. New York: Harper and Row, 1935.

Nixon, E. B., ed. *Franklin D. Roosevelt and Conservation, 1911–1945*. Hyde Park, N. Y.: General Services Administration, 1957.

Noble, L. L. *The Life and Works of Thomas Cole*. Cambridge: Oxford University Press, 1964.

Norris, F. *The Octopus, A Story of California*. New York: Doubleday, Page, and Company, 1901.

Norton, J. "Field Notes: Southwest." *Wilderness* 51(Summer 1988): 55.

Oka, J. N. *Sedimentation Study, Glen Canyon Dam, Colorado River Storage Project*. Salt Lake City, Utah: U. S. Bureau of Reclamation, 1962.

Outdoor Recreation Resources Review Commission. *Wilderness and Recreation: A Report on Resources, Values, and Problems*. Washington, D. C.: Government Printing Office, 1962.

O'Connor, H. *The Guggenheims: The Making of an American Dynasty*. New York: Covici, Friede, 1937.

Palmer, W. C. *Meteorological Drought*. Washington, D. C.: National Oceanic and Atmospheric Administration, Research Paper 45, 1965.

Park, C. F., Jr. *Minerals and the Political Economy*. San Francisco: Freeman, Cooper and Company, 1968.

"*Parker v. United States*: The Forest Service Role in Wilderness Preservation." *Ecology Law Quarterly* 3(1973): 145–50.

Patric, W. C. *Trust Administration in the Western States*. Denver: Public Lands Institute, 1981.

Peffer, E. L. *The Closing of the Public Domain: Disposal and Reservation Policies, 1900–1950*. Stanford, Calif.: Stanford University Press, 1951.

Pererson, H. J. "Wyoming." In *Rocky Mountain Politics*, ed. by T. C. Donnelley, 122–30. Albuquerque: University of New Mexico Press, 1940.

Perkins, C. A., Nielson, M. G., and Jones, L. B. *Saga of San Juan*. Monticello, Utah: San Juan County Daughters of Utah Pioneers, 1968.

Pettigrew, R. *Triumphant Plutocracy: The Story of American Public Life from 1870 to 1920*. New York: Academy Press, 1921.

Pinchot, G. *Breaking New Ground*. Reprint. Seattle: University of Washington Press, 1972.

———. "How the National Forests Were Won." *American Forests and Forest Life* 36(1930): 615–19.

Pinkett, H. T. *Gifford Pinchot: Private and Public Forester*. Urbana: University of Illinois Press, 1970.

Polenberg, R. *Reorganizing Roosevelt's Government: The Controversy Over Executive Reorganization, 1936–1939*. Cambridge, Mass.: Harvard University Press, 1966.

Powell, J. W. *Canyons of the Colorado*. Meadville, Pa.: Flood and Vincent, 1895.

———. *Report on the Lands of the Arid Region of the United States, with a More Detailed Account of the Lands of Utah*. Washington, D. C.: U. S. Geological and Geographical Survey of the Rocky Mountain Region, 1878.

———. *Tenth Annual Report of the Director of the United States Geological Survey: Part 2, Irrigation*. Washington, D. C.: U. S. Department of Interior, 1889.

———. "The Non-irrigable Lands of the Arid Region." *Century Illustrated Magazine* 39(1890): 915–22.

———. *Eleventh Annual Report: Part II: Irrigation*. Washington, D. C.: U. S. Geological Survey, 1891.

———. *Thirteenth Annual Report*. Washington, D. C.: U. S. Geological Survey, 1893.

———. "Ownership of Lands in the Arid Region." *Irrigation Age* 6(1894): 143–49.

Price, R. *History of Forest Service Research in the Central and Southern Rocky Mountain Regions 1908–1975*. Ft. Collins, Colo.: U. S. Forest Service, General Technical Report RM-27, 1976.

Proceedings of the Twelfth Trans-Mississippi Commercial Congress Held at Denver, Colorado, August 16–21, 1909. Denver: Commercial Congress, 1909.

Puter, S. A. D., and Stevens, H. *Looters of the Public Domain*. Portland, Oreg.: Portland Printing House, 1908.

Pyne, S. J. *Grove Karl Gilbert, Great Engine of Research*. Austin: University of Texas Press, 1980.

Rabbitt, M. C. *John Wesley Powell: Pioneer Statesman of Federal Science*. U. S. Geological Survey Professional Paper 669-A, Washington, D. C.: Government Printing Office, 1969.

———. *Minerals, Lands, and Geology for the Common Defense and General Welfare: Vol. 1, Before 1879*. Washington, D. C.: U. S. Geological Survey, 1979.

———. *Minerals, Lands, and Geology for the Common Defense and General Welfare: Volume 2: 1879–1904*. Washington, D. C.: U. S. Geological Survey, 1980.

Rachford, C. E. *Range Appraisal Report*. Washington, D. C.: U. S. Forest Service, 1924.

Raymond, R. W. *Report of the General Land Office*. U. S. House of Representatives, Executive Document 207. 41st Cong., 2nd Sess., 1877.

Recreational Resources of Federal Lands. Washington, D. C.: National Conference on Outdoor Recreation, 1928.

Reitan, C. H. "Trends in the Frequencies of Cyclone Activity over North America." *Monthly Weather Review* 107(1980): 1684–88.

Richard, T. A. *A History of American Mining*. New York: Appleton, 1932.

Richardson, E. *The Politics of Conservation: Crusades and Controveries, 1897–1913*. Berkeley: University of California Press, 1962.

Robbins, R. *Our Landed Heritage: The Public Domain, 1776–1936*. Princeton, N.J.: Princeton University Press, 1942.

Roberts, P. H. *Hoof Prints on Forest Ranges: The Early Years of National Forest Range Administration*. San Antonio: Naylor Company, 1963.

Robinson, G. O. *The Forest Service: A Study in Public Land Management*. Baltimore: Johns Hopkins University Press, 1975.

Robinson, M. C. *Water for the West: The Bureau of Reclamation*. Chicago: Public Works Historical Society, 1979.

Rogers, A. D. *Bernhard Edward Fernow: A Story of North American Forestry*. Princeton, N. J.: Princeton University Press, 1951.

Ronald, A. *The New West of Edward Abbey*. Albuquerque: University of New Mexico Press, 1982.

Roosevelt, T. *Hunting Trips of a Ranchman*. New York: Putnam, 1885.

———. "First Annual Message." In *A Compilation of the Messages and Papers of the Presidents*, vol. 15. New York: Bureau of National Literature, 1901.

———. *Autobiography*. New York: Charles Scribner's Sons, 1913.

Roth, F. "Administration of the U. S. Forest Reserves." *Forestry and Irrigation* 9(1902): 191–93, 241–44.

Rudnick, L. P. *Mabel Dodge Luhan: New Woman, New World*. Albuquerque: University of New Mexico Press, 1984.

Russell, C. P. "100 Years in Yosemite." Yosemite National Park: Yosemite Natural History Association, 1968.

Sale, R. D., and Karn, E. D. *American Expansion: A Book of Maps*. Lincoln: University of Nebraska Press, 1962.

Sampson, A. *The Arms Bazaar: From Lebanon to Lockheed*. New York: Viking Press, 1977.

Sannon, F. A. *The Centennial Years, A Political and Economic History of America From the Late 1870s to the Early 1890s*. Garden City, N. Y.: Doubleday, 1967.

Saway, O. *Bernard DeVoto*. New York: Twayne Publishers, 1969.

Schenck, C. A. *The Biltmore Story: Recollection of the Beginnings of Forestry in the United States*. St. Paul, Minn.: Minnesota Historical Society, 1955.

Schurz, C. *The Reminiscences of Carl Schurz*. 2 vols. New York: McClure Company, 1907.

Searle, R. N. *Saving Quetico-Superior: A Land Set Apart*. St. Paul, Minn.: Minnesota Historical Society Press, 1977.

Sellers, W. D., and Hill, R. H. *Arizona Climate 1931–1972*. Tucson: University of Arizona Press, 1974.

Shaforth, J. A. "Real Conservation and Federal Ownership." *Continental Magazine*, May 1912, 112–13.

Shanks, B. *This Land Is Your Land: The Struggle to Save America's Public Lands*. San Francisco: Sierra Club Books, 1984.

Shannon, F. A. "The Homestead Act and the Labor Surplus." *American Historical Review* 14(1936): 637–51.

Sharpe, C. F. S. *What Is Soil Erosion?* Washington, D. C.: U. S. Department of Agriculture, Miscellaneous Publication 286, 1938.

Shepard, J. *The Forest Killers—The Destruction of the American Wilderness*. New York: Weybright and Talley, 1975.

Shephard, W. "The Handout Magnificent." *Harper's Monthly* 163(1931): 594–603.

Sherman, E. A. "Thirty-Five Years of National Forest Growth." *Journal of Forestry* 24(1926): 129–35.

"Sierra Club Financial Report." *Sierra* 73, no. 2(1988): 77–79.

Smith, H. H. "Rain Follows the Plow: The Notion of Increased Rainfall for the Great Plains, 1844–1880." *Huntington Library Quarterly* 10(1947): 169–93.

Smith, W. O., Vetter, C. P., and Cummings, G. B. *Comprehensive Survey of Sedimentation in Lake Mead, 1948–1949*. Washington, D. C.: U. S. Geological Survey, Professional Paper 295, 1960.

Smyth, W. E. *The Conquest of Arid America*. New York: Harper and Brothers, 1889.

Sparks, W. A. J. *Annual Report, General Land Office*, Washington, D. C.: Government Printing Office, 1885.

Steen, H. K. *The U. S. Forest Service: A History*. Seattle: University of Washington Press, 1976.

Stegner, W. *Beyond the Hundredth Meridian: John Wesley Powell and the Second Opening of the West*. Boston: Houghton Mifflin Company, 1953.

———. *The Uneasy Chair: A Biography of Bernard DeVoto*. New York: Doubleday, 1973.

Sterling, E. W. "The Powell Irrigation Survey, 1888–1893." *Mississippi Valley Historical Review* 27(1940): 421–34.

Stevens, W. B. "When a Missourian Forced a Special Session of Congress." *Missouri Historical Review* 23(1928): 46–47.

Stewart, W. H. *Reminiscences*. Washington, D. C.: Neale Publishing, 1908.

Sundquist, J. L. *Politics and Policy: The Eisenhower, Kennedy, and Johnson Years*. Washington, D. C.: Brookings Institute, 1968.

Sutton, S. B. *Charles Sprague Sargent and the Arnold Arboretum*. Cambridge, Mass.: Harvard University Press, 1970.

Swan, H. *History of Music in the Southwest, 1825–1950*. San Marino, Calif.: Huntington Library, 1952.

Swanberg, W. A. *Citizen Hearst: A Biography of William Randolph Hearst*. New York: Charles Scribner's Sons, 1961.

Swift, T. T. "Date of Channel Trenching in the Southwest." *Science* 63(1925): 70–71.

Taft, W. H. "Address to the National Conservation Congress in St. Paul, Minnesota, September 5, 1910." In *A Compilation of the Messages and Papers of the Presidents*, vol. 16, 7555–74. New York: Bureau of National Literature, 1910.

Tanner, F. M. *The Far Country: A Regional History of Moab and La Sal, Utah*. Salt Lake City, Utah: Olympus Publishing Company, 1976.

Tarlock, A. D., and Tippy, R. "The Wild and Scenic Rivers Act of 1968." *Cornell Law Review* 55(1970): 707–39.

Taylor, R. W., and Taylor, S. W. *Uranium Fever, or No Talk Under $1 Million*. New York: Macmillan, 1970.

Teele, R. P. *Irrigation in the United States: A Discussion of Its Legal, Economic and Financial Aspects*. New York: D. Appleton and Company, 1915.

Thornthwaite, C. W. *Climate and Accelerated Erosion in the Arid and Semi-Arid Southwest, with Special Reference to the Polacca Wash Drainage Basin, Arizona*. Washington, D. C.: U. S. Department of Agriculture, Technical Bulletin 808, 1942.

Trefethen, J. B. *Crusade for Wildlife: Highlights in Conservation Progress*. Harrisburg: Stackpole Company, 1961.

Trijonis, J., and Yuan, K. *Visibility in the Southwest: An Exploration of the Historical Data Base*. Research Triangle Park, N. C.: U. S. Environmental Protection Agency, 1978.

Trimble, S. W. *Man-Induced Soil Erosion on the Southern Piedmont 1700–1900*. Ankeny, Iowa: Soil Conservation Society of America, 1974.

Urdang, L. *The Timetable of American History*. New York: Simon and Schuster, 1981.

Utah Department of Natural Resources and Energy. *Project BOLD: Alternatives for Utah Land Consolidation and Exchange*. Salt Lake City: State of Utah, 1982.

U. S. Bureau of the Census. *Compendium of the Eleventh Census: Part I, Population*. Washington, D. C.: Government Printing Office, 1890.

———. *Statistical Abstracts*. Washington, D. C.: U. S. Department of Commerce, 1953.

U. S. Bureau of Land Management. "A Comparison of Grazing Fees on National Forest and Grazing Districts. Unpublished paper, National Advisory Board Council, Bureau of Land Management, 1950.

———. *Draft Environmental Impact Statement, Livestock Grazing Management on National Resource Lands*. Washington, D. C.: Government Printing Office, 1974.

———. *Public Land Statistics*. Washington, D. C.: Government Printing Office, 1978.

———. *Public Land Statistics*. Washington, D. C.: Government Printing Office, 1986.

U. S. Congress. House. *Establishment of Public Land Law Review Commission*. H. Rept. 88–1008, December 7, 1963.

———. *Grazing on the Public Domain*. Hearings on H. 11816. 72d Cong., 1st sess., 1932.

———. *National Wilderness Preservation System*. H. Rept. 88–1538, July 2, 1964.

U. S. Congress. Senate. *A National Plan for American Forestry*. S. Doc. 12, 73d Cong., 1st sess., 1933.

———. *A National Plan for American Forestry*. S. Doc. 12, 73d Cong., 1st sess., 1933.

U. S. Department of Agriculture. *Agricultural Statistics*. Washington, D. C.: Government Printing Office, 1980.

U. S. Department of Commerce. *Historical Statistics of the United States, Colonial Times to 1970*. 2 vols. Washington, D. C.: Government Printing Office, 1975.

U. S. Forest Service. *Highlights in the History of Forest Conservation*. Washington, D. C.: Government Printing Office, Agricultural Bulletin 83, 1952.

———. *Forest Service Manual: Section G, Wilderness Areas*. Washington, D. C.: Department of Agriculture, 1955.

———. *Roadless Area Review and Evaluation: Final Report*. Washington, D. C.: Government Printing Office, 1973.

———. *RARE II: Final Report*. Washington, D. C.: Government Printing Office, 1979.

van Alstyne, R. W. "International Rivalries in the Pacific Northwest." *Oregon Historical Quarterly* 46(1945): 221–50.

Van Dyke, J. C. *The Desert: Further Studies in Natural Appearances*. New York: Charles Scribner's Sons, 1901.

Vanoni, V. A., ed. *Sedimentation Engineering*. New York: American Society of Civil Engineers, 1975.

Voigt, W., Jr. *Public Grazing Lands: Use and Misuse by Industry and Government*. New Brunswick, N. J.: Rutgers University Press, 1976.

Waters, F. *The Colorado*. New York: Holt, Rinehart and Winston, 1946.

Watkins, T. H. "The Terrible-Tempered Mr. Ickes." *Audubon* 82(1984): 93–111.

Watkins, T. H., and Watson, C. S., Jr. *The Lands No One Knows*. San Francisco: Sierra Club Books, 1975.

Webb, W. P. *The Great Plains*. New York: Grosset & Dunlap, 1931.

Wellman, J. D. *Wildland Recreation Policy: An Introduction*. New York: John Wiley & Sons, 1987.

Wells, P. P. "Phillip P. Wells in the Forest Service Law Office." *Forest History 16(1972): 22–29.*

Weyerhaeuser, F. K. *Trees and Men*. Princeton, N. J.: Newcomen, 1951.

Who Was Who in America with World Notables. Chicago: Marquis Who's Who, 1985.

Who's Who in American Politics, 1987–1988. New York: R. R. Bowker Company, 1987.

Wilber, C. D. *The Great Valleys and Prairies of Nebraska and the Northwest*. Omaha, Nebr.: Daily Republican Printer, 1881.

Wilbur, R. L., and Hyde, A. M. *The Hoover Policies*. New York: Charles Scribner's Sons, 1937.

"Wilderness Society, Annual Report, 1986." *Wilderness* 50(Spring 1987): insert, 30–31.

Wiley, P., and Gottlieb, R. *Empires in the Sun: The Rise of the New American West*. New York: G. P. Putnam's Sons, 1982.

Williams, R. L. *The Loggers*. Alexandria, Va.: Time-Life Books, 1976.

Wiltse, C. M. *Aluminum Policies of the War Production Board and Predecessor Agencies*. Washington, D. C.: War Production Board, 1945.

Winter, C. E. *Four Hundred Million Acres: The Public Lands and Resources: History, Acquisition, Disposition, Proposals, Memorials, Brief, States*. Casper, Wyo.: Overland Publishing Company, 1932.

Wolf, R. "New Voice in the Wilderness." *Rocky Mountain Magazine* (March/April 1981): 29–31.

Wooley, L. "Federal vs. State Control of the National Forests of the United States." B. A. thesis, University of Utah, 1911.

Wyant, W. K. *Westward in Eden: The Public Lands and the Conservation Movement*. Berkeley: University of California Press, 1982.

Young, O. E. *Western Mining*. Norman: University of Oklahoma Press, 1970.

Zahniser, H. C. "How Much Wilderness Can We Afford to Lose?" In *Wildlands in Our Civilization*, ed. by D. R. Brower, 33–39. San Francisco: Sierra Club Books, 1964.

Zelinski, W. "North America's Vernacular Regions." *Annals of the Association of American Geographers* 70(1980): 1–16.

Index

H

I

R

S

T

U

V

W

Y

Z